DINOSAUR MEMORIES II

POP-CULTURAL REFLECTIONS on DINO-DAIKAIJU & PALEOIMAGERY

Allen A. Debus

Foreword—Mike Bogue

Text Copyright – 2017 by Allen A. Debus. All Rights Reserved.

ISBN:

Cover Design by Todd Tennant.

ISBN-13: 978-1976543593

ISBN-10: 197654592

For Diane … my co-conspirator of many years

Vincent Lynch's fantasy illustration of the "Gigantosaurus" (later described as Brachiosaurus) parading along a New York City street. The art appeared in the Nov. 28, 1914 Scientific American, illustrating an article by paleontologist W. D. Matthew, "The Largest Known Dinosaur." See Chapter Eighteen for more.

Table of Contents

FOREWORD BY MIKE BOGUE 1

A PALEO PREFACE 3

ONE 8

    LOST HORIZONS 10
    CHAPTER ONE—PREHISTORIC MONSTERS OF SIR ARTHUR CONAN DOYLE'S 'LOST WORLD' 13
    CHAPTER TWO—THREE GIANT VENUSIAN DINO-MONSTERS: ARE THEY DAIKAIJU? 33
    CHAPTER THREE—TIME STEPS ASIDE 39
    CHAPTER FOUR—REVISITING THE LOST CONTINENT 46

TWO 52

    CHICAGOLAND DINOS 54
    CHAPTER FIVE—CHICAGO ATTACKED BY A SWARM OF GIANT DINOSAURS STORMING FROM CANADA 56
    CHAPTER SIX—GIANT DINO-MONSTERS & 'KONG' INVADE CHICAGOLAND 62
    CHAPTER SEVEN—MY 'GORGOSAURUS' 76

THREE 82

    PALEOIMAGERY VERSUS PALEOART 84
    CHAPTER EIGHT—PREHISTORIC SCENES 88
    CHAPTER NINE—NEW-MILLENNIAL PALEOIMAGERY 94
    CHAPTER TEN—IN SEARCH OF THE OLD MASTERS 107
    CHAPTER ELEVEN—ON THE QUESTION OF HAWKINS' SPIRITUAL NATURE 112
    CHAPTER TWELVE—GOING 'POSTAL' OVER NEAVE PARKER 115

FOUR 124

    SCIENCE FICTIONAL DINO-MONSTERS 126
    CHAPTER THIRTEEN—LOST WORLDS OF SCIENCE FICTION 130
    CHAPTER FOURTEEN—GET REAL: DINOSAUR MASQUERADE 137
    CHAPTER FIFTEEN—INVENTION OF TOHO'S FIRST TWO FLYING PALEO-MONSTER SPECIES 149
    CHAPTER SIXTEEN—SLEUTHING THE GIANT MYSTERIOUS MONSTER 158
    CHAPTER SEVENTEEN—SPOILER ALERT! DINOSAURS IN FANTASTIC FICTION—"EXTRAS" 163
    CHAPTER EIGHTEEN—BIG, FIERCE, EXTINCT ... RADIOACTIVE: GODZILLA'S ESSENTIAL FORMULA 188

FIVE 202

    STATUESQUE DINOSAURS 204
    CHAPTER NINETEEN—PREHISTORIC MONSTER MEMORIES 206
    CHAPTER TWENTY—CALGARY'S PREHISTORIC ZOO 210

CHAPTER TWENTY-ONE—OF PREHISTORIC (SCULPTED) CREATURES GREAT & SMALL 217

SIX 226

UNSUNG PALEO-MONSTERS 228
CHAPTER TWENTY-TWO—A TRIO OF UNLIKELY PREHISTORIA..................231
CHAPTER TWENTY-THREE—NEW WINGS ON THE PALEO-PERCH - (A 1996 PERSPECTIVE)..245

SEVEN 258

MYTHIC MONSTROSITIES 260
CHAPTER TWENTY-FOUR—REFLECTIONS OF DOOMSDAY: WHEN DID DINOSAURS BECOME APOCALYPTICAL? 261
CHAPTER TWENTY-FIVE—TOWARD A UNIFIED THEORY OF DINOSAURIAN KAIJU: A PACIFIC RIM-INSPIRED, 'WHAT IF? 266
CHAPTER TWENTY-SIX—"THINGS IN THE MIST": STEPHEN KING'S FORAY INTO CREEPY GIANT DINO-MONSTERS 270

EIGHT 274

PLANET OF APES VERSUS DINO-MONSTERS 276
CHAPTER TWENTY-SEVEN—TWO MYSTERIOUS MONSTERS OF EARLY SCIENCE FICTION LITERATURE...278
CHAPTER TWENTY-EIGHT—FROM "INCOGNITUM" TO ODO: PREHISTORIC ROAR 283
CHAPTER TWENTY-NINE—GODZILLA VS. KING KONG: SKULL ISLAND'S 'TRUE' (ALTERNATIVE) TALE 289
CHAPTER THIRTY—MODES OF SURVIVAL: GODZILLA 'VERSUS' KONG 304

NINE 310

ASTRO-PALEONTOLOGY 312
CHAPTER THIRTY-ONE—METEOR CRATER'S "IMPOSSIBLE BUT TRUE" 313
CHAPTER THIRTY-TWO—GOING FOR THE CYCLE: DARK MATTER, MASS EXTINCTIONS, AND THE RESURGENCE OF PALEONTOLOGICAL PERIODICITY 316

TEN 324

BELLICOSE BEHEMOTHS 326
CHAPTER THIRTY-THREE—WHEN DINOSAURS ATTACK! 328
CHAPTER THIRTY-FOUR—PLATED PARAGON: STEGOSAURUS' GREATEST MOVIE MONSTER BATTLES 337
CHAPTER THIRTY-FIVE—IMHO: VERY BEST OF THE PRE-1960S GIANT PSEUDO DINO-MONSTER BATTLES 347

ELEVEN 356

PERSONAL PETROGLYPHS 358
CHAPTER THIRTY-SIX—ONE EARTH FOR US 362
CHAPTER THIRTY-SEVEN—GODZILLA, THE REAWAKENING 368
CHAPTER THIRTY-EIGHT—THE ORIGINAL DINOSAUR SCRAPBOOKS........374

INDEX	**383**
ABOUT THE AUTHOR	**389**
ABOUT THE COVER	**389**

Foreword by Mike Bogue

There's no doubt in my mind that Allen A. Debus would have been as at home with Dinosaur Town as Reptilicus would have been with Monster Island. Dinosaur Town was a long-running saga my older brother Frank and I played as kids – it seemed a radioactive meteor had enhanced the IQs of nearby Marx plastic dinosaurs, which subsequently evolved over a two-year period from cave dwellers to spaceship builders. (Naturally, they fled a dino-dying earth to live on Venus.) This fantastic scenario features all the elements with which Allen is acutely conversant – dinosaurs, evolution, dinosaurs, mutations, dinosaurs . . . well, I think you get the point. Allen probably knows more about dinosaurs than most practicing paleontologists, but he isn't content to let the borders of science confine his passion. Quite the contrary, he delights in lost worlds, fantastic fiction, giant monsters, and (yes) Japanese maninsuitasauruses.

In July 2014, at G-Fest (an annual convention for Godzilla fans), I had the pleasure to serve on a panel with Allen. The topic was "Atom Age Connections," which explored the differences between Japanese and American atom age movies. The night prior to our panel session, Allen, fellow G-fanatic Mark Matzke, and I discussed filmic mutants, monsters, and mushroom clouds well into the wee hours of the morning. Our conversation cemented the respect I already had for Allen, for unlike some scholars whose upturned noses bump against the stratosphere, Allen remains a fan at heart, never putting on airs. Yes, Allen is fluent in both fanspeak and scholarspeak, an exceeding rarity, and his love of both the real and the imaginary, of both science and storytelling, shines through every page of Dinosaur Memories II, the book you now hold in your hands.

Indeed, I know of no Monster Kid as knowledgeable as Allen in the multiple fields of paleontology, paleo-art, paleo-literature, paleo-parks, and paleo-film, and in this anthology, he covers each area with style and intelligence – and also more than a little nostalgia. After all, Allen speaks of his exuberance as a kid when he spied the 1963 newspaper movie ad for *King Kong vs. Godzilla*, and his utter rapture upon seeing the film, which branded itself indelibly upon his formative brain. And this fandemonium encompassed (and still encompasses) both stop-motion animation and suitmation; both Harryhausen and Tsuburaya; both Rhedosaurus and

Rodan. And let's not forget those famous dinosaur masqueraders strutting their befrilled, lizardy stuff in 1940's One Million B.C. and 1960's *The Lost World* remake. Unlike many genre fans, Allen accords these faux dinos some respect – his childhood memories of seeing 1960's Lost Word on a vast movie screen make his six-year-old shock and awe palpable!

At times, Allen's tongue is planted firmly in cheek, as in his droll account of Chicago attacked by a horde of Canadian dinosaurs which, coincidentally of course, occurred on April 1. Even in his mostly serious Dinosaur Memories II entries, wit peppers his prose like sugar suffusing sweet bread. He also, like me, has a predilection for puns – Gaw-d-zilla, anyone?

Allen also scrutinizes fantastic fiction dealing with all things prehistoric. He thoroughly examines Arthur Conan Doyle's trend-setting novel *The Lost World*, as well as Ray Bradbury's equally trend-setting short story "A Sound of Thunder," a tale that begat a wealth of time travel follow-ups. But although *Lost World* and "A Sound of Thunder" may shine as Allen's personal favorites, he also turns the spotlight on the prehistoric tale-spinning of Jules Verne (*Journey to the Center of the Earth*) and Edgar Rice Burroughs (*At the Earth's Core, The Land That Time Forgot*).

Of course, Allen the scholar speaks of fossil finds, extinction events, birds as dinos (or is it dinos as birds?), and evolution periods. And Allen the art critic gives a long, fond look at paleoimagery and paleoart. So, what's the difference between the two? Read Chapter 3 and find out! But suffice to say, Allen is a big fan of the prehistoric vistas of Charles R. Knight and Neave Parker.

If you have followed Allen's work for the past two decades in Scary Monsters Magazine, G-FAN, *Prehistoric Times*, or Mad Scientist, then you know with this book that you're in for a treat. And if you've never read Allen's work, uncharted territories lie ahead! For just as Allen would have been at home with my boyhood saga of Dinosaur Town, you will be at home with his nostalgia-laden tome of Dinosaur Memories II.

A Paleo Preface

In recent decades there've been numerous pop-cultural books published intended for laymen enthusiasts concerning science and technical aspects of dinosaurs and other prehistoric animals or distinctive facets in the history of paleontology and paleoart. However, there have been relatively few publications celebrating *popular culture* interest in prehistoric life—especially reflecting the 'prehistoric monster' perspective. Donald F. Glut's *The Dinosaur Scrapbook: The Dinosaur in Amusement Parks, Comic Books, Fiction, History, Magazines, Movies, Museums, Television* (illustrated with more than six hundred photographs, drawings, cartoons and posters) may be the first in the latter category, inaugurating the 'genre' four decades ago, in 1980. In the intervening period, there have been slightly more than a handful of books, with the two most recent entries (known to me) being Ulrich Merkl's *Dinomania: The Lost Art of Winsor McCay, the Secret Origins of King Kong, and the Urge to Destroy New York* (2015), and my own *Dinosaurs Ever Evolving: The Changing Face of Prehistoric Animals in Popular Culture* (2016). One of my favorite books in this vein was another of Glut's: *Jurassic Classics: A Collection of Saurian Essays and Mesozoic Musings* (2001), which inspired my own (with Diane Debus) analogous *Dinosaur Memories: Dino-trekking for Beasts of Thunder, Fantastic Saurians, 'Paleo-people,' 'Dinosaurabilia,' and other 'Prehistoria'* (2002).

This brings us to the volume at hand – *Dinosaur Memories II*. "II"? 'Oh, hell—no!' some of you may mutter, as many did upon reading one of those *Sharknado* movie sequel subtitles. What's this one about, why was it made, and how *original* might it be?

To a certain extent, Glut acted as a mentor to me when it came to dinosaur writing during those heady days (~1996 to 2001), producing the now extinct *Dinosaur World* fanzine—felled by world shattering events. But here I seek not to emulate Glut's *JC* (let alone his *other* achievements—an impossible task!), but to follow as a mere 'disciple'—in his once forged & fabled fossil tracks. While Glut's *Scrapbook* and *JC* were highly original for their time, I claim little originality in *DM II*.

I cannot recall exactly where or how this crazy urge to write on such topics originated. But I spent many weeks of my 1970 summer vacation following my sophomore year in high school writing an unpublished (and no doubt unpublishable!) 'book' of well over one hundred hand-written manuscript pages—replete with illustrations, concerning the history of life through geological time. It rested in my bedroom closet at my parent's house for many decades. The following summer, I authored a short essay titled "The Fate of the Dinosaurs," exploring the mystery of dinosaur extinctions, mirroring concepts popularized by Dr. Edwin H. Colbert (whom with I corresponded briefly over a decade later). My three-page, hand-written copy is dated August 1971. A concluding sentence reads, prophetically: "Something happened 70-million years ago on our Earth which resulted in a severe and rapid biological change. Only by studying evidence he has uncovered from that time can Man hope to discover what actually did occur during this relatively brief, but important primeval episode." Amen—release the hounds!

My first printed avocational fanzine and newsletter articles began appearing in the late 1970s. Although I strove for originality, most of this early production wasn't very good and it took nearly two decades before I learned how to write material marginally suitable for a 'real' (peer-reviewed) magazine or edited book. Writing various kinds of reports, letters and other documents for employers didn't help much in this regard. I had to step out on my own, researching and writing on the topics of most interest to me for the proper 'succeed or fail' experience. Once I got my bearings, by the later 1990s a steady outpouring of material funneled into a number of fanzines and magazines of general interest to dinosaur groupies, dino-monster/sci-fi fans and assorted fossil paleo-philes. These days, Mike Frederick's magazine, *Prehistoric Times*, remains the key treasure trove for ready/steady information on a wide, disparate range of topics pertaining to interests in paleo-pop-culture; over 120 issues so far & counting. You will see that many of the chapters included in this book originally appeared in Frederick's brainchild 'zine. Likewise, J. D. Lees' *G-Fan* magazine, in which many of the present chapters originally appeared, is a goldmine of info on dino-*daikaiju*.

All selections herein have a 'prehistorical' bent: the primary theme. I also scanned published pages of additional printed material—*much* more than included in this book—onto a "Compendium" DVD in 2015 (with a 2016 update) for a handful of friends and family. I consider these scans from the original publications as a form of electronic "tear sheets." I decided to attempt this self-published "CreateSpace" online book too, incorporating a small subset of this scanned (paleo-themed) material, edited or corrected as needed.

Fanzines typically are not found in library collections; they quickly go out of print, passing into prized private ownership where they are considered "collectible," hence inaccessible to most. *DM II* permits availability of my original articles in a more permanent, accessible form for those relatively few eclectic souls who may wish to read them, obviating any need to track down so many old, rare fanzines, prying them—with tempting bushels of money—from private hands, say, at a comic-con. In so doing, the old articles (transformed into chapters here) won't become 'extinct' quite so readily. One drawback: I included only a very small fraction of the pictures and images that originally accompanied the printed fanzine articles.

Of the seven prior books I have authored, not including the aforementioned *Dinosaurs Ever Evolving*, three of my McFarland publications — *Paleoimagery* (2002); *Dinosaurs in Fantastic Fiction* (2006); *Prehistoric Monsters* (2009) — collectively dealt with diversified paleontological aspects in popular culture. An additional two of the seven were paleosculpting 'instructional' books, but meaty Chapter Two in *Dinosaur Sculpting: A Complete Guide, 2nd ed.* addressed historical themes in paleoimagery. Arguably, several of these were 'unique' for their time. My first book, *Dinosaur Sculpting: A Beginner's Guide from Hell Creek Creations and Dragon Attack!* (1995) – coauthored with Bob Morales and Diane Debus – was self-published. It sold over a thousand copies (every copy printed), and ranks as the very first 'how-to' book on the sculpting of table-top prehistoric animals. *Paleoimagery* was the first such book outlining paleoart's history, spanning the genre's history up through the new millennium. *Dinosaurs in Fantastic Fiction: A Thematic Survey* (2006) was the first book delving into science fictional literature concerning dinosaurs and other prehistoric animals. Other books mentioned in my McFarland 'series' have expanded upon some of the early ideas.

But the pairing of *Dinosaur Memories* titles I and II are, for lack of a better handle, a loose combo of 'essay collections.' As in the case of the 2002 *DM* volume, all chapters included herein were previously printed in fanzines, as indicated in short introductions. (I retain copyright protection for all textual material.) Although I strove to avoid repetition of material that was printed in any of my previous books, due to the nature of this beast, some degree of overlap was unavoidable. Each major section concerns a paleontological theme — something of personal interest — reflected in a few subsequent chapters therein. The style and organization of this book thus mimics Glut's treasured *Jurassic Classics*, of which I remain a fan. This is *not* intended as a scholarly booklet. Accordingly, I've decided to omit a handful of my 'dinosaur/paleo-' articles that have appeared in journals which are more heavily peer-reviewed. I've simply selected a number of paleo-themed titles that seemed to flow together. In this go-around (and as opposed to the original *DM*), I've dispensed with interviews conducted with various paleontologists and other 'paleo-people.' Scans of these items may be found on a forthcoming (2018) version of the aforementioned DVD.

'Predictably'—for those of you know me or are familiar with my writings—most of the present chapters (at one time published as fanzine articles) explore the "intersection" of real dinosaurs with *kaiju*, certain dinosaurian displays and theme parks we've trekked to, science fictional dinosaurs both in film and literature, certain paleoartists, mass extinctions, dinosaurabilia, and what is for me the 'time-honored' theme of Godzilla vs. Kong, so deeply imprinted on my young psyche.

Readers may detect a personalized undertone—an intention. However, unlike in *DM*, I won't lapse into nostalgia in this Preface. Instead, there's a section here that delves into aspects of my personal paleo-history (or as Glut referred to in his *JC* volume, his "Personal Petroglyphs"), and that should suffice for this rodeo. Yes—for me, dinosaurs and paleontology are a 'genetic' phenomenon, sort of a legacy. My grandmother took geology courses at Northwestern University during the early 1920s. She was the one who indoctrinated my Dad with a love of 'prehistoria,' as documented today in "prehistoric monster" scrapbooks, consolidating clippings from newspapers and magazines, which she began compiling for him in 1935. And he passed on the love—nay, the family's mission—to me.

Growing up, I've always been a 'dinosaur paleo-person' but have sought interrelated ties with paleo-fiction in its many sundry forms. Thus, many of my works meld facets from geology and the paleontological sciences with dino-monster and dino-*daikaiju* (pseudo-dinosauria that are gigantic, menacing and otherwise mysterious) lore, fiction & film. However, for those of you who may wonder, my academic, 34-year professional career was staked out in chemistry (i.e. my major) and several geochemistry and geology courses (including one in invertebrate paleontology). Later, it was bolstered by post-graduate work at two U of I academic institutions and environmental science work experience—spanning a handful of jobs and positions. In order to appreciate popular culture of dino-monsters/daikaiju, prehistoria and real dinosaurs, one must first make valiant strides into understanding the science of paleontology, and not be dismissive of it. Then science may be melded with fantasy.

For me, writing is a lot like sculpting. You do your research, sketch out an idea into a plan, outline or design. As the work progresses, you find there are numerous ways to correct and edit. There's usually a spark of creativity at the beginning when you conceive the idea, and also at numerous stages along the way as you resolve problems. Along the journey, a stray thought might be captured while walking in the woods, having a crazy waking dream, or while showering. That lightning 'Eureka' moment may spur means of improving, say, a sculpture by adding that certain blob of Super Sculpey to a particular body surface—thus resolving a nagging problem. Likewise, with analogous inspiration, maybe a better word, sentence or phrase could be inserted to enhance a key paragraph. Also, for me writing is usually slow and methodical. A sentence or a couple of paragraphs here and there, written each and every day can add up to a rock pile of words over an extended period of time – from beginning to (a hopefully coherent) end. Ultimately (and again, hopefully!) you'll be pleased with your end product. Likewise, I hope you all shall be pleased with this product – *Dinosaur Memories II*, representing paths once explored and fondly remembered.

Here I must acknowledge the following individuals: Todd Tennant—who designed the illustrious book cover, Jason Croghan, Ryan Dennis, Wendell Ricketts, Tim Paxton, Mike Bogue, Dennis Druktenis, Mike Fredericks, J. D. Lees, John Lanzendorf, Lisa Debus, Gary Williams, Lynne Clos, Mark Berry, Martin Arlt, Jack Arata, and Don Glut. And finally, special thanks to Kristen Dennis of *Full Proof Editing*, who performed a yeoman's task in patching this remarkable volume together, performing many invaluable and considerable editing tasks along the way to make it a far more palatable paleo-feast portion!

A.A.D.
Hanover Park, IL
August 2017

ONE

Illustration by Malcolm Smith showing a Venusian dino-daikaiju named the "Grimalkin" appearing in G. H. Irwin's science fiction tale, "Lair of the Grimalkin." Fantastic Adventures (April, 1948). See Chapter Two for more on this topic. (A. Debus collection; copyright 1948, Ziff-Davis Publishing Co.)

Lost Horizons

There was a rustling in the walls, or so it seemed. A mouse perhaps? Had a burrowing chipmunk penetrated the 'fortress of solitude' of my home office? Groaning, I crouched down to peer underneath the desk where a succession of dogs used to nap peacefully. I saw something in the corner! Two—no three little black legs leaning upside down on the nearby wall in a hard-to-reach location. Changing angle of sight, I looked down to verify that I was indeed looking at some sort of animal's 'paws.' Why was the intruder so still though? I found a metal rod and boldly jabbed downward so as to release the little creature's apparent grip from painted plaster. In a rushing flash I saw black fur, tiny feet, and ... white gleaming *tusks*? Not a rodent after all; no need for traps! It was a plastic, four-inch Woolly Mammoth that had fallen behind my desk, perhaps decades ago, only to become disinterred because of the strange sounds I'd heard. I'd never thought of my own office as a 'lost world' before—until now. Strangely, the peculiar under-the-desk sounds disappeared after my long lost (Invicta) Mammoth was restored to its rightful place atop a book shelf.

In December 2011, Dennis Druktenis (then editor of *Scary Monsters Magazine*) asked whether I could write an article for his forthcoming "Scary Summer Special 2012, no.4" issue on the silent *Lost World* film. I had been writing articles for his magazines since early 2002, and we'd met and spoken several times during the annual summer Godzilla "G-Fest." Although he was a handful of years older than me, it was intriguing that Dennis lived only a few miles from where I'd grown up during my high school years, in Deerfield, Illinois. Dennis knew I'd written about Arthur Conan Doyle's novel and classic film adaptations that had followed for other 'zines, so he'd come to the right person for the job. I spirited the work along and several months later, a handsome soft-cover volume arrived in my mailbox. My feature article led off the book.

Dennis also included a comprehensive "film-book" inside his Scary Summer Special no. 4, the movie script and more as well. Because there were only 200 copies of this rare 'collectible' printed, pricing was a lofty $30. But you now have my contribution to it, represented here as Chapter One.

During the early mid-20th century, while astronomers and popular writers pondered a possibility that planet Venus' supposedly steamy, hot, and once conceivably swampy (and therefore ancient) terrain might host creatures mimicking those out of Earth's prehistory, a number of fanciful sci-fi tales emerged concerning man's discovery of such monsters. How strange to view another planet of our solar system as an active, 'living' lost world, thus prompting a question—if 'exo-dinosaurs' existed there, could some of them conceivably be regarded as "*kaiju*"? Venusian dino-monsters hold a long, time-honored history in science fiction stemming back to the 1890's however, and I encourage readers to also consult Chapter Six of my 2006 book, *Dinosaurs in Science Fiction* and Russell Hawley's short interesting article, "Dinosaurs On Venus" in *Prehistoric Times* no.116 (Winter 2016) for illuminating insights on this topic. In the ensuing decade since material represented here as Chapter Two was written, numerous additional extrasolar planets have been discovered, at least several of which might have aqueous oceans. Although it's now known that Venus cannot sustain dinosaurian organisms, could one of these other worlds instead be in a 'Mesozoic' stage of life?

Or wouldn't it be stranger still to travel to an actual moment lost in prehistory on a dinosaur safari—with the fate of the world in your grasp. Ray Bradbury didn't write too much 'dinosaur' sci-fi but at least two of his paleo-monster tales rank among the greatest ever. In Chapter Three I investigate a key influence behind his "A Sound of Thunder" short story, involving 'safaris' to lost places of the globe. Bradbury wasn't the first to connect a time travel device with dinosaurs either; but his 1952 tale is arguably the best of its kind.

It should be mentioned that Bradbury's short story was continued decades later in a number of novels written by Stephen Leigh (e.g. *Ray Bradbury Presents Dinosaur World* 1992, and *Ray Bradbury Presents Dinosaur Planet* 1993), although these lacked the brain-blasting imagination of the 1952 classic.

1954's *Gojira* wasn't the first movie to tie dino-monsters to radiation; as you shall soon see, 1951's *The Lost Continent*—discussed here in Chapter Four—was.

> These chapters originally appeared as articles in the following publications:
> Chapter One—*Scary Monsters Magazine,* Scary Summer Special 2012 no.4 (June 2012), pp.5-19.
> Chapter Two—*G-Fan* no.96, Summer 2011, pp.18-22.
> Chapter Three—*Prehistoric Times* no.80, Winter 2006, pp.20-22.
> Chapter Four—*G-Fan* no. 114, Winter 2017, pp.24-27.

Chapter One—*Prehistoric Monsters of Sir Arthur Conan Doyle's 'Lost World'*

When early scientists considered the nature of fossils over two centuries ago, unwittingly, they were entering a mysterious and primeval "lost world" that would forever shift perspectives of reality.

If history has taught us anything, it is that in popular view our godlike 'prehistoric' monsters reside neither on Mount Olympus nor in Valhalla, but instead on obscure, forbidding and highly metaphorical 'islands,' insulated from modernity. Our term "prehistoric monsters," as in the 2012 titular event "Prehistoric Monster Bash" (Butler, Pennsylvania), has convoluted origins. Unlike supernatural vampires and werewolves, prehistoric monsters were once really alive and continue to exist in museum halls and imaginations, or in the form of fossil reconstructions. Yet in fictional accounts (movies and science fictional/horror tales), they live again, anachronistically threatening modernity and fragile civilizations, with awful, foreboding circumstances. As with matter and antimatter, the 'living' prehistoric and modern worlds should never, *ever* join. For the prehistoric world is a misty, mythical, forbidding place, infested with sinister monsters—symbols of bad omen, Frankensteinic creatures of absurdly joined body parts that instead should have remained separated and long dead. Beware—for prehistoric monsters walking the Earth again are harbingers of doom! Prehistoric monsters may not be supernatural, yet they are eternal. Because they did and do exist, reminding mankind of his folly, his arrogance, and our deepest psychological fears (i.e. of our evolution from beasts and impending extinction), they're much more difficult to destroy!

Arthur Conan Doyle's *The Lost World* remains perhaps the most singularly known and beloved, if not quintessential tale concerning mankind's fatalistic brush with prehistoric animals—several of which are portrayed as hideous monsters. This novel, collectively with Jules Verne's *A Journey to the Center of the Earth* (1864/67 eds.) and Edgar Rice Burroughs' *The Land That Time Forgot* (1918) and two "Caspak" sequels, represents the prototypical, classic triumvirate of stories involving living prehistoric life discovered in remote, prehistoric settings. Many modern science fictional tales (think of the *Jurassic Park* novels and film trilogy, for example), concern encounters with prehistoric life in obscure, prehistoric 'islands.' And although not isolated by an ocean, Conan Doyle's lofty "lost" plateau "Maple White Land" is also an 'island' conforming to an ecologist's definition of an environment whose indigenous 'prehistoria'—with few exceptions—are essentially restricted from modern organisms found below in the surrounding South American Amazonian wetland.

Are the dinosaurs and other prehistoric animals depicted in the novel and 1925 silent film truly "monsters"? Well, naysayers should know they receive their due in Donald F. Glut's *Classic Movie Monsters* (1978), included in discussion of Kong, along with chapters devoted to fellow 'prehistoria'—Godzilla and the Gillman.

I've previously outlined intriguing aspects of (both) Conan Doyle's 1912 story, and First National's famous 1925 silent film, *The Lost World*—see my *Dinosaurs in Fantastic Fiction: A Thematic Survey* (McFarland, 2006), and *Prehistoric Monsters: The Real and Imagined Creatures of the Past That We Love to Fear* (McFarland, 2010), respectively. As paleoartist Mark Rehkopf stated recently, "Dinosaurs are the perfect science fiction ... parts of them are actual science fact, and the remainder is imagined fiction based on science speculation." No wonder writers such as Conan Doyle found such 'prehistoria' to be perfectly suited for a fresh kind of science fiction story, melded with elements of adventure and horror. Here, we'll outline Arthur Conan Doyle's novel fabrication and introduction of the *idea* of the archetypical, resurrected "prehistoric monster," whose metaphorical, primeval 'descendants' still thrill and haunt us today in film and fiction!

Had Conan Doyle conceived of a lost world-ish tale today, it would have probably involved trendy 'Zombie-saurs.' But by the early 1910s, dinosaurs were very trendy. Paleontologists had divulged a considerable wealth of knowledge about the 'prehistoric world'; many museum displays and skeletal reconstructions of a host of dinosaurs and other prehistoric 'monsters,' including fossils of early men, were known to science. For decades prior, science popularizers were hell-bent on introducing dinosaurs in lavishly illustrated books to an adoring public, which, coupled with the most eye-dazzling painted and sculpted life restorations imaginable, fueled pop-cultural fascination over the science of dinosaurs and cavemen. One such volume, H. N. Hutchinson's 1892 *Extinct Monsters,* which went through several subsequent editions, equated the extinct with somehow being 'monstrous' in nature too. Yes, dino-"monsters" would have seemed entirely relevant then and, especially thanks to Victorian and Edwardian debates concerning evolution and inspiring British Museum fossil displays, all very much in public consciousness.

Yet before 1912, there were very few science fiction or fantasy tales concerning prehistoric life or monsters derived from prehistoric worlds. So—besides Sherlock Holmes—Conan Doyle's most brilliant achievement was in setting prehistoric creatures on a firm science fictional foundation in popular literature leading to filmic depiction. At a time when, generally speaking, monsters still weren't so popular, perhaps with exception of Mary Shelley's "Frankenstein," in his *The Lost World,* Conan Doyle succeeded eminently. The character of Professor Challenger would also appear in several other tales penned by Conan Doyle, including *The Poison Belt*. Interestingly, Conan Doyle had by then recently written a paleontologically-themed short story titled "The Terror of Blue John Gap," (1910), concerning a gigantic evolutionary descendant of the extinct Cave Bear that terrorizes the British countryside.

So, what were the scientific, paleontological influences behind the novel, leading to the prehistoric monsters encountered on the prehistoric plateau? There were several.

Perhaps Conan Doyle's interests extrapolating into the paleontological realm may have stemmed from a strangely coincidental discovery and contemporary accounts of live, 'giant,' 9-foot long reptilian 'dragons' found in a prehistoric island of Komodo in Indonesia. Also, there were reports of other, allegedly 'prehistoric' beasts living both in the Amazon as well as in the Congo, which Conan Doyle *may* have learned of. In the case of the former (as reported in the Jan. 11, 1911 *New York Herald*), the intrepid explorer Franz Herrmann Schmidt and his party were attacked by a 35-foot long, "bullet-proof" plesiosaur. In the Congo case, on the basis of dubious folk tales, wildlife collector Carl Hagenbeck attempted to bring a brontosaur back alive! While Hagenbeck's expedition proved unsuccessful, at least his "zoological garden" in Hamburg profited, as it was being decorated with an impressive array of interacting, life-sized dinosaur statues designed and sculpted by Josef Pallenberg. And then came the discovery of Great Britain's sensational fossilized "Piltdown Man," conveniently (and rather suspiciously) found only a fossil stone's throw away from Conan Doyle's manor. The romantic idea of prehistoric monsters living in obscure, prehistoric places, and then having humans going off to find them (at a time when remote places still *really* existed), was at least certainly in play, – a cryptozoological *zeitgeist*. True, there were other influential factors—both scientific as well as sociopolitical—involved besides the appeal of cryptozoological elements, but here we must move on to the story itself.

Furthermore, just as Verne relied on a visually replete factual paleontology book as a template for his imaginative *Journey* (i.e. Louis Figuier's *Earth Before the Deluge*, chock full of Edouard Riou's magnificent illustrations of fossils and prehistoric animals and flora), Conan Doyle also relied on contemporary paleontological volumes, perhaps principally British evolutionist E. Ray Lankester's (1846-1929) amply illustrated *Extinct Animals* (1905).

Conan Doyle (1859-1930) was a trained physician, and he took exceptional interest in natural history. But fiction writing was his 'forte'. In the case of *The Lost World*, he intended to pen a story to delight "… the boy who's half a man, or the man who's half a boy." While the prehistoric creatures populating Conan Doyle's novel (initially published as a serial in *The Strand Magazine*), remarkably differ from those animated in the 1925 feature silent film, they do rather closely conform to those outlined in Lankester's popular paleo-book (restorations for which were evidently illustrated by Lankester himself). Lankester, in fact, may have been one of the three historical figures from whom Conan Doyle forged the persona of his fictional Professor Challenger.

Okay! We generally know the basic story and premise. While probably some readers have seen the silent film and assuredly the 1960 Irwin Allen 'remake,' considerably fewer may have enjoyed reading the novel. While here we'll focus on dino-monsters and other prehistoria appearing in the 1925 film—with its amazing stop-motion animated effects (and to lesser extent, the 'in-its-own-way' impressive, fun 1960 production)—for proper perspective we also need to know a bit about what Conan Doyle originally wrote.

I happen to own three copies of *The Lost World* novel and have eagerly pored through all the original pages as printed in copies of *The Strand Magazine* (April to November, 1912). My collection consists of a circa 1925 edition published by A. L. Burt Company, containing several Plate photos from the silent film and two annotated volumes. Of these, the most informative is the 2002 *The Lost World of Arthur Conan Doyle - Collector's Anniversary Edition*, offering a considerable wealth of material prefacing the reprinted novel contributed by John R. Lavas, a zoologist, artist and 'Lost World' scholar (only 500 copies were printed). Lavas introduces readers to Conan Doyle's life and influences preceding his writing of the novel, underscoring that this was a great age of exploration and discovery, a time when many intriguing zoological discoveries were being made around the world, intriguing news and information that Conan Doyle most certainly kept abreast of. As he wrote in his autobiography, "This old world has got some surprises for us yet." The more available "The Annotated Lost World" (1996) is carefully referenced by Roy Pilot and Alvin Rodin.

I'll now describe the story as simply as possible while also encouraging you all to read the novel, which I find to be one of sci-fi's absolute finest and most fluent. Understandably, considering the time when it was written, a significant theme of the novel is organic evolution, or rather a curious lack of it upon the 'lost' plateau. The slightly 'mad,' often angry scientist, Professor Challenger, who leads the charge from London back to Maple White Land in the Amazon jungle, is an ardent evolutionist. He will not tolerate those who 'challenge' his ham-fisted ideas. Meanwhile, journalist Mr. Malone, seeking to impress his girlfriend, Gladys, opts for a romantic 'get away' adventure and joins the expedition. Also joining the expedition is Professor Summerlee, an expert in comparative anatomy, and wild game sportsman, Lord John Roxton. Their mission—which they've all accepted—is to witness the prehistoric creatures documented in notebook illustrations and photographs by the deceased explorer Maple White with their very eyes, and possibly to capture specimens as well. And so off they go to the forbidding land of "curupuri." Roy Pilot deduced that their year of departure must be 1908.

When they reach the base of the plateau, they realize that they, much like Maple White before, must ascend vertically some 600 feet to reach the summit. Things look gloomy indeed. During their continued survey of the cliff from below, they spy a flying pterodactyl with a 20-foot wingspan swooping out of the night sky toward their camp, confirming Maple White's records. Summerlee apologizes to Challenger for his skepticism, and they scale a perilous ascent to the top.

Now stranded upon the plateau, they find a three-toed dinosaur track, which "puzzled a worthy Sussex doctor some ninety years ago" (i.e. Gideon Mantell). Conan Doyle had even found such a track at a nearby ironstone quarry. Almost immediately, they encounter an *Iguanodon* family (two adults and three juveniles), looking like monstrous kangaroos, munching on foliage. Fossilized remains belonging to this type had been discovered in southern England during the 19th century. Next, they stray into a pterodactyl rookery, where, with rifles ablaze, the party staggers from the nasty, leathery-winged, dive-bombing brutes.

On the third night atop the plateau, they're awakened by an awful, ear-splitting sound, likened to a screeching railway engine, although "deeper in volume and vibrant with the uttermost strain of agony and horror." In the morning they discover carnage; iguanodonts had been slaughtered by a large, carnivorous creature, which Challenger proclaims an *Allosaurus*, while Professor Summerlee suspects *Megalosaurus*. (Think of this latter genus, perhaps, as England's version of North America's allosaur, or a lightly built tyrannosaur.) Whatever it was—the dino-monster remains menacingly at large in the forest! Later, after catching glimpses of a hideous ape-man observing their party, Malone observes a plated, lumbering *Stegosaurus*. Maple White had sketched this creature in his notebook and it was also depicted by Harry Rountree in *The Strand Magazine* so very similarly to the restoration appearing in E. Ray Lankester's book!

While evidently "time stands still" on Maple White Land, in reality, climatic conditions coupled with forces of natural selection and evolution have frozen or "suspended" the environment of an earlier geological age—the Jurassic. Challenger reasons that following sudden volcanic uplift of the plateau, Mesozoic creatures swept up on the mountain continued to thrive without facing extinction as they did elsewhere. But only in certain, exceedingly rare, occasions passages opened within the mountain, permitting certain organisms such as early anthropoids to filter through, immigrating to the top. Incidentally, the surface area of the plateau approximates the size of Sussex – Conan Doyle's own stomping grounds. As John Lavas demonstrated, a map of Maple White Land curiously conforms to Sussex's geology and geography.

Then in a riveting, hair-raising passage, Malone is chased by the megalosaur/allosaur dino-monster through the dark forest. The flesh-eater with the "toad-like face" absurdly 'hops' after Malone, in a fashion consistent with early views of how bipedal dinosaurs moved that remained more or less conventional until the 1930s. Malone is only saved after falling into a deeply dug pit, with a huge wooden spike in the middle, evidently used by natives on the plateau to capture and kill large prey! Meanwhile, the savage, primitive ape-men capture the rest of the expedition party, save for Roxton.

There's quite a bit of paleo-anthropology at issue in the novel. More or less as in H. G. Wells's *The Time Machine* (1895), it turns out there are *two* kinds of anthropoids living on the plateau – a shaggy, 'primitive' red-haired variety (possibly "Pithecanthropus" or possibly "Dryopithecus" as our professors debate) referred to as a race of "missing links," and a more modern 'indian' type who inhabit a labyrinthine system of caves. The two races have long remained in conflict with each other. Ironically, with his flowing "Assyrian" beard and stout, barrel-shaped body, Challenger even resembles those missing links. However, in an astonishing scene conceptually reminiscent of Jack London's 1907 novel *Before Adam*, with rifles blazing, Challenger, Roxton and the others later assist the Indian tribe in decimating the crude missing links. Thus, Challenger triumphantly declares self-servingly, "Man was always the master." In sci-fi and horror tales, man must slay the metaphorical dragon – the monster. So, are the anthropoidal 'links,' no longer 'missing,' in their own right … 'monsters'? To the extent that they represent man's primitive traits and worst nature – resoundingly, yes!

It should be mentioned that in the September *Science 83* article written by J. H. Winslow and A. Meyer, Conan Doyle, who was renowned for his belief in 'channeling' through mediums, or supernatural spiritualism, was accused of actually perpetrating the infamous Piltdown Man fossil forgery. Additionally, the article lambasted *The Lost World* as an elegantly cryptic yet conceivably decipherable ruse, mocking Lankester's anti-spiritualism views. But the 'trap' was never sprung. Yes, Conan Doyle retained adequate knowledge for insidiously doing such a misdeed, and did even include a curious—if not rather startling—reference to the relative ease with which one might fake a fossil in *The Lost World* novel. However, the 'missing link' ape-men described in *The Lost World* depart from the Piltdown type, or the "earliest Englishman" (which was supposed to have been a more evolutionarily 'advanced,' yet still geologically older, type). While I do not concur with their thesis and absolve Conan Doyle, recounting the affair at Piltdown here would be a bit tangential for present purposes.

The plateau's central lake was inhabited by primitive, fresh-water aquatic reptiles—ichthyosaurs and long-necked Plesiosaurs. Conan Doyle describes a scene where two fearsome flesh-eaters attack the Indian tribe at the mouth of their cave network, however the dino-monsters are overcome by poisoned arrows (recalling a scene in 1940's *One Million B.C.*). Professor Challenger is also chased by a large flightless bird – *Phororhacos* (later animated by Ray Harryhausen for 1961's *Mysterious Island*), before Roxton dispatches it with rifle rounds. They also encounter a docile *Toxodon*, described as a 10-foot long guinea pig.

Analogously, as in Verne's *Journey* (i.e. recalling Arne Saknussemm), the explorers 'decode' the means of descending through the cave network and passing through the old volcanic plateau, escaping to civilization. Returning to London, Challenger triumphantly displays a live baby pterodactyl at the next Zoological Society meeting, which after terrifying the audience ultimately escapes into the night sky. Malone muses,

"If its homing instinct led it upon the right line, there can be no doubt that somewhere out in the wastes of the Atlantic the last European pterodactyl found its end."

In homage to Conan Doyle's masterpiece, several other novels have continued, been based on, or otherwise inspired by the story of *The Lost World* . Perhaps the most widely known is Michael Crichton's *The Lost World* (1995). Crichton's novel brings the horror of genetically resurrected tyrannosaurs and slashing "raptors" to Isla Sorna in the Pacific – where certain death awaits. In Greg Bear's *Dinosaur Summer* (1998), intended as an indirect sequel, Willis O'Brien, Merian C. Cooper and Ray Harryhausen visit the 'real' lost world plateau in Venezuela—in the year 1947—of Conan Doyle's novel. They intend to return aging circus dinosaurs captured years before, while during their visit also witnessing other living dinosaurs that evolved considerably since the Mesozoic Era. Nicholas Nye's *Return to the Lost World* (1991) was intended as a direct sequel, involving Conan Doyle's characters who revisit the plateau in 1912. Ultimately, Professor Challenger perpetrates the infamous Piltdown forgery. And then there is Don Glut's *Frankenstein in the Lost World* (1976, revised and reprinted in 2002).

Now on to the 1925 movie, which Conan Doyle greatly admired! I probably first saw this movie with my dad in the mid-1960s on the Chicagoland television series, "The Toy That Grew Up," which featured silent films – although the version I saw was probably severely chopped. The 1925 film, filled with wonderful prehistoric monsters and miniature 'puppets' animated through the magic of stop-motion photography, is THE film that all other dino-monster movies should be compared with, in the context of its time. Predating *King Kong*, it is a seminal work. Originally 104 minutes in length, film footage running time was cut in half. Recently, however, a more complete version of the movie was discovered in Prague. The reliable DVD version by Image Entertainment, nearly fully restored, is approximately 94 minutes long.

There are several significant differences in the plot between novel and silent film that should be clarified before describing those incredible stop-motion animation scenes. In the novel, American explorer Maple White dies prior to opening events. But in the 1925 film, the expedition is, in part, intended as a mission to rescue Paula White's father, who as we later learn, perished on the plateau. Furthermore, while no ladies venture along the journey in the novel, First National chose to add a young beauty who would serve as love triangle interest both for Malone as well as the older, honorable Roxton. Also, while Conan Doyle inhabited the plateau with two tribes (ape-men and Indians), in the 1925 film there is but a single, hairy ape-man convincingly played by actor Bull Montana, who along with his chimpanzee 'sidekick,' (a Tarzan/Cheetah-like combo!) represent the novel's anthropological leanings. Then, in the novel, the living dinosaurs represent a suspension of evolutionary forces, while in the film Malone refers to the dinosaurs as creatures that died out "10 million years" ago, although on the plateau, these are "living descendants" of those creatures that are supposed to have been dead for millions of years (implying more sensibly that at least some minimalistic measure of evolution has taken place).

Finally, prehistoric animals populating the plateau in the film conform more closely to fossil vertebrate discoveries made in North America and also to those which had been restored magnificently in paintings by Charles R. Knight. And of course, rather than a baby pterodactyl, Challenger manages to raft a huge sauropod *Brontosaurus* back to London, which escapes into the nighttime city streets; the first 'realistic,' filmic rampage, dino-monster-on-the-loose-in-civilization scene, emulated frequently in subsequent decades (e.g. Kong, Rhedosaurus, Godzilla, Gorgo, Paleosaurus, Reptilicus, Gappa, Rodan, *Jurassic Park II's Tyrannosaurus*, etc.).

Perhaps one of the *key* early set-up scenes in the film shows paleontological displays mounted inside London's 'Zoological Hall,' where we spy skeletal reconstructions of a brontosaur and an allosaur; a panel mount of a more slender theropod—possibly *Struthiomimus*, not appearing later as an animated creature—adorns a rear wall. There is even a large, flightless (ratite) bird skeleton visible, which although not animated for this film, recalls Challenger's encounter with the *Phororhacos* in the novel. Here, the director, Harry O. Hoyt, is attempting to suspend our disbelief by showing creatures that will be caught 'live' on camera really did (and do) exist, that is, as known to science through their genuine fossil bones (which, by the way in these scenes are not real skeletons but fabrications), as well as on the plateau. Similar or analogous scenes showing dinosaur skeletons in prehistoric monster movies also appear in films such as *The Beast From 20,000 Fathoms* and *Godzilla, King of the Monsters*. ('Artificial' dinosaur skeletons, including a *Triceratops*, were perhaps first seen in a faux museum setting in a 1923 silent film, *Adam's Rib*.)

But what we see in *The Lost World* is a visual extrapolation of what Conan Doyle thought was an important element in the novel. For in the novel, Challenger shows Malone pages in Maple White's diary that most intriguingly 'prove' the existence of 'living' prehistoric animals. First, Challenger shows Malone White's drawing of a *Stegosaurus*, allegedly drawn from life conforming perfectly to a drawing in Lankester's *Extinct Animals* book. Then, to underscore the probity of the sketches, Challenger hands Malone a hefty, dinosaur-sized bone, not fossilized, but 'fresh.' In the novel, Challenger has more tricks up his sleeve, as he produces a photograph of a pterodactyl and a section of wing from a specimen that he shot with his rifle, while once more referring to a substantiating restoration in Lankester's book. This scene is recreated in the 1925 movie, although instead of a stegosaur, Malone is shown Maple White's drawings of a brontosaur and an allosaur, both of which figure prominently in special effects scenes. After all, 'seeing is believing,' so if Malone can be verily convinced then also will the reading and, likewise, viewing audience. So, in the 1925 film, we proceed from lifeless museum skeleton props, to Maple White's live 'flesh' drawings, and on to the real (animated) animals - prehistoric movie monsters!

The stop-motion puppets were the marvelous handiwork of paleo-sculptor Marcel Delgado. Special effects master, Willis O'Brien had utilized the stop-motion technique years before for other shorter films that he created prehistoric animal recreations for, such as 1919's *The Ghost of Slumber Mountain*. But for a feature film such as *The Lost World*, with so many planned animation scenes which had to be executed excellently and convincingly, considerably more skill was needed. Hence, O'Brien offered Delgado (who was then working as a grocery clerk while taking art classes part time) a position as assistant; the team succeeded magnificently, not only in this film, but also with their later magnum opus – *King Kong*.

Delgado's puppets were constructed using armatures that could be realistically moved using a jointed internal ball-and-socket system. This allowed for the many steady movements that would be photographed, tediously, one infinitesimal segment of a step or turn at a time until – when seen altogether in sequence as continuous footage—it seemed as if the puppet was actually walking, leaping, flying, eating, growling or fighting. Delgado filled in the body cavities, covering metal armatures with sponge rubber for "muscles" and in some cases (for realism) added a balloon bladder that he could laboriously pump up and then release with air to simulate a dinosaur 'breathing' stop-motion effect. Over the bodies, latex skin, plates, horns or other integument and protuberances were molded and applied. To Delgado's credit, this truly was a herculean effort. His individual dinosaur puppets were typically about 18 inches long. Delgado sculpted on the order of 50 puppets altogether, whose appearances were based on Charles R. Knight's paintings. Dinosaurs seen on camera include *Allosaurus, Tyrannosaurus, "Trachodon," Stegosaurus, "Brontosaurus,"*

Brachiosaurus, Pteranodon, Triceratops and *Agathaumas*. Of note, Knight had never painted a life restoration of *Brachiosaurus*; the astonishingly huge skeleton of this new dinosaur genus—until then the largest known—was undergoing reconstruction in Germany.

Flora, trees and bushes in the plateau setting were made sturdily. Leaves on trees were cut from sheet metal, so that they would remain in place while animators climbed into the setting to move the creatures from one sequence to the next. Nighttime scenes were tinted bluish, while for daytime a sepia tint was used. Lush jungle settings, matte paintings, even the awakened volcano that would become cliché' in so many subsequent dino-monster films (originally conceived for Jules Verne's *Journey*), worked together splendidly.

Beyond battling dino-monsters, audiences must have been absolutely thrilled by scenes in which dinosaurs and members of Challenger's party appeared together in the same shot, a 'split screen' illusion achieved by combining separately filmed sequences shot 'in the can' that was still quite uncommon for the mid-1920s. While in real life, the actors were large relative to the size of Delgado's puppets, using his trademark trick photography O'Brien reversed this reality most convincingly. Roy Pilot relates that script writer Marion Fairfax was asked to write two versions. The second, unused version eliminated any plot elements pertaining to or otherwise incorporating stop-motion dinosaurs – that is, in case O'Brien couldn't pull things off. When he found out afterward there was a 'backup' contingency plan in place, O'Brien was indignant.

To a large extent, seemingly so alive on the plateau, the dinosaur recreations take center stage, relegating actors Wallace Beery (Prof. Challenger), Lloyd Hughes (Ed Malone), Lewis Stone (Sir John Roxton) and the rest amidst their plight and plot lines to the back burner. This may be the movie's only flaw; the dinosaur recreations are simply too good for their time, refocusing interest and attention to the props rather than human acting! This is because O'Brien, Delgado and other technical staff had their latex and sponge 'players' acting so convincingly, like never seen before. However, in his *Stop-Motion Filmography* (1999), Neil Pettigrew shares his perspective, that "… Harry Hoyt was a poor choice as director: He shows no concept of pace or excitement, presenting each scene in a flat, uncinematic way."

Winged *Pteranodon* is the first stop-motion animated creation seen on the plateau. We see it feeding, flapping its wings and gliding realistically (although a sudden jerkiness indicates that someone with wayward fingers misaligned the wires). Then as stalwart team members chop down a tree that will serve as a bridge over the deep gorge, an inquisitive *Brontosaurus* munching on vegetation notices the silent film's tree-cutting sound. We see a composite scene of miniature actors observing the giant dinosaur; then (like Kong would 8 years later) a brontosaur knocks the felled log into the ravine, thus apparently blocking the expeditions' only means of escape. In fact, so many plot analogies can be made between *King Kong* and *The Lost World* that the latter film has been considered a filmic template for *King Kong*, where O'Brien's ideas, skills and craftsmanship were perfected. (But rather than listing these conceptual plot analogies here, why not enjoy spotting these yourselves on a weekend afternoon!)

Now, aboard the magical plateau, audiences are treated to "special effects-land" marvels! A pesky *Allosaurus* takes center stage through several impressive scenes. First, the carnivore attacks a duck-billed "Trachodon" that is spied chomping foliage. With "Trachodon's" throat gripped tightly within the carnivore's cruel jaws, the deceased duck-bill rolls into a lake, which according to Pettigrew was actually not water, but liquid glycerin that would hold its position between camera shots. This scene is derived from an aforementioned scene in the book where the alleged 'megalosaur' dispatched iguanodonts (evolutionary ancestors to Late Cretaceous duckbills, like "Trachodon") in the night. But here, audiences were treated to the carnage, witnessed 'live.' We then see the allosaur ripping gobbets of flesh from the dinosaur corpse.

Next, there's a scene featuring a *Triceratops* mother and baby, fortified with a third adult horned dinosaur; the *Triceratops* family is menaced by an allosaur. This scene is no doubt inspired by Charles R. Knight's famous 1906 painting, completed for display in New York's American Museum, showing a *Triceratops* 'family' threatened by a tyrannosaur striding in for a fresh kill. Knight's static painting appealed to viewers' imaginations, instilling a sense of primeval wonder. But O'Brien and Delgado showed us more—graphically illustrating how these horrific scenes of prehistoric combat happened, and which of these great animals lived or died in their struggle for existence. The allosaur, gored by a three-horn, briefly ponders the outcome then wanders off-screen, while the baby greets its mother affectionately. Given that Knight's oft-repeated *Triceratops* vs. tyrannosaur (or in this case – allosaur) combat theme is, pop-culturally speaking, dino-paleontology's most recognized symbolic portrayal, this sequence must have been intended as a 'piece de resistance.' In fact, this Knightian metaphor even translated into Japanese monster movies, such as 1955's *Gigantis, the Fire Monster* (and 1977's *The Last Dinosaur*).

At nighttime, in a scene inspired from the novel, an allosaur—possibly the one which was already stabbed?—threatens Challenger's jungle encampment. Rifle shots and a burning brand ward off the dino-monster. Then the picture switches to another combat scene, this time involving another horned dinosaur, made famous in an 1897 Knight painting—the *Agathaumas* (sometimes incorrectly referred to as "*Monoclonius*"). Clearly, the horned vs. carnivorous dinosaur theme was a major inspiration underlying *The Lost World*'s special effects. *Allosaurus* circles the large *Agathaumas*, leaping onto its back, but is thrown off. *Allosaurus* succumbs, impaled by *Agathaumas*' prominent nasal horn. But wait, there's more! Much more.

Representing a period when nature reigned, red in bloody tooth and claw, there is no rest for the weary on Maple White Land. Otherwise peaceful *Agathaumas* can barely enjoy its next meal, when a larger, presumably hungrier Tyrannosaur enters the scene, disturbing what passes for relative tranquility on this tumultuous plateau. Leaping onto *Agathaumas*' prickly back, the stronger carnivore dispatches the horned dinosaur with relative ease, going for its jugular then gulping down raw, blood-stained flesh. The tyrannosaur, evidently not sated with this huge kill, now jumps, snatching a winged scavenger pterosaur right out of the sky; the pterosaur is greedily devoured!

An epic battle is about to ensue between another snarling allosaur and its intended prey, a long-necked *Brontosaurus* that will become, literally, a 'cliff-hanger.' As Professor Challenger observes, a feeding brontosaur is suddenly disturbed by the allosaur. The two behemoths stare each other down momentarily before brontosaur lunges in defensively for a throat hold on the predator. But allosaur turns the tide, seizing brontosaur's long neck in its toothy jaws. As brontosaur backs up toward the cliff edge, it loses its footing and soon tumbles over. Instinctively, allosaur takes a swift bite of air, as its huge 'meal' plummets into oblivion. But, landing in soft river mud far below, the wounded brontosaur is spared death. We can see it 'breathing' – although this time not through stop-motion photography. This 'specimen' will prove highly important later on after the expedition team returns to London.

The chaotic dinosaur stampede amidst the volcanic eruption is a magnificent achievement! For this scene, a much larger table jungle landscape milieu was built, measuring 75 feet by 150 feet long. A record number of puppets were used simultaneously in this sequence; even the smoldering volcano—always a risky effects element lest it look fake—looks reasonable. Herds of stegosaurs, brachiosaurs, brontosaurs, Triceratopses, and ravenous allosaurs, intent on feasting even during the midst of the melee', flee the eruption, with Ed Malone looking on dangerously close to the action. Volcanic smoke was superimposed after stop-motion scenes were filmed. Next, there's a (more placid) 'morning after' river bank scene, showing five allosaurs devouring a corpse, mingling with brontosaur and stegosaur survivors. According to Pettigrew, O'Brien—who filmed most of the effects so incredible for their time, himself—must have sought considerable help from stage technicians in rendering these dramatic scenes. Overall, the effort to produce stop-motion scenes for *The Lost* World required 14 months of intensive labor; at most only 20 seconds of "straight-forward scene" screen time could be accomplished daily.

The lone, scary-looking ape-man, wounded earlier by a gunshot, vengefully menaces the party as they escape from a cave atop the plateau by climbing down a rope ladder. Roxton kills the monstrous ape-man who is yanking the rope upward with Malone dangling perilously high above the jungle floor. Roxton thus spares Malone, who will eventually claim Paula as his lover. The missing link species suffers extinction—much as the turn of events in the novel. Anglo-Saxon man is the conqueror.

If this wasn't enough in itself, the shocking, never before seen, dinosaur-on-the-rampage climax certainly startled audiences. After an alarmed Malone notifies Challenger that his prize specimen—the one that fell off the plateau, landing safely in river mud—has escaped its cage into the nighttime London streets, we see the behemoth rounding a street corner. The brontosaur wreaks havoc along a city block, and here we see near seamless displays of terror stricken people fleeing the monster, as well as smashed buildings and city structures (accomplished using a traveling matte process). Some scenes relying on full-sized props such as the monster's foot and tail menacing intended victims were spliced into the action to heighten the sense of reality. Unlike some of the previous jungle scenes, where, for example, a dinosaur simply eats foliage, here O'Brien pulls out all the stops; there's considerable mayhem in the London set.

The brontosaur nearly crushes a woman, topples a statue, singes its snout on a street lamp and rears on hind limbs against a multi-story building. In a scene cut from the original print, the brontosaur then inserts its head through a window, frightening those inside. Next, the brontosaur demolishes the building. The exterior London scenes foreshadow what audiences would see the "prehistoric ape," Kong, perform eight years later.

But the idea of a large dino-monster sizing itself up, anachronistically, against modern city structures had been visually conveyed decades before. One such picture first published in an 1886 French volume by Camille Flammarion showed a godzillean-sized, horned *Iguanodon* pressing up against the height of a 7-story city building. Another similar and more relevant picture appeared a dozen years later on the cover page of a November 1898 issue of the *New York Journal*. Here, readers treated to a lead story concerning dinosaur discoveries were astonished to see an illustrated brontosaur rearing up facing a skyscraper, peering through an 11th story window! Today, it's impossible to know whether O'Brien and Delgado had ever seen or remembered these images. But surely *The Lost World's* climax must have inspired Katharine Metcalf Roof's little known short story "A Million Years After," published in *Weird Tales* (Nov. 1930), in which a brontosaur hatches from a petrified egg. The issue's striking cover art featured the fully grown bronto-monster advancing upon a metropolitan area.

Finally, *The Lost World's* brontosaur, intent on escaping turmoil, strides toward Tower Bridge. However, the creature is too immense, and when it rears up once more, its crashing weight collapses the structure in an impressive, well-timed sequence where an automobile also tumbles toward the river. Later we see the dino-monster swimming off to sea from the stricken area. Pettigrew muses, "… *The Lost World* set many standards for monster-on-the-loose films, but the cliché of the arrival of the military wasn't one of them."

Many movie stills were taken showing ape-man and dinosaur puppets in various poses, in some cases with actors posing in a split screen effect camera view, although not all of the dino-monsters (e.g. *Brachiosaurus* and *Tyrannosaurus*) were photographed for posterity as stills. Also, about 13 minutes of masterfully done footage—these were either test reels or outtakes—from *The Lost World* highlighting special effects featuring 'live' dinosaur puppets has been found. This footage is included as an 'Extra' on the Image Entertainment DVD. Given how arduous it was to produce even a few minutes of stop-motion animation, it's incredible to see that 13 minutes of material could be excised from the final production!

Even at an alleged production cost of $1 million, a *Variety* reviewer stated, "… no matter what the cost, either in labor or money, the results justify the expenditure." Due to obvious time limitations and other complexities, the film remained in production for many months. Finally, in 1922, a test reel was prepared, which Conan Doyle showed to the Society of American Magicians on a night when Harry Houdini performed. In this test footage, magicians witnessed a pair of tyrannosaurs attacking herbivores. Afterwards, the tyrannosaurs battle to the death (a dramatic pose imaginatively conceived by paleontologists Henry F. Osborn and Barnum Brown for New York's American Museum). According to Pettigrew, the tyrannosaur "…victor is set upon by a triceratops, then flees and unsuccessfully attacks a stegosaurus. Conan Doyle, relishing the opportunity to astonish his audience, refused to answer any questions as to whether what he had shown was reality or trickery."

Vintage advertising for the movie shows a gigantic, incorrectly 4-fingered allosaur tumultuously overturning a rail trolley car with a domed capitol city building visible in the distance. In the foreground, terrified people flee from the allo-monster — a godzillean sort of scene that would be reenacted decades later, independently by Max Fleischer (*Superman – "The Arctic Giant"*), Ray Harryhausen (*The Beast From 20,000 Fathoms*) and Eiji Tsuburaya (*Gojira*). The Image Entertainment DVD includes a nice reproduction of the original souvenir program. According to this program, one or several of the long-necked sauropods were intended to be not *Brontosaurus*, but instead *Diplodocus*. *Brachiosaurus* (with arms proportionally larger than its hind limbs) isn't mentioned in this brochure, but this genus is clearly seen during the dinosaur stampede. Therein are some interesting anecdotes concerning Wallace Beery (Prof. Challenger), who at one time lived in the Chicago north shore area. My father recounted hearing from my grandparents that occasionally they would see Beery out driving in his automobile, possibly along Green Bay Road.

For the most part, O'Brien's and Delgado's dino-monster puppets were state-of-the-art, based on contemporary scientific knowledge. Of course, anyone who has seen the *Jurassic Park* movies now knows that dinosaurs didn't look or move as seen in the 1925 film. However, even stranger looking dino-monsters were featured in Irwin Allen's 1960 remake of *The Lost World*, at a time when his special effects team and monster designers should have known better. In fact, these titanic creatures were so odd looking and for their time known to be anatomically inaccurate, given what they were supposed to have represented, that they've been dubbed by Mark F. Berry not as 'dinosaurs,' but 'Irwinosaurs.' Regardless, to me—then a young lad seated with my grandparents in a downtown Chicago theater watching this grand spectacle in July 1960—these Irwinosaurs seemed very effective indeed!

Besides the kinds of monstrous prehistoria and means of making them come alive on screen, significant plot and character development differences may be noted between the 1925 and 1960 film versions of *The Lost World*. For our purposes, Maple White is now "Burton White" – who, blinded, has survived, stranded upon the plateau. Procrastinating John Roxton (Michael Rennie) is a notorious ladies' man, while Paula White's (Bessie Love's) role is switched to eye candy Jill St. John's quite humorous "Jennifer Holmes" (no relation to Sherlock, however, yet homage to Conan Doyle) and her poodle. A cute native girl's cameo in the silent version has been expanded to the Native Girl role played by Vitina Marcus. Besides Challenger's (Claude Rains) pompous pursuit of knowledge and fame, now diamond discoveries form a financial basis for mounting the plateau. While Harry O. Hoyt cast only a single primitive ape-man to represent races described in Conan Doyle's novel, instead Irwin Allen scripted a tribe of essentially modern humans. But while the novel's indians were friendly to Challenger's party, Irwin Allen's natives are cannibalistic.

Rather than climbing up the plateau, Gomez (Fernando Lamas) flies them up there in a helicopter. Conan Doyle's evolutionary angle is considerably downplayed, with Challenger only once declaring before the Zoological Institute that the land of curupuri is "… a land where monsters live … so isolated as to be insulated from laws of evolution." Additionally, no dinosaur is seen to be returned to London, although at the end, before their very eyes, a pilfered dinosaur egg hatches a strangely-horned gecko that Challenger incongruously pronounces a 'tyrannosaur,' (the gecko probably will be out selling car insurance before it ever terrorizes London). 'Shame,' Professor; 'shame,' Irwin! At least, like the 1925 film, Irwin Allen's version displays impressive production values.

Agreed – 'gigantic' creatures appearing in the 1960 film do not in the least resemble any self-respecting dinosaurs. And I can sense the pangs that stop-motion animator experts Ray Harryhausen and Willis O'Brien must have felt when Irwin Allen decided to bypass their remarkable talents, opting to instead use "gagged-up lizards" for his production. However, here I must differ with the gist of Berry's critique. (See Mark Berry's entry for *The Lost World* 1960, in *The Dinosaur Filmography*, 2002.) While I cannot disagree with his chiding analysis of the movie, well, I simply liked it! And I enjoyed its crazy looking, scary dino-monsters. It 'works' for me. Why?

I saw this film on one of those old-time, long movie screens (unlike typical, mid-sized stages in today's movie theaters where 30 screens are all scrunched together in the same lot). To me, *Lost World* 1960's dino-monsters were HUGE; their roaring sounds reverberated awesomely throughout the theater, penetrating my soul. I was totally scared %&*$-less! Having probably just come from a Sunday visit to the nearby Field Museum of Natural History, I probably realized—even at age 6—that the pseudo-"brontosaur" shown in the picture (with its rubbery ceratopsian neck frill and sporting stegosaurian plates adorning its spine) didn't at all look like the long-necked fossil skeleton we'd just seen in the museum. But I was completely spellbound, scared out of my wits at the sound of trees splintering and crashing to the ground. And what kind of titanic monster could smash and destroy a helicopter!? Then I cowered in my seat when "Jennifer" chases her stray poodle into a clearing, encountering the peculiarly gigantic, horned iguana which began roaring directly at ME. In short, abject fear endeared me to this film, imprinting these scenes upon my definition of "prehistoric dino-monster." Later, it was also televised regularly on Chicago's *Family Classics*, hosted by WGN's Frazier Thomas, although seeing it broadcast on a black and white portable fell far short of that initial full-screen experience!

Clearly, Irwin Allen was trying to emulate or capitalize upon successes of 20th Century-Fox's 1959 *Journey to the Center of the Earth*, starring James Mason, Pat Boone and Arlene Dahl. Applying rubbery prosthetics (horns, sails, plates, etc.) to reptiles and lizards with glue to create the illusion of prehistoric monsters on screen had already been done tolerably well for Hal Roach's *One Million B.C.* (1940). And L. B. Abbott certainly concocted impressive fin-backed *Dimetrodonts* from iguanas for 1959's *Journey*. It would not have been possible to achieve similar, anatomically *convincing* successes with the assortment of reptiles selected for Irwin Allen's 1960 dinosaurian extravaganza. Anatomically, the reptiles just didn't conform to any real dinosaurs, in itself an excusable flaw. Unfortunately, fuddy duddy Challenger, who should have known better, repeatedly insists that the huge monsters are in fact living dinosaurs. (And even if these *were* monstrously mutated descendants of original dinosaurs, then Challenger's claim about the suspension of evolutionary forces upon the plateau is incorrect.) 'Shame,' Professor Challenger. What kind of zoologist are you anyway?

Certain scenes in the 1960 film truly captivate viewers. For instance, there is the titanic struggle between the cool looking caiman alligator with recurved 'Angilas-like' brow horns, lateral spines and a vertebral sail, versus the aforementioned pseudo-brontosaur (actually a monitor lizard). As contrived as it all was and horrific, I simply could not turn away! While Ed Malone (David Hedison) and Jennifer watch nervously from behind a rock in a split screen, the two 'enormous' dino-monsters wage a fight to their death. As in the 1925 version, the bronto-monster, battling for its life, falls over the edge of the cliff, although this time in the 1960s film, its assailant—the predatory alligator, or "alliga-saurus"—tumbles as well. This rather conventionalized 'dinosaurs fighting to the death, only to plummet together over a cliff edge' sequence had recently been choreographed for Toho's *Gigantis, the Fire Monster* (1957). Footage from the 1960s *Lost World* scene's cliff edge battling dino-monsters sequence was excerpted and spliced into another Irwin Allen production—a 1964 *Voyage to the Bottom of the Sea* episode, "Turn back the clock," in which David Hedison assumed his role as the *Seaview's* 'Captain Crane.' This particular *Voyage to the Bottom of the Sea* episode also borrowed plot elements from another dino-monster flick, 1957's *The Land Unknown*, in which dinosaurs are paradoxically discovered in a verdant Antarctic setting.

Speaking of 'fire monsters,' Irwin Allen incorporated a spectacular creature dubbed thusly for *The Lost World's* climax. This was a second monitor lizard, whose added prosthetics rather resembled the aforementioned caiman alligator's, sporting a huge sail on its back, twin forward-pointing horns and a third, shorter nasal horn. The Fire Monster was evidently aquatic, slumbering within a lake bed adjacent to a lava flow, until the explorers clumsily awaken the beast (and the volcano) while escaping cannibals along the most perilous "Wipeout" course ever insidiously devised.

Because a clutch of eggs is found along the path nearby the Fire Monster's lava lair—one of which hatches into Challenger's 'tyrannosaur,' presumably and inexplicably, this dino-monster distortion must also be a (sigh!) tyrannosaur! Embarrassed for Challenger, even I—a mere 6-year old—scorned his wacky conclusion, given that I had just seen the Field Museum's real *Gorgosaurus* skeleton on display (which resembles a smaller version of *Tyrannosaurus*). Thankfully, the giant horned iguana startled by Jennifer Holmes' poodle went unnamed. Incidentally, I've had the pleasure of meeting the last reptile to appear as a faux prehistoric dino-monster, Don Glut's pet tegu lizard, Neecha, appearing in his *Dinosaur Valley Girls*, 1996.

Not counting the famed 1997 *Jurassic Park* sequel, in his *Dinosaur Filmography* Mark F. Berry lists four additional 'Lost World' films completed between 1992 and 2001, which probably most of you have never seen. While some of these are well done and fortified with special effects that in many respects may surpass those achieved by O'Brien, Delgado and, later, L. B. Abbott, why is it that these two older 'classic' versions are those which audiences typically associate "The Lost World" movie title with, instead of the latest, slick versions? Well, perhaps we admire the older artists for their verve. Many of us, now 'oldsters,' enjoy and appreciate those early experimental attempts to recreate realistic prehistoric monsters on the silver screen, even if we realize that the very latest forms of trick photography, now relying on copyrighted computer software packages—with CGI effects flowing so seamlessly and with considerably more accuracy—can outdo the pioneers. Additionally, in the case of O'Brien and Delgado, we appreciate the painstaking, labor of love dedication needed to put dozens of stop-motion puppets through their paces, tediously making miniature dinosaur models seem so alive on camera. *King Kong* (1933) is especially heralded among readers, and *Lost World* (1925) was a sensational, historically significant check-in point leading to their ultimate masterpiece. Nostalgia also enters the equation. For the fantastic movies we acclimated ourselves to, for one reason or another, during our respective youths are those which we find most endearing and feel most allied to in adulthood.

Conan Doyle's tale remains timeless, and doubtless it will be redone many times over. But part of the thrill and enjoyment of the 1925 and 1960 versions—neither of which was entirely 'faithful' to the 1912 novel—is that both were made at times when a great sense of mystery for 'lost' crypto-places of the planet and their indigenous fauna—some of which, when discovered, conceivably could be 'missing' links belonging to bygone ages—still lingered.

Chapter Two—*Three GIANT Venusian Dino-Monsters: Are They Daikaiju?*

In a 1964 *Science* article, "The Non-Prevalence of Humanoids," eminent paleontologist George Gaylord Simpson summarized, "We do not really know [what the atmosphere of Venus is like], and we are thus not severely limited in our conclusions." Therefore, for many decades, in glimpsing Venusian life, science fiction writers were left unbounded as to the limits of imagination. Scientists pondered theories as well, which, given today's factual knowledge, might seem implausible or peculiar. For instance, during the late 19th century, astronomer-popularizers suggested that compared to the other planets in the Solar System, Venus was "most likely to be inhabited by beings with affinities to earthly organisms."

Eventually, scientists considered that Venus' rain-swept, swampy terrain must be geologically younger than Earth – just as Mars was older – because Venus formed from the solar nebula millions of years *after* Earth and Mars. So, it wasn't too far of a stretch for sci-fi writers to populate Venus with Mesozoic-like forms. Accordingly, during the mid-20th century some writers authored stories about strange dinosaurian creatures, conceivably *daikaiju*[1] of our sister planet, three 'species' of which will be presented here.

Henry Kuttner's April 1940 short story, "Beauty and the Beast," (*Thrilling Wonder Stories*, pp.67-76) is first rate sci-fi out of the classic pulp era. An astronaut returning from a mission to Venus crashlands in Earth's remote country wilderness. Then, farmer Jared Kirth recovers items from the cindered spacecraft. Kirth unscrupulously reclaims a diary with photographs taken by the astronaut, a strange gleaming jewel (which is an egg!) and pocketfuls of seeds – derived from Venus. Apparently, there was intelligent life on Venus – beings capable of building immense primeval looking structures, now in ruins. Now, only fields of orchid-like flowers grow on the surface. Save for the plant life, Venus appears dead.

Not reporting the deceased spaceman or whereabouts of the wreckage, nor the relics retrieved from within, Kirth returns with the 'loot' to his farm. The egg hatches in an incubator—a lizard-like form emerges that resembles a "miniature kangaroo," a reptile with a long tapering tail. Due to Earth's more ameliorative surface conditions, the lizard grows exponentially, soon outstripping his living quarters, an outside sty pen. Now known as "Beast", the voluminous Venusian is carnivorous ("or at least omnivorous"), fed with slops and swill. Kirth intends to sell the gentle Beast to a circus. Meanwhile, planted Venusian seeds are growing, reflecting a horticultural "riot of color" from his garden.

As an aside, for several decades paleontologists believed large theropodous dinosaurs leaped like kangaroos. In fact, author Kuttner (pp. 124-125 of the cited *TWS* issue) envisioned his tragic, misunderstood heroic creature as an intelligent pseudo-dinosaur.

Soon, Beast's ancient racial memories begin to stir as he deciphers his alien origin. Although he cannot speak, Beast develops an understanding of human language. With the awakening of his vast mind and intellect, Beast further recognizes the deadly flowers Kirth has grown, comprehending inherent danger. Certain planetary doom! For on Venus, millennia before, the flowers pollinated a plague that destroyed the world of his ancestors.

But when intelligent, logical Beast tries to communicate the danger to Kirth, the stupid man doesn't understand. Instead, he wounds Beast with a shotgun. So, Beast stumbles off into the mountains, remaining hidden while devising a plan for aiding mankind, before it is too late. Physiologically stimulated by the Sun's "actinic rays," (here, unblocked by Venus' dense atmosphere), Beast grows to Godzillean proportions. Now fully mature, Beast approaches Washington DC. Military planes bomb him, while buildings topple. A science fictional "heat ray" situated atop a derrick fires a deadly blast upon Beast as its paw clumsily smashes the Capitol Building's dome (i.e. the dramatic scene captured by artist Howard V. Brown on the April 1940 *TWS* fanzine cover). Beast lies dying even as pollen within the pretty Venusian flowers—the story's titular "Beauty"—germinates. The pollen carries a virus deadly to all life, which as the Beast frustratingly fails to communicate with a dying gasp, will destroy mankind and all other animal life on Earth.

Author Kuttner conceived "Beauty and the Beast" after visiting a friend's greenhouse where he was warned not to touch a pretty flower's petals that secreted a toxic chemical. Kuttner wrote, "Somehow the human race has always associated ugliness with evil and beauty with good… in a certain California museum there's a restoration of a dinosaur, and … I've always felt a sneaking liking for the fellow. He's ugly as sin, but he's a brontosaurus, and they never had much of a chance with the ferocious carnivores of prehistoric times. I began to wonder what might be the experiences of an *intelligent* [my italics] dinosaur set down in modern civilization. Does man value superficial beauty above intrinsic worth?"

Given its size, 'mysterious nature' and brain power, Beast is a probable *daikaiju*, per J. D. Lees' definition.[2] Arguably, Beast was a conceptual precursor for Toho's later cinematic creations, as well as Ray Harryhausen's "Ymir" Venusian creature—featured in *Twenty Million Miles to Earth* (1957)—which also emerges from an egg as a tiny lizard-like hatchling. *"Twenty's"* movie script was written by Bob Williams and Christopher Knopf (and later novelized by Henry Slesar in 1957). Here, however, I cannot prove that Ymir was intentionally derived from Beast.

Before moving on to our next candidate *daikaiju*—(the "Whip")—let us feast a bit more on the distinguished career of Howard Vachel Brown, who painted the titular "Beast" dino-monster in its precociously iconic scene on the *TWS* April 1940 cover. Brown (1878-1945), a Chicago Art Institute alumnus, was a well-known illustrator, specializing in cover art for numerous magazines, trade journals and sci-fi fanzines. His artistry graced covers of *Scientific American, Radio News, Science and Invention,* and *Everyday Engineering*, for example, as well as many *TWS* and *Astounding Stories* issues. As stated on a website, Brown "…probably had as much to do with shaping our concept of the future as any three science or science-fiction writers you could name. What the futurists thought, Howard put into pictures so we could immediately see what it could be like."

Extraterrestrial dinosaurs are staple sci-fi characters and traditional props today, although the Venusian variety enjoyed their heyday over half a century ago. Before "Beast's" 1940 entry, *TWS* readers were treated to Arthur K. Barnes' "The Hothouse Planet" (Oct. 1937, pp. 12-29). Even today, Brown's cover illustration of the story's most ferocious monster—a long-eared, pseudo-theropod, rather resembling "Gorgo," menacing men fleeing back into their spaceship—certainly seems 'thrilling.' This tale was inspired partly by professional zoologist/adventurer, Frank Buck, who decades ago gained notoriety as a big game wildlife trapper. Except in "Hothouse Planet," the interplanetary animal catcher is a woman – Gerry Carlyle, who doesn't fully understand the ramifications of the Venusian ecosystem.

Unfamiliar with Venus' challenging dense, foggy atmosphere, adverse climate and swamps, 'catch-'em alive' Carlyle relies on Tommy Strike in planning her excursion to trap 'murris.' True to her 'catch' phrase, she wants to bring them to London's Interplanetary Zoo. Murris are simian-like creatures who are exceptionally tricky if not impossible to remove from their hives, and along the way, Carlyle, Strike and other safari team members also encounter peril from Venusian 'gill-men.' But it is the dinosaurian "Whip" which concerns us here.

When the long-eared Whip attacks, Carlyle also wants to catch this dangerous specimen alive. But the Whip is the "…most terrifying of all products of Venusian vertebrate evolution. …Fully fifty feet the monster towered into the mist, standing upright on two massive legs reminiscent of the extinct terrestrial *Tyrannosaurus* rex. A set of short forelegs were equipped with hideously lethal claws; the head was long and narrow like a wolf's snout, with large ears and slavering fangs. Everything about the nightmare creature was constructed for efficient annihilation…"

Whip earned its name from its 50-foot long, 'razor-edged' ropy tongue, which it can unleash lightning-fast to seize prey. In fact, one team member is slashed and ends up bleeding to death! Shortly after this misfortune, the Whip is subdued. As Barnes explained in the 1937 *TWS* issue, "As in the past, I have tried not to invent things that are too far removed from life as we know it on this Earth. The weird 'whip,' for instance, has its Earthly counterpart not only in the ant-eater, but in the sphinx-moth, which sports a hose-like tongue a good deal longer than its body." The Whip is sufficiently large to be *daikaiju*, although, critically, it lacks intelligence beyond what might be expected of any self-respecting Venusian 'dinosaur' – it is merely controlled by instincts.

Twelve years later, "Hothouse Planet" was reprinted in *Startling Stories* (Sept. 1949); this time the Whip featured on the cover, illustrated by Earle Begley, looks more like a terrestrial tyrannosaur. Indeed, in 'Carey Rockwell's' 1954 novel, *The Revolt on Venus*, Tom Corbett and his *Polaris* pals risk their lives in hunting a vicious Venusian Tyrannosaur (which is not a *daikaiju*). The Venusian dino-theme persists to present day, as in S. M. Stirling's 2006 novel, *The Sky People*. Author Barnes' Gerry Carlyle tales were recently compiled in an e-book, *The Complete Interplanetary Huntress: The Adventures of Gerry Carlyle* (Dec. 2007).

Finally, we have in 'G. H. Irwin's' "Lair of the Grimalkin," (*Fantastic Adventures*, April 1948, pp. 8-53), a true monster quest. (Note – 'Irwin' was a pseudonym for Richard S. Shaver.) What begins as a humdrum geological expedition to verify a vast platinum ore reserve in the forbidden jungle recesses of Venus, builds to a battle for human survival against an enormous 3-eyed winged dragon – the Grimalkin (expertly illustrated by Malcolm Smith). It turns out one of the ore seekers (a man named Farrar) all along intended to hunt this legendary creature – if it actually existed, which of course it does.

Rodan-sized Grimalkin is the last of its kind (on Venus), although in one curious passage there's indication that it may have arrived there eons ago after having "winged across worlds," (like Toho's Ghidorah). Winged Grimalkin possesses an uncanny telepathic sense (known as "yarva") allowing it to command a stunted and subservient bipedal race of Venusian amphibi-humanoids known as the "Bunae." After so many millennia, the Grimalkin and Bunae relationship has become symbiotic (if not parasitical, favoring Grimalkin's control).

Irwin's tale borrows themes from (both) *The Lost World* and *King Kong*. First, we have the idea of the intrepid explorers rafting along a dark Amazonian-like jungle river, leading through uncharted wilderness toward a dreadful destination inhabited by a monster. Along the way, hideous aquatic reptiles and snakes must be avoided or killed. Secondly, we immerse ourselves in the 'Kong-ish' theme of the beautiful maiden, held captive by the native Bunae. She is destined to be sacrificed to (and eaten by) monstrous Grimalkin. Thirdly, not unlike both novel (pterodactyl) and movie (brontosaurus) 'Lost World' versions and as in *King Kong,* the vengeful monster tracks fleeing explorers back to (Venusian) civilization, where it wreaks havoc upon mankind – at the human colonists' domed cities built near the industrial mines.

Grimalkin is personified evil incarnate. Its mighty roar and hurricane force gust winds generated by beating of its titanic wings seem hellish. Its regenerative powers seem unbounded, as it even heals following a vampiric impaling engineered using an enormous stake fashioned from a tree trunk (a scene illustrated by Smith)! And, accentuating its nightmarish nature, Grimalkin is an intelligent creature, capable of telepathically mind-controlling Venusian sentients (Bunae), as well as – to some degree – humans, using yarva.

So, would Grimalkin, like Kuttner's Beast, qualify as *daikaiju*? My only quibble about naming both Beast and Grimalkin as bonafide *daikaiju* concerns their abilities to withstand human military might. In Kuttner's case, Beast seems invulnerable to ordinary rifles and (pre-atomic) bombs dropped from military planes (i.e. on Earth). In fact, puny humans must resort to a science fictional heat ray laser blast to destroy the tragic monster. Thus, Beast probably is *daikaiju* (even though it never appeared, per se, in a Japanese sci-fi flick). Conversely, while Grimalkin shrugs off high-caliber bullets, and survives impaling of an almost unimaginable scale, it succumbs to a single bomb dropped from a plane. (We don't know exactly what kind of bomb, however.) I think its yarva powers outweigh its relative degree of vulnerability to human weaponry. Therefore, to me, both Beast and Grimalkin are Venusian *daikaiju*, whereas Whip decidedly is not.

'Exo-dinosaurs,' sci-fi dinosaurs from other worlds, remain popular denizens of fantasy fiction and film. It's hard to draw the line in such cases as to what's a 'dinosaur' vs. not (or maybe a mere 'exo-reptiloid') because exo-dinosaurs are inherently alien. Few would label Ghidorah as a dinosaur, although it probably is, well, an exo-planetary 'drago-dinosauroid.'

The fact that so many planets beyond our Solar System have already been discovered (429 and counting, as of 2012!) suggests that life may be more common in the universe than skeptics might believe.[3] No doubt, as sci-fi writers and dreamers learn more about these and other exo-planets as they come to light, non-terrestrial dinosaurs and other prehistoria will populate these extraterrestrial ecosystems too. Shall the prehistoric *daikaiju* phenomenon spread beyond Earth and Venus, as once imagined by 'Alexander Blade' (pseudo. for Edmund Hamilton) in his whimsical essay "Stories of the Stars ... Aldebaran," illustrated by Frank R. Paul (*Fantastic Adventures*, Dec. 1945, pp. 178 and back cover)? What would mankind discover should we trespass upon these far flung, forlorn worlds?

So, there you have it! Prototypical Venusian variants of Godzilla, Rodan and even a kind of 'Gorgo'– all there in print and illustriously depicted years before Toho's *Gojira*.[4]

Notes: (1.) "*Daikaiju*" is a Japanese sci-fi term meaning "giant mysterious beast." (2.) In *G-Fan* no.78 ("What is a *Kaiju*?"), J. D. Lees defined *kaiju* as creatures that are very large, menacing in aspect, and intelligent beyond mere animalistic tendencies. (3.) However, for a great modernistic science-fictional perspective of this concept, see p. 93 in the July 2010 issue of *Asimov's Science Fiction*, in Robert Reed's story titled "A History of Terraforming." Furthermore, since astrophysicists have recently proposed that evolution of complex life on planets lacking an axial-stabilizing satellite, as manifested in our Earth-Moon system, would be less likely, the odds of finding 'exo-dinosaurs' on other worlds may not be so high after all. (Jennifer Heldmann televised in "Earth Without the Moon," *Nat. Geog.*, 2010) (4.) Although it was not here intended to discuss giant Venusian dino-monsters portrayed in other media, such as film, it may be of interest to mention that dinos also appeared in the 1962 Russian *Planeta Burg*, as well as the NBC television series, *Tom Corbett, Space Cadet* (e.g. 4/15/1952 episode). Exo-dinosaurs were featured in other *Tom Corbett* episodes too.

Chapter Three—*Time Steps Aside*

Jurassic Park aside, eventually, most dinosaur time travel adventures wind up in the Late Cretaceous period. Why? Partly cuz' that's where *Tyrannosaurus rex*, three-horned *Triceratops* and even more demonic creatures known collectively as 'raptors' dwell! But mainly because that's where Ray Bradbury staged the most influential dino- time travel tale ever written—"A Sound of Thunder" (1952).

Although H. G. Wells's 1895 novella *The Time Machine* is regarded as the landmark time travel tale, Ray Bradbury's beloved "A Sound of Thunder" became the launching point for most time travel stories involving prehistoric life. Bradbury's threefold legacy is apparent because following "Thunder," (1.) the usual Mesozoic temporal destination is the Late Cretaceous (as opposed to, say, the Jurassic) (2.) *Tyrannosaurus* or an equivalent apex predator, dino-monster is often encountered by the time travelers, and (3.) there is often a 'low tech' time safari mission objective which backfires in extraordinary fashion. While developments in modern physics and recent paleontological discoveries may have altered the scope of dino-time travel fiction since "Thunder's" first publication in 1952, in homage, core elements of Bradbury's classic "A Sound of Thunder" are often cleverly retained.

As a young man, Bradbury absolutely delighted in dinosaurs, having attended the 1933/34 Chicago World's Fair with his parents where he saw life-sized mechanical restorations of *Tyrannosaurus, Triceratops* and other dinosaurs. Besides the writings of Edgar Rice Burroughs and Jules Verne, who incorporated prehistoric fauns into such novels as, respectively, *The Land That Time Forgot* (1918) and *Journey to the Center of the Earth* (1864), Bradbury also was undoubtedly excited by Wells's *The Time Machine*. At an even younger age, he'd feasted his wondering eyes on a silent movie, *The Lost World* (1925), offering "metaphors … so powerful that they changed my life."

So, wouldn't one naturally presume that Bradbury's "A Sound of Thunder" was naturally derived from Wells's 1895 classic? Despite outward, peripheral similarities—such as use of a machine to move through time and a confrontation with horrible monsters—Wells's and Bradbury's aims couldn't have contrasted more. But machines and monsters are where apparent similarities end, and where Bradbury began to fashion a 'time-honored' legacy of his own.

Wells perceived *The Time Machine* as an "assault on human self-satisfaction." He wrote, "In my student days we were much exercised by talk about a possible fourth dimension of space; the fairly obvious idea that events could be presented in a rigid four-dimensional space time framework had occurred to me, and this is used as the magic trick for a glimpse of the future that ran counter to the placid assumption of that time that Evolution was a pro-human force making things better and better for mankind." For as the time traveler discovers in the far-off future, 802,701 years hence, the course of evolution has split humanity into two species—the hideous subterranean Morlocks, and the gentle, childlike Eloi, who are subjugated by the Morlocks. And at an even more distant future, Man ceases to be recognizable. Surely, for its time, this was thinly veiled social commentary on Victorian class structure.

Unlike Wells and despite "Thunder's" prehistoric setting, Bradbury's theme was quite apart from 'evolutionary.' Instead we find that our eternal conflict with dinosaurs, evident in time safari tales, is a reflection of Man's self-destructive tendencies and disregard for the environment.

In a sense, dinosaurs embody one of science fiction's most conventional themes—time travel. By luring men's fantasies to misty, 'lost' primordial worlds, traditionally, dinosaurs have represented deep geological time. Yet, from any dinophile's perspective, arguably, what good are time machines anyway, unless one may rely on them to travel backward in time to witness a primeval bestiary?

Curiously, time travel involving dinosaurs and other prehistoric animals weren't common for many decades. Partially this is because the prospect of finding dinosaurs alive in the present-day world, although in remote, primeval settings seemed more romantic, if not downright plausible. Certainly, it must have seemed easier to suspend disbelief simply by chartering a voyage to some unexplored region of the world, akin to Kong's 'Skull Island,' Burrough's 'Caspak,' or Arthur Conan Doyle's lost Amazonian plateau where dinosaurs and other prehistoric fauna may be encountered, as opposed to finding the wherewithal of building a seemingly impossible device—a Wellsian time machine. After all, what could be *more* improbable than escaping into the past?

Many of you would recollect that in Bradbury's "Thunder," intrepid hunters commission guides on a safari to the Cretaceous Period where they can shoot a *Tyrannosaurus rex*. But the travelers cannot interfere with natural surroundings and must confine their footsteps to a stasis 'Path' leading from their time machine out into the primeval environment. For straying from this metaphoric Path could result in changes which would "… add up infinitesimally. A little error here would multiply in sixty million years, all out of proportion." Therefore, every precaution is taken to prevent disruption of the past to preserve the future. Soon, out of the jungle strides the mightiest, most horrifying Mesozoic monster—magnificent *Tyrannosaurus*.

"It came on great, oiled, resilient striding legs. It towered thirty feet above half the trees, a great evil god, folding its delicate watchmaker's claws close to its oily reptilian chest. Each leg was a piston, a thousand pounds of white bone, sunk in thick ropes of muscles sheathed over in a gleam of pebbled skin like the mail of a terrible warrior … Its armored flesh glittered like a thousand green coins. The coins, crusted with slime, steamed. In the slime, tiny insects wriggled, so that the entire body seemed to twitch and undulate, even while the monster itself did not move. …The Monster … lunged forward with a terrible scream."

Petrified, one hunter slips off the Path, crushing a single butterfly, whose silent scream—(ironically) the tale's real 'sound of thunder'—echoes down through the corridors of time to an irrevocably altered future. In Bradbury's story, safari guides know when to hunt the monster only because they've already observed its natural death during a prior trip. However, they're prohibited from ever encountering themselves in that setting on subsequent voyages because "Time steps aside," a key point we shall return to.

Primarily remembered for its 1960 projection onto the silver screen, Wells's *The Time Machine* is perhaps the most famous time travel tale involving the remote future, while Bradbury's precocious "Thunder" is the most beloved story coupling the deep past with dinosaurs. Despite the perfection of Wells's classic 1895 tale, most dinosaur time travel tales owe their genesis to Bradbury's brilliance. According to Bradbury, "A Sound of Thunder" was little more than a serendipitous 'what if' experiment. "What if we could travel in time and run back to hunt the prehistoric beasts? This last was an experiment I tried in 1950. I simply sat down to my typewriter one morning with no idea where I would wind up, and hammered together a Time Machine, and shot my hunters back a few million years to see what would happen. Three hours later, after a butterfly had been stepped on, making it one of the first, and unconscious, ecology storied, the tale was done, the beast slain, and all political history changed forever." Imagine—Bradbury composed his timeless time travel story in only three hours!

Wells, striving for a convincing level of technological realism, directed time travel toward Man's future, whereas Bradbury, writing in a more enlightened period with respect to modern physics, pointed to the nostalgic abyss of deep time. By introducing the idea of the time safari, Bradbury—who never specialized in technological or scientifically (including evolutionary and paleontologically) founded fiction—illustrated mankind's technological folly. Wells's *The Time Machine* was written before the Einsteinian revolution in physics of the early 20th century, when there was much hand-wringing over downright impossible, paradoxical scenarios that would result from hypothetically traveling backward in time. However, for reasons that go beyond the present scope, physicists finally accepted the plausibility of time travel into the past as theoretically possible, by 1949.

As opposed to dealing with 'hard' science fiction, Bradbury confessed that, instead, many of his early writings probed the 'psychological aura of loneliness,' as in his 'other' famous dinosaur story, "The Fog Horn" (1951). Furthermore, in the early 1950s, Bradbury admitted "I don't like what science is doing to the world …I think science is a good thing to escape from." As summarized by Isaac Asimov in 1980, "Bradbury… created moods with few words. He wasn't ashamed to tug at the heartstrings and there was a semi-poetic nostalgia to most of those tugs. He still views science with intense suspicion, but supports the space program enthusiastically, largely because … he finds it poetic." While Bradbury may have seemed rather anti-scientific to some readers, Wells reveled in exploring the consequences and impact of scientific advance upon human society.

Beyond technological hurdles, stepping into prehistory is both foreboding and forbidden. Traveling through time into the deep past signals misery and death; metaphorically, the past is already dead and buried. It's like crossing the River Styx into Hades, or into the shadowy Underworld. Moreover, in the Mesozoic, coprolite happens! We see this illustrated in stories written by Frederick Pohl ("Let the Ants Try" (1949), Arthur C. Clarke's "Time's Arrow" (1950) and Asimov's "Day of the Hunters" (1950). Which collectively perhaps, paved the way to Bradbury's classic. By degree, dinosaur hunting of one kind or another increasingly became an undercurrent theme of time travel to the deep past, although it was Bradbury who put it all together, inaugurating the idea of the time safari, and coupling dino-time travel with paradox.

Whereas Pohl's, Clarke's and Asimov's stories were printed in the traditional home for sci-fi tales—pulp magazines—Bradbury's "A Sound of Thunder had wider exposure, first appearing in the slick *Collier's* (June 28, 1952). Thereafter, "A Sound of Thunder" has been reprinted numerous times, including *Playboy* (June 1956), and as the dawn of the modern age of dinosaur fandom peaked, his story also appeared in numerous anthologies including Bradbury's *Dinosaur Tales* (1983). Such a rich publishing history pattern is highly unusual for a science fiction short story. Why has it aged so well? Asimov suggested "People who didn't read science fiction, and who were taken aback by its unfamiliar conventions and its rather specialized vocabulary, found that they could read and understand Bradbury." Perhaps also its precautionary, culturally relevant message—to interfere with or destroy our heritage or harm nature often has dire consequences for our future—really dug home. After all, at least, "A Sound of Thunder" is an ecological story.

But evidently, there was another key influence as well behind "Thunder's" creation, that is, besides ecology and dinosaurs. For another clue rests in another Bradbury statement. "One day… I said to myself, 'This is the day. I'm going to sit down and write myself a story about dinosaurs. Let's form a safari … I've always been interested in people like Frank Buck and Martin Johnson and his wife, when I was a child, and going to Africa. So, let's make up a safari in time and send them to hunt a dinosaur."

Who were Frank Buck and the Johnsons?

Older film buffs may recall a vintage generation of wildlife movies and natural history serials from the 1920s through the 40s, with titles like *Jungle Menace* (1937), *Among the Cannibals of the South Pacific* (1918), *Jungle Adventures* (1921), *Simba* (1928) and *Congorilla* (1932)—the first 'talkie' made in Africa, or even a comedy co-starring Abbott and Costello—*Africa Screams* (1949). Such film fare captured the heady atmosphere of exploration on the "dark continent" and other relatively 'uncivilized' places of the planet. Frank Buck (1884-1950), who produced *Jungle Menace* and appeared in the Abbott and Costello film also wrote books about his travels, such as *Bring 'Em Back Alive* (1930), a phrase which became his motto. Buck transported African animals to zoos and exhibitions across the United States, such as his Jungle Camp at the 1933/34 Chicago World's Fair. Osa (1894-1953) and Martin Johnson (1884-1937), who produced the other films listed above, popularized the 'camera safari.' Osa's best-selling, *I Married Adventure*, an account of their travels abroad, was published in 1940. In retrospect, the Buck/Johnson legacy was wildlife conservation, a theme which must have impressed itself on the young Bradbury's fertile mind.

So, whether or not Bradbury's "Thunder" resulted from a serendipitous vision or, instead, from fond, vicarious recollections of wild African adventures, the master poet had profound influence on his contemporaries as well as a later generation of dinosaur writers. One of the first to tackle the time safari theme following Bradbury was L. Sprague de Camp. If you recall, in Bradbury's "Thunder," Time 'steps aside' when an individual attempts to revisit himself in the past. For according to the laws of physics as understood before 1949, it would have been impossible for any guide to observe the exact moment and place of the tyrannosaur's demise and then return exactly to those space-time coordinates to shoot the same dinosaur without encountering himself—(unless when time "stepped aside" they had entered an alternate Mesozoic universe).

L. Sprague de Camp admitted how "… in 1954, I read a story by a colleague, telling of a hunt for dinosaurs by time machine. I had been a dinosaur buff since the First World, Kaiserian War … Fate denied me the lifetime role of paleontologist; but I still thought I saw egregious scientific errors in my colleague's story. These irked me to the point of writing my own story of dinosaur hunting by time machine, "A Gun For Dinosaur," trying to show how it should be done." So, instead, in de Camp's 1956 classic tale, concerning a tyrannosaur hunt recounted by safari guide Reginald Rivers, revisiting a slice of time already occupied by oneself in the past results in a horrific explosion of space-time forces. In contrast to Bradbury's space-time stabilizing "Path," de Camp's explosion resolved paradox, preserving a 'well-run universe' without adverse consequences for the future.

Bradbury's legacy? Following Bradbury's "Thunder," time travelers began wending their way to the Mesozoic and other geological time periods more so than ever before. According to physicist Paul Nahin, after publication of D. R. Daniels's 1935 story, "The Branches of Time,"—which didn't involve dinosaurs—"…the splitting time line concept quickly became part of standard science fiction lore, and writers could use it without having to offer a lot of explanation." Bradbury's precocious 1952 tale, however, was the first to meld the time safari idea with dinosaurs and multiple realities, where time steps aside.

In Bradbury's stead, for example, writers Clifford D. Simak (*Mastodonia*, 1978), Robert J. Sawyer (*End of An Era*, 1994), Geoffrey A. Landis ("Embracing the Alien," 1992), Will Hubbell (*Cretaceous Sea*, 2002), Steven Utley ("Walking in Circles," 2002), Stephen Baxter (*The Time Ships*, 1995), David Drake (*Tyrannosaur*, 1993), and Diane Carey and Dr. James I. Kirkland (*Star Trek: First Frontier*, 1995) have all in one way or another paid homage to "Thunder's" unheralded dinosaurian 'thunder.' Another time safari story, David Gerrold's *Deathbeast* (1978), featuring its titular, demonic tyrannosaur, inspired a film, *Planet of Dinosaurs* (Cinema Dynamics, 1978). True—while some of these titles were published during the post-*Jurassic Park* (1990) era, such stories themselves arguably owe more to Bradbury's genius than Michael Crichton.

Through my detailed researched of science fiction's dinosaur/time travel theme, it would seem as if the motion picture industry—milking the *Jurassic Park* dino-DNA cash 'cowasaurus' for all its worth—has been challenged by the dizzying variety of dinosaur time travel ideas evident in science fiction literature. Yet in film, we have only a handful of stories to reflect upon, such as Karol Zeman's *Journey to the Beginning of Time* (1956) and a classic *Twilight Zone* episode written by Rod Serling, "The Odyssey of Flight 53," first aired on February 24, 1961, in which an airline jet is mystically transported to the Jurassic Period. However, time travel elements in such filmed fare aren't nearly as profound as "Thunder's" remarkable premise. Even *Planet of Dinosaurs* was transformed into an outer space thriller rather than the time travel adventure Gerrold envisioned.

With respect to science fiction literature, "Thunder" has been more influential than Bradbury's 'other' famous dinosaur story, "The Fog Horn" (1951)—projected onto film as *The Beast From 20,000 Fathoms* (1953). Furthermore, "A Sound of Thunder" arguably is as famous as Wells's 1895 novella, known chiefly through a 1960 George Pal film adaptation—*The Time Machine*. Readers may recall that Bradbury's "Thunder" was televised for an episode of *Ray Bradbury Theater,"* first aired on August 11, 1989. In August 2005 the disappointing *A Sound of Thunder*, directed by Peter Hyams, was released taking extreme liberties with Bradbury's story.

Bradbury's "A Sound of Thunder" is perhaps *the* dinosaur science fiction 'cult' time travel tale most recognized by readers, possibly the only one generally recalled. Regrettably, and rather incredibly, no 'thunderous' movie production was ever based on the essence of "Thunder" during that marvelous era of 1950s and 1960s dino-monster B filmdom.

Ironically, by directing his time machine into prehistory, Bradbury beckoned to the future of sci-fi writing. That Bradbury surely is an 'illustrative man.'

Chapter Four—*Revisiting the Lost Continent*

"… in 1951 the only genre subjects with proven commercial appeal were dinosaurs and rockets. What could be more natural than to devise a script combining them – an updated 'lost world' … with Space Age trappings." (William Fogg in *Filmfaxplus* 105, Jan/March 2005, p.109)

No, this article isn't about the mythical "lost" continent of Atlantis. Nor will this story outline Marc Cerasini's audiobook, *Godzilla and the Lost Continent* (1999). Instead, it's about a misguided atomic rocket, and a strange island inhabited by real (or 'reel') dinosaurs. Here we're delving back to a 'pre-Godzilla' era when people were learning to cringe in "… fascination and fear of the atom." (Mark F. Berry, *Dinosaur Filmography* 2002, p.237). I first saw this film during the early 1960s, before I could fully appreciate its (perhaps not too) subtle messaging.

Something very odd happened to the traditional "lost world" story during the early 1950s, as signified in Lippert Pictures' 1951 movie, *The Lost Continent*. Decades prior, in 1912, Sir Arthur Conan Doyle conceived his remarkable, quintessential tale, *The Lost World,* involving an expedition to the fabled lost Amazonian plateau infested with prehistoric animals ("curupuri"). It was conceived as an engaging adventure story intended for "… the boy who's half a man, Or the man who's half a boy." However, during the mid-20th century, following the scourge of nuclear weapons development and proliferation during the onset of the geopolitical Cold War, *Lost Continent*'s appeal instead promised this odd juxtaposition: "Thrills of the atomic-powered future! Adventures of the prehistoric past." Yes, quite an emphasis shift in the oft-told tale of discovering a lost land where anachronistic, living prehistoric relics dwell!

As zoologist and "Lost World" booster John Lavas aptly opined:

"Today, as in 1912, the concept of a 'lost world' still holds the same fascination for readers as do unsolved mysteries, hidden cities, and as yet undiscovered animals. Our curiosity will always be aroused not only by the unknown but perhaps even more by the unknowable… these scenes still prove evocative because they were based on the real travelogues of explorers as they ventured into the uncharted territories of the New World for the very first time. *The Lost World* will, without doubt, continue to entertain readers of all ages for generations to come." (p.lvi)

So, what remains the significance of 1951's largely forgotten *The Lost Continent* today? How and why did Conan Doyle's original satisfying, armchair-reading concept evolve into something far more dire along the way? Rather than left feeling "lost," let's delve into it now.

Of course, through the decades since, there have been numerous remakes (inspired novels *and* films) capitalizing from Conan Doyle's perfected 'Lost World' theme, including 1933's *King Kong*. Yet, in striving for its sinister originality, 1951's *The Lost Continent* ("*TLC*") reflects early ripples of a new wave stemming from the old lost world theme, introducing cold war specter, nuclear age angst. Intriguingly, *TLC* established ties between dinosaurs/dino-monsters and nuclear weaponry and the Cold War, years before other directors and script writers made such connections, as in 1953's seminal *The Beast From 20,000 Fathoms* and 1954's *Gojira* (e.g. presuming that we may loosely refer to the Rhedosaurus and Godzilla as filmic, albeit 'dinosaurian' mutations). However, while TLC's dinosaurs have been exposed to natural radiation, they do not vengefully (metaphorically so) invade civilization.

What may be said of *The Lost Continent*[1]? For starters, here's an interesting notation found in George E. Turner's (et. al.) reference book, *Spawn of Skull Island* (2002): "*The Lost Continent* (1951) – Cesar Romero toplines this *Lost World* knock-off from the low-budget producer Robert Lippert. The dinosaur footage *may* represent a salvage job from an unrealized project called *The Lost Atlantis.*" (p.53) To me, that review sounds too harsh, and we'll return to that "Lost Atlantis" remark later. Even more mocking criticism may be found in Volume 2 of Bill Warren's *Keep Watching the Skies* (McFarland, 2010). However, it would be more proper to state that this film is an interesting reflection of its time.

Accordingly, I'll *briefly* outline signature plot elements, emphasizing essentials central to my thesis. When a launched test rocket, powered by an atomic-powered engine, crashes on a South Pacific island, a team of scientists are sent out to retrieve components of said engine—of course, for purposes of top level national security. This particular island has two unusual twists though. First, its geological, volcanic bedrock is highly radioactive due to presence of concentrated uranium. Fortunately, because uranium radiation is "mysterious," the rocket's guidance system is presumed to have been somehow "inevitably" attracted toward a high-level radioactivity field emanating from a plateau on the island, before running out of fuel. The search airplane's instruments conveniently go haywire just as they near this plateau, referred to by a native islander as a "sacred mountain." And the island is about to blow due to the strange volcanic vapors force interaction with radiation, ultimately unleashing an energy equivalent to a "stockpile of hydrogen bombs." Potentially, this explosion would then seem to be on the order of Krakatoa's.

Five of the six crash team members survive a really tough climb made along narrow rock ledges, surmounting the island plateau (e.g. the titular "continent"), cast in an eerie green radioactive glow. (No, they haven't arrived in Oz.) Here they discover living representatives of the Mesozoic Era – an assortment of (non-avian) dinosaurs, surviving in "…a world that hasn't existed for millions of years… as if time forgot this place." (That's partially a nod to Edgar Rice Burroughs' 1918 novel, *The Land That Time Forgot*.) Fortunately, the radiation isn't considered dangerous to human health, Rostov remarks, "…but if the uranium in these fields were *refined*, we'd have been dead hours ago." A Geiger counter registering radioactivity facilitates tracing of their wayward rocket inland. The missile is eventually located, although another team member is killed by a roving *Triceratops*.

One team member, the sinister-seeming "Rostov" played by John Hoyt, happens to be Russian, thus inducing a spattering of paranoiac Cold War insinuations. At one point "Major Joe Nolan" (Cesar Romero's character) implies that buttons could be pushed to launch more rockets so as to punish those who have corrupted countries. (The prospect of pushing those buttons on certain kinds of rockets indiscriminately then instilled fear.) Nolan refers to how nuclear bomb testing had altered the atmosphere, continuing, how man might "… destroy the one (i.e. world) we've got … nightmares." So here we have a peculiar dinosaurian denizen juxtaposition, melded with a strange 'natural' interaction of volcanic and radioactive forces, the latter two constituting a quasi-hydrogen bomb upon which dinosaurs—soon to become vaporized, extinct—live. Conan Doyle couldn't have conceived anything quite like this!

In his *The Stop-Motion Filmography* (1999), Neil Pettigrew focuses on the film's animation effects, stating, "Although an interesting independent attempt to create stop-motion effects, *Lost Continent* ultimately only serves to emphasize the genius of (Willis) O'Brien." (p.424) Let's discuss these effects a bit further. First, there's a *Brontosaurus* which chases one poor soul up a tree, perhaps in homage to a similar scene in RKO's *King Kong* (1933). Then there's the highly combative *Triceratops* duo whose expressions have been criticized as "cute," and a winged pterodactyl briefly appears too that's quickly dispatched by a rifle shot. We must wait patiently (nearly an hour into an 83-minute picture) before arrival of the dino-monsters. But, in my opinion, the wait is fairly rewarded. Despite what was implied in promotional advertising, there isn't a theropod/tyrannosaur featured in *TLC*.

The dino-monster animation has been described as amateurish and choppy. Furthermore, as stated by Berry in 2002, "The identity of the animator, as well as the maker of the models, seems … lost to history." (p.237) Yes, dark mystery remains as to who actually performed *TLC's* special effects, although this circumstance hasn't deterred some from theorizing. As early as 1999 (and again in 2002), George E. Turner suggested that some of *TLC's* stop-motion dinosaur effects might have been borrowed from a reel prepared a few years earlier for an unfinished, aborted film titled *The Lost Atlantis* (1938-1940). The intriguing idea was supported in 'definite-maybe' fashion later by Mark F. Berry in 2006, and further reported by myself in 2009. Yet through personal e-mail communications received (most recently in 2016) from Don Glut, there seems little credence for a possibility that any dino-monster footage prepared for "Lost Atlantis" (e.g. either *Triceratops* or theropod dinosaur) later wended its way into *TLC*.[2] As an aside, however, footage of one of TLC's *Triceratops* was spliced into the infamous *Robot Monster* (1953).

TLC's script, written by Richard Landau – based on a story by Carroll Young, is pretty good, and throughout the acting is solid, engaging, performed by a talented cast (also) including Hugh Beaumont, Whit Bissell, Sid Melton and Hillary Brooke.

As Mark F. Berry notes in his *Dinosaur Filmography* (2002, pp.235-236), the year 1951 was still in the height of a very frustrating war in Korea. Berry adds, "…the Korean conflict and other forces were causing the patriotic, post-World War II victory glow to dwindle, and all through the film runs an unmistakable thread of contempt for the military." Furthermore, the Committee on Un-American Activities led by Senator Joseph McCarthy sought to root out communism in Hollywood and elsewhere in the United States. Many were unfairly "blacklisted" in the resulting "Red Scare," reflected in many films of the time. To place matters further in perspective, the Soviet Union had the atomic bomb (in 1950), while the USA was making steady technological advances toward its first successful hydrogen fusion bomb test (conducted in 1952).

Whereas most dino-monster philes know of Conan Doyle's classic (written over a century ago), few recall *The Lost Continent* 1951 movie. Despite this, however, William Fogg (in 2005), claims there are those who would suggest that *TLC* is the "greatest dinosaur movie ever made." While I disagree, if there is any truth to Fogg's claim, I wonder if much of *TLC's* appeal could be the rather subliminal yet seminal ties made between the beginning of global proliferation, nuclear weapons angst and dinosaurs—then increasingly (yet vaguely) becoming metaphoric for mankind's worsening plight in popular culture for the *idea* of *extinction*, which has increasingly cemented itself further into today's pop-culture as mankind faces its everyday challenges. In one symbolic scene we see the two *Triceratops*, the *Brontosaurus* standing around the crashed rocket in the midst of the uranium field. Yes, *TLC's* animated dinosaurs are fun and interesting to see (even today), but even then, they were clearly running secondary to the prospect of cold war paranoia.

The Lost Continent is the first film to ally ideas of nuclear bombs with prehistoric life (i.e. dino-monsters). *Gojira* would follow as the third such film only 3 years later. Thus, *TLC* is a 'precursor' of what would then come shortly, and what became so prevalent in monster movies during the late 20th century.

Notes: (1.) C. J. Cutliffe Hyne's 1899 same titled novel wasn't a source material for the film.

(2.) George E. Turner, et. al., *Spawn of Skull Island* (Luminary Press, Balt. MD), 2002, pp.53, 221-222; William Fogg, "Rediscovering Robert Lippert's Lost Continent," *FilmFaxplus*, no. 105, Jan/March 2005, pp.108-112, 122; Mark F. Berry, "The Lost Atlantis: Forgotten Dinosaur Epic of the Thirties," *Prehistoric Times*, no.79, Fall 2006, pp.22-23; Allen A. Debus, "Eerie Eyre: 'War Eagles – Beyond the Test Reel," *Mad Scientist*, no.20, Fall 2009, pp.24-35. Donald F. Glut, pers. Comm., 2016; "Models that Save Lives," *Popular Mechanics*, Sept. 1939, vol. 72, pp. 392-395, 130A-131A; Letter from Robert Skotak in *FilmFaxplus,* no. 106, April/June 2005, p.132.

TWO

Illustration by Prehistoric Times editor Mike Fredericks made especially for this book showing a Gorgosaurus about to devour its prey—Lambeosaurus. Inspired both by paleoart by Zdenek Burian as well as the Field Museum of Natural History's skeletal display, "Predator & Prey." See Chapter Seven for more. (copyright 2017, Mike Fredericks, used with permission)

Chicagoland Dinos

From a young age with my first viewing of RKO's *King Kong* in the late 1950s, thanks to my father and, in turn, his parents—my grandparents especially, I was raised *destined* to love fossils, paleontology, and dinosaurs. And while our family moved around at times during the late 1950s and 1960s for extended periods, including overseas, basically the Chicago 'burbs have always been my 'home turf.' In no small way because of the relative proximity of the world class Field Museum of Natural History and its wonderful geological exhibits, this contributed to my fascination with dino-science, and no doubt science in general.

By the late 1990s, the 'dinosaur invasion' of our Windy City was still in full swing as it had been for over a century, as we may glean from the following chapters delving into aspects of Chicago's special hold on dinosaurian imagination. In my mind, the Chicago region will forever remain a hub of paleo-intellect, for the many dino-displays shown here, its paleontology programs at major institutions—the University of Chicago especially, or even because of the annual Godzilla dino-monster convention held annually near O'Hare airport. During the 1990s and 2000s my parents and my wife, Diane, and I attended several of Dr. Paul Sereno's and Robert Bakker's lectures (and others), the Burpee Museum's annual "Paleofest," and Navy Pier's "Dino-Fest." Additionally, there were *numerous* other local dinosaur displays, several of which I channeled my journalistic self to cover for fanzines of the time.

When we casually think of dinosaurs, many of us, especially children, tend to think of them as angry, stomping fighting machines—we'd all better run and hide if they ever reach civilization! We also view their prehistoric world as somehow being more savage and fierce than whatever struggles might be witnessed in modern times (most likely a misguided opinion). Such perspectives certainly borrow from knowledge of the African plains where great animals struggle in riveting

'Nature red in tooth and claw' competition for food, water, breeding rights and wallowing places. And so, analogously, our images of dinosaurians battling to the death persist. For their titanic struggles and their potential to destroy other races of creatures (or, in a sci-fi sense, vengefully versus *Man*!) surely must have been epic—as captured via a variety of popular 20th century media!

So, I trust the following three chapters will offer some sense of the 'fantastic' concerning the time-honored, popular grasp of dinosaurs on Chicagoland history—especially my own.

These chapters originally appeared as articles in the following publications:
Chapter Five—*G-Fan*, no. 112, Spring 2016, pp.42-45.
Chapter Six—*Scary Monsters Magazine*, no.92, April 2004, pp.107-119.
Chapter Seven—*Fossil News: Journal of Avocational Paleontology*, vol. 14, no.4, April 2008, pp.12-16.

Chapter Five—*Chicago Attacked by a Swarm of Giant Dinosaurs Storming From Canada*

Did this really happen, or not? After all, it was recorded in the papers as 'breaking news' once and usually that means its just gotta be true – right? Not fake news? It happened so long ago though that maybe it was just generally forgotten, until the story recently resurfaced.[1] I don't recall my grandparents mentioning anything about it and they lived in Chicago then. So, what's this all about? A *kaiju* invasion that 'evidently' took place in 1906, almost half a century before Godzilla raided Japan, half a century before I was born. Sheesh – incredible. At least no nuclear bombs were involved. Army men solely relied on conventional weapons.

Okay, let's move on to the details, those alleged 'facts' as reported over a century ago. The story broke on April 1st, 1906. Hmmm. Sounds a little suspicious, I know. But just eyeball those riskily taken on-the-scenes photographs (see Note at end), proving veracity of the calamitous event. There were no computers back then, meaning you couldn't easily doctor up pictures like these, right? Not in the case of all the news that's fit to print.

Strangely enough, the story wasn't even the day's headline, as details of the dino-monster invasion began on p. 28 of *The Chicago Sunday Tribune*. But the feature's title carried punch: "Chicago Invaded by Hordes of Prehistoric Monsters Dealing Death and Destruction." A two-page spread, no less. So, what's the how, when, where and why of it all? Everything points to a genuine *kaiju* attack. As reported then, this was the "…most terrifying enemy that ever threatened the destruction of mankind. … Never in 6,000 years of history had so terrifying" a calamity "been forced upon a people."

At first, Chicago newspapers described early reports coming from Canada, flippantly. The dinosaurs evidently had trekked southward across Canadian forests while marine reptiles infiltrated through lakes to the shores of Lake Michigan. One of the earliest sightings was of a "sea serpent" near Duluth. Then there was a dramatic encounter between a plesiosaur and a tugboat in Lake Michigan waters. Then "batlike" birds began manifesting themselves near Lake Winnipegosis in Canada, causing terrified refugees to flee aboard Canadian Pacific rails. As later reports arrived, some of the 'birds' were later observed to have had wingspans up to 120 feet across! These bat-like birds—no, pterosaurs—were witnessed swooping over Evanston (north of Chicago where I later attended college), and had heads shaped "like monstrous pelicans." A tugboat was sunk by a pterosaur. But when the growing masses of dino-monsters reached Racine, Chicagoans had reason to fear what was about to ensue! And our town would bear the brunt of the full attack, one that challenged our military might.

As the *Tribune* article documented, "Hordes of gigantic beasts swarm down from the north and overwhelm the city, leaving it a mass of wreckage, tangled iron, piles of brick and stone, killing tens of thousands of the inhabitants. Parks turned into deserts, great skyscrapers leveled by blows from the tails of the monsters, and entire populations threatened in the panic and pestilence that follows the invasion." The Dearborn Street Bridge (no doubt the one which I've crossed many a time was rebuilt) became an early target, as a massive *Diplodocus* caused it to collapse when it attempted to venture across on its way toward the great post office building. (This bridge collapsing phenomenon seems to have been borrowed later for the climax of 1925's *The Lost World*.)

Furthermore, we read in summary, "Awful combat between gigantic dinosaur and huge tyrannisaurus (sic) on Grant Park results in the destruction of the immense buildings along Michigan Avenue. National government sends warships and troops to battle with the invading hordes of supposedly prehistoric beasts on the lake and in the streets of the city. Terrifying scenes in the stockyards when 25,000 cattle, sheep and hogs are eaten by pterodactyli. Art Institute destroyed and people driven to refuge in cellars."

It may have seemed at first, however, that things weren't going too badly, as many of the dinosaurs – such as the sauropods by temperament lacked ferocity, and partly due to their small-brains seemed rather docile. Some brave souls even mingled about the herd, even in the midst of a 39-foot-long *Triceratops*. But then *Tyrannosaurus rex* (which had then only recently been described in the scientific literature) arrived, hellbent on causing death and destruction after feasting within the stockyards area.
Pictures often tell a thousand words, and several attest to exactly how it all went down. It was clearly a massive attack, not unlike that left later in the wake of a Godzilla or Rodan. One shows a sauropod standing in tripodal pose leaning upright against a building on Clark Street (perhaps not far from where I once worked) smashing its exterior. The image is not unlike a similar imaginative image printed a decade before on the front page of an

1898 newspaper, the *New York Journal*, in which the most "colossal animal ever on Earth"—the "Brontosaurus"—had been allegedly discovered in the American west. (An earlier similar influential illustration dating from 1886, showing a gigantic *Iguanodon* rising up tripodally, bracing itself against an early 'skyscraper' conjures similar fantasies.) Except in our case – that of Chicago's – printed photos documenting the invasion and the instance were real!

Another photo shows destruction of the Art Institute, while yet another captures citizens fleeing from the gigantic brutes. But perhaps the most poignant image is that of the *Diplodocus* and *Tyrannosaurus* in a death duel, mounting the Montgomery Ward building tower along Michigan Avenue, at the climax of a horrific, epical combat which was also vibrantly described in the story. (Today's giant monster movie lovers are familiar with the strange phenomenon of how towers often are toppled whenever *kaiju* attack.) This 80-foot tall *T. rex* moved (e.g. hopped) in 100-foot long kangaroo-like strides. A true monster, reporters documented its bloodthirsty attack upon the *Diplodocus*, which had been grazing in Grant Park! "Everyone who saw the gigantic beast and noted the ponderous frame and its powerful jaws with their double rows of teeth, realized at once that it was an enemy of the diplodocus in the park and that a battle such as civilized man had never before witnessed was about to take place." And so, with a bellow of rage, the carnivore leaped upon *Diplodocus*' back. When the tower crashed, both dino-monsters plunged to their death, where their carrion flesh was feasted upon by "vulture-like" pterodactyls and presumably other reptilian brutes. I wonder how many souls lost their lives during the melee, those who were riveted by the ensuing primeval turmoil yet simply could not turn away in time to save themselves on the afternoon of that first invasion day.

The carnage became rampant. Pterodactyls devoured thousands of cattle and sheep in the stockyards. During the wanton destruction of the Lincoln Park area, zoo animals were attacked by giant dinosaurs, at first curious of the strange caged creatures. While the lions and tigers put up a good fight, according to the account, elephants "fell easy prey," circumstances recalling filmic projection of a scene in 1969's movie, *Gwangi*. By the second day of the mass invasion, city streets had crumbled beneath the dinosaurs' massive bulk, and the Elevated train structure was smashed in up to a dozen places around the Loop.

And then of course, what happens in the movies evidently did actually happen then in 1906, as scientists were called in to testify and, as things turned out – pontificate, as to the nature of the great attacking beasts.

So, while Professor Dryasdust's "learned disquisition" apparently revealed his utter fascination with the huge monsters, and Professor Oldashills conjectured as to why and where they originated, the Mayor was disinterested in the ivory tower pursuit of such knowledge. He simply wanted the dino-monsters who were tearing apart his city dead. Our only hope, or so it seemed, lay in a massive cannon counterattack.

By this time, as even more ugly carnivorous brutes arrived, the dino-monsters had come to recognize mankind as their bitter enemy. And so they fiendishly engaged the military called upon to stop them in their massive tracks. This military defense of a major city against rampaging dino-monsters may have been the first in recorded *daikaiju* history. The first assault failed, however, as the dreaded dinosaur charge caused stalwart soldiers to flee. In fact, this dramatic account may remind readers of that 1980s Topps collector 55-card set *Dinosaurs Attack!* which also carried those phony newspaper headlines of their global invasion on the backs of each graphically illustrated card.

Now hundreds of gigantic dino-monsters roamed within city limits, causing officials to question how many more would come. Fort Sheridan, situated north of Chicago along Lake Michigan, had been destroyed, so that when fresh military units were dispatched by the Secretary of War, they had to travel from Omaha and St. Louis. While it may have seemed like trying to hit the side of a barn, given many of the monsters' enormous size, hovering pterodactyls proved especially difficult to shoot down due to their aerial maneuverability and armor-plated hide. Thousands of troops were deployed, while many 'amphibious' dino-varieties kept emerging along the lake shore. Ammunition must have been running low as the carnage continued. But soldiers were aided by a swiftly organized citizens corps, who began hurling improvised "dynamite bombs" in the dino-monsters' paths.

The battle raged within Lake waters as well. Fish-lizard 'ichthyosaurs' were on the loose in Lake Michigan, so the President ordered submarines and torpedo boat destroyers to engage. Because all the aquatic reptilian varieties were scared of the submarines' electric headlight, these dino-monsters could be rooted out of the Lake, accordingly chased toward the surface where they were dispatched by torpedo boat machine gun fire. Then their floating carcasses were tugged to the Indiana Sand Dunes where they were cremated, thus minimizing extent of water contamination due to putrefaction. Finally, blessedly, reports came that the swarm of dinosaurs stemming from Canadian woods had finally ceased!

The final extent of the damage was severe yet not 'apocalyptic.' That is, many skyscrapers were stripped of their facades, and several other buildings were mostly or partially destroyed—like the Post Office and the Montgomery Ward Building. Michigan Avenue was a total loss though, and tons of debris lay piled up to ten feet deep along main downtown thoroughfares such as Clark, Monroe, LaSalle, Madison and Washington streets—my old haunts. Most downtown bridges were demolished.

Professors Dryasdust and Oldashills presented their theory on the origin of the attacking dino-monsters. They made astute connections to the lost city of Atlantis (yes, that "long lost prehistoric city") which (may) "have held these great monsters in safety against the warring elements of time… secure from the destroying march of the ages, the dinosaurs and other monster reptilians had been conserved." Hmmm, where might they have got that peculiar notion? Curiously, C. J. Cutliffe Hyne's novel *The Lost Continent* (a tale of Atlantis also incorporating prehistoric creatures), had only then recently been published, in 1899. According to the two professors, mighty earthquakes and convulsions may have thrown Atlantis toward Earth's surface within the Arctic Circle near King William's Land, near the north magnetic pole. Those "antediluvian" monsters would have instinctively sought warmth and nourishment available in more southerly climes, hence the fierce southward migration through the Canadian wild toward more highly populated regions. As conjectural as it may seem based on today's science that explanation sufficed to explain this dai*kaiju*-ish matter at the time.

But today is there cause for skepticism? Well, like Mulder, I just want to believe. Can this ever happen again? Who knows, but it does appear that Atlantis has sunk again, so for now, we may be safe. Although not of flesh and blood variety, many dinosaurs have since "invaded" Chicagoland's museums, Fair grounds and shopping malls since that first 1906 *daikaiju* dino-war. However, all you 2016 G-Fest goers should remain vigilant and beware. Cuz' there's really no telling when Atlantis might suddenly reemerge from way up north. And those bat-winged nightmares travel ever so quickly.

Note (1.): Spoiler alert! Despite what may or may not have been implied herein, yes, there really was such an (anonymously written) April Fools fictional account printed in the cited 1906 newspaper. I discovered this oddity in an interview with Ulrich Merkl (who recently published a book titled *Dinomania*). The interview, printed in the December 2015 issue of *Monster!* was conducted by Stephen R. Bissette. A reproduction of the two-page 1906 newspaper spread illustrated the *Monster!* interview although one cannot read the very diminutive type. Fortunately, this 1906 newspaper is available online, accessible via a Google search, so you can view actual 'photos' of the alleged invasion. The draft of this/my article was forwarded to J. D. Lees just in time for April Fool's Day 2016.

Chapter Six—*Giant Dino-Monsters & 'Kong' Invade Chicagoland*

Okay ... well, not exactly "giant" dino-monsters; they're just normal sized. But my headline rings true because, as I write this, computerized, animatronic prehistoric denizens *have* invaded Chicagoland .... *again*! Which is strange, because usually don't they attack Tokyo, London or New York instead? Okay—they're currently on display at Brookfield Zoo (until October 27th 2013), scattered around the premises, menacingly lying in wait. But those are NOT the manifestations I intend to focus on here. For, instead, giant-dino-monsters as well as a *bona fide* 'King Kong' did invade Chicago, possibly for the first time during the Chicago World's Fair of 1933/34, also billed as "A Century of Progress Exposition." These weren't building-smashing brutes, or allegorical manifestations of nuclear bombs—which hadn't been invented yet anyway. They weren't even alive but, rather, life-sized model recreations, several of which did move (robotically, mechanically) while engaged in titanic struggles. They established their primeval "beachhead" along Lake Michigan, in the vicinity of where the McCormick Place stands now, south of Soldier Field. My father and grandparents witnessed them "in the flesh" (yes—for real this time). And they survived to tell the tale. (I have seen several, but don't call me bonkers, or mock me with silly pejoratives like "mokele mbembe"!)[1]

By the 1930s, sculptural representations—both life-sized and miniature—had become all the rage! Prehistoric animal "paleoart" was an accepted staple at most museum institutions featuring paleontology displays. And paleoimagery increasingly wove its extraordinary spell upon the American public. Perhaps Charles R. Knight's famous models and paintings (i.e. his then recently completed Chicago Natural History Museum murals especially!) had much to do with it, coupled with New York paleontologist Henry F. Osborn's evolutionary ideologies. But also, psychologically, quite possibly the art of making miniature stop-motion sculptural "puppets" of dinosaurs built with moveable armatures seem absolutely gigantic on the silver screen was even more influential!

Mid-1920s movie makers had been experimenting with stop-motion animation special effects for about a decade, but in creating dinosaurs which appeared in First National's sensational silent film, *The Lost World* (1925), and later RKO's *King Kong* (1933), an unstoppable wave was set in motion. (Knight's powerful artistic influence on Marcel Delgado and Willis O'Brien – masterminds behind those two sensational films – is well documented.) Here was a new, innovative way to enjoy *animated* prehistoric life, and the key to success began with sculpting *miniature* scale dinosaurs! Besides Knight who on his own couldn't possibly satisfy demands from every museum globally, during the early 20th century, other paleo-sculptors began testing their talented hands in the art of miniature dinosaur model- and diorama-making either for museum display or the film industry, including (among others) Vernon Edwards, Charles Whitney Gilmore, Willis O'Brien, Marcel Delgado, one "Prof. Amalitzky of Warsaw," Erwin S. Christman, Richard Swann Lull and the young Ray Harryhausen (who later in his career relied on Arthur Hayward's talents).

By the early 20th century, few people had seen sculptures or statues of life-sized dinosaurs. Rare glimpses were possible, however at Sydenham, England, outside London where Benjamin Waterhouse Hawkins' life-sized concrete models of "antediluvians" have stood since the mid-1850s. And during the pre-World War I period, Joseph Pallenberg sculpted life-sized dinosaurs for the Hamburg Zoo, several of which have not stood the test of time. There were a handful of other cases cropping up by the late 1920s, but most people had not stared in person and in awe in the presence of life-sized dino-monsters, let alone any which could startlingly move, swing their tails, eat foliage and growl!

Sure—by 1930, museum displays sometimes featured fully reconstructed skeletons of prehistoric animals, and occasionally miniature sculptures offering glimpses as to how they appeared in life. But while miniature sculptures of individual prehistoric animals were enormously captivating, static, sculptural and statuesque *life-sized* prehistoric creatures proved even more mind-boggling. Better yet—why not make the life-sized prehistoric sculptures move *on their own*, growl, flash their teeth, "feed," swivel their extended necks, roll their eyes, and then unveil them logically and, traditionally - that is, as understood by an adoring public—exhibited in a format certainly suggestive of a quasi—"living" life-through-time context! Now here was an eyeful experience the public would never forget!

Although perhaps not directly inspired by First National's sensational 1925 silent film, *The Lost World*, based on Arthur Conan Doyle's 1912 novel, George Harold Messmore and Joseph Damon's "The

World A Million Years Ago" ("WMYA") motorized, animatronic display of life-sized prehistoric animals, which opened at the Chicago World's Fair in late May, 1933, certainly stoked the imaginative fire and interests of those who'd witnessed dinosaurs "living" at larger than life-size on the silver screen in major motion picture productions. In the world of the 'cine-fantastic,' it was a heady time: the spirit of King Kong and his dinosaurian foes reigned! And Messmore and Damon's (M&D's) electronic, "robotic" dinosaurs were the latest invention for resurrecting prehistory to life. For their grandest exhibition, at the Chicago World's Fair, they presented 150 different kinds of waterproofed papier-mache' and cloth-skinned animals, about 15 of which represented extinct genera. Additionally, there were four species of cave- or ape-men and a giant prehistoric gorilla (likely inspired by RKO's *King Kong*, which opened three months earlier) to boot! There was also a second "giant" gorilla squatting outside the WMYA Globe, over the entrance, which was billed in promotional ad newsreel footage as "King Kong."

The most famous WMYA dinosaur, however, was their 49- foot long *Brontosaurus*, made during the mid-1920s, but erroneously attributed to the Triassic Period in a souvenir postcard description. Of course, Messmore and Damon's prehistoric animals didn't all live exactly "one million years ago," but the moniker simply 'worked,' much as did the apt title "Jurassic Park" decades later, wherein most of the featured dinosaurs in that 1993 film actually hailed from the Cretaceous Period instead. Although their display was arranged more or less chronologically, certainly the WMYA title fostered a popular misconception that dinosaurs, cavemen and all other prehistoric animals lived together in prehistory, not unlike the later "Flintstones" mentality, or what was seen shortly thereafter in Hal Roach's One Million B.C (1940), or in those 1950s Marx 'Prehistoric Times' playsets mixing plastic cavemen with dinosaurs and other examples of prehistoric life.

For several years, Joseph Damon was a butcher out of Mt. Vernon, IL who also later studied art and animal sculpting for 4 years while in St. Louis, thus learning anatomy; George Harold Messmore (nicknamed 'Harold' or just 'George') was a gifted scene painter who had a knack for electrical devices and mechanical intuition. They fortuitously met in Newark at an Elks convention in 1916, and decided to collaborate. And so by 1918 they opened their prehistoric monster-making factory on West 27th Avenue in New York.

A journalist in 1933 described the process of designing one of these prehistoric creatures. After sketching how the animals should appear, "The sculptor first has a skeleton of wood and metal made in the proper size and proportion. In the case of the giant ground-sloth, the man-made skeleton was about nine-feet high. Then he made a clay statue of the beast, carrying out the drawing in every detail. When the statue was finished it was covered with a thick coating of plaster which, when it had hardened, was carefully removed in two half sections. These sections were, of course, a 'negative' mold of the statue. Skilled workers in papier-mache then took the halves of the cast and lined them layer upon layer of wet paper. When the paper was built up to the proper thickness, the casts were put in a huge oven and baked until the papier-mache was both firm and flexible. The paper lining was stripped from the plaster and put together to form a 'positive' figure of the nine-foot ground-sloth, a figure exactly like the solid clay statue, but light in weight and really only a shell. This shell was … dissected…" Then the most inventive steps began when technicians constructed the mechanized metal armature and motor systems that allowed the neck, jaw, eyelids, tail, and limb hinge joints to realistically rotate in menacing lifelike fashion. Motors rhythmically expanding and contracting the abdominal region effectively simulated "breathing." Old silent black and white film footage, however, demonstrates how *Brontosaurus* skin was made of sewn-together burlap swatches.

Actually, Messmore & Damon had a small menagerie of 'prehistoria' already built by early 1929, as they were exhibiting a "dinosaur, mammoth, etc…." in the Detroit showrooms of the Richards-Oakland automobile company in the spring of that year. Mr. Richards, President of the company, claimed that in March, 1929, 350,000 patrons stormed through their showroom, a month during which they sold 1,200 automobiles! Dino-monsters had become "…one of the answers to a salesman's prayer…"

Interestingly, there was a lawsuit filed for copyright infringement over the workings of the innards of America's first mechanical dinosaur with a trial taking place in a New York City courtroom in January 1934. Messmore and Damon had built their first such 49-foot long dinosaur named "Dinah" in 1919 for a $30,000 cost; afterward it made appearances as an animated parade float. A caption to a photo of Dinah in the Dec. 1927 Express Messenger vol. X, no.6, indicates that 'she' "… played a part in a motion picture," title unstated. (Dinah contained ten electrical motors inside.) And so their first dinosaur, *Brontosaurus*, was rented by a Midwest firm (Fanchon and Marco), who used it for a stage act in which the dinosaur "devoured" a show girl pinned within its sturdy jaws.

But they had a rival in theatrical showman Earl Carroll, who by August 1931 had constructed his own *Brontosaurus* nicknamed "Sarah" (although the newspapers called her "Dinny"). Carroll's dino-monster facilitated a stage act called "Dance of the Dinosaur," in which an actress/dancer gyrated and writhed in "Sarah's" flexible mouth while technicians operated the flexible neck mechanism. Although a number of popular magazine articles suggested how dinosaurs like 'Dinah' and 'Sarah/Dinny' were operated from behind the scenes, accompanying illustrations therein were a ruse, intentionally. George H. Messmore had filed a patent, wishing to protect said rights, granted on Feb. 21, 1933. Therefore beforehand, it was paramount to protect engineering details as to how those gears, rotors and cables in the neck and head, etc., truly functioned and moved, etc. Also, Carroll had toured Messmore's and Damon's construction facility in the fall of 1930, and so the latter claimed Carroll had occasion to steal ideas as to how (i.e. specifically the neck of) a mechanical dinosaur could be built and operated. The lawsuit, however, was dismissed by Judge Schmuck. (Yes – that was his name.) Whereas Carroll's mechanical brontosaur faded into obscurity, Messmore and Damon's dino-monsters were enjoying the height of their popularity at the Fair, exhibits which took 4 years to accurately build! Of course, a small army of talented artists, technicians and model-makers were poured into that effort (as well as in other projects).

The "Official Guide Book of the Fair: 1934 with Supplement" (p. 125) certainly whetted patrons' appetites for all manner of things prehistoric: "It is hard for us to conceive of a world inhabited by monsters other than those of industry. But when we cross the broad plaza at Twenty-third street to a spherical building on the hillside by the lagoon, we see examples of prehistoric creatures that would, in the flesh, terrify the bravest man. Step onto a platform, in motion, and you will be transported through 'The World a Million Years Ago.' You are carried past a series of six dioramas displaying the animals of the ice age and 'man' before the dawn of history. Then you enter the main arena. Here gigantic, prehistoric beasts are brought to life—a platybelodon, a huge hairy mammoth, a 9-foot tall giant gorilla, saber-tooth tiger, and ground sloth are seen as if nearly interacting. Also, the glyptodont, triceratops, pterodactyl, the massive brontosaurus, and the "Vernops" and finback *dimetrodon* in a death struggle are represented in their natural habitats—seem to be alive, breathing, uttering cries, and moving." The WMYA globe stood near Admiral Byrd's Polar Ship and not many yards from the (rival) Sinclair Refining Dinosaur display.

My Dad (Allen G. Debus) was a young boy then, escorted numerous times to the Fair with my grandparents, both in 1933 and 1934. He was particularly fascinated by the WMYA dinosaurs, in 1997 recalling, "… I can almost hear the mechanized screams of these animals even today. For me this exhibit was far more powerful than the Sinclair dinosaurs because it was inside and relatively dark. This made it seem more lifelike than the Sinclair exhibit which was filled with tourists in the sunlight. The moving walkway –so common today in airports—was then a real novelty and it kept the people at a distance from the reconstructed animals. … I recall the head of the *Brontosaurus* moving over the people on the walkway. Outside once more it was possible to buy postcards and small cast-iron replicas of the prehistoric animals. …"[2]

Messmore's and Damon's dino-creations were housed inside a 150-foot diameter metal half-globe building representing Earth and showing continents and oceans on the exterior. During 1933, to keep audiences flowing, 700 at a time, patrons moved past jungles and animals 'wallowing' in swamps, constituting several insular dioramas, along a motorized beltway or concourse. Technicians craftily hidden in adjacent quarters operated a switchboard controlling each of the creatures' motorized movements in such a way that the animals appeared to "interact" with patrons. The WMYA exhibit had the capacity of carrying 84,000 people every twelve hours along the winding ramps. As of June 1932, it was planned to provide earphones along railings so that spectators could listen to recorded growls played on disk records by behind-the-scenes technicians and hear educationally oriented descriptions of the animals in each display, but it is logistically difficult understanding today how this might have been arranged while continually moving along a beltway, especially when the speed of the beltway was adjusted faster or slower depending on how many people waited in line for admission. Besides *Brontosaurus*, chief attractions witnessed within the globe were a terrifically detailed restoration of strange-mouthed *Platybelodon* (a "shovel-tusked" Pliocene elephant), *Tyrannosaurus rex* posed adjacent to its traditional nemesis *Triceratops*, furry mammoths, a woolly rhino and *Stegosaurus* a Jurassic dinosaur erroneously deferred to the Cretaceous Period in company advertising.

Recently a DVD compiled by Jeff Quinn was offered for sale through *Prehistoric Times* magazine. This curiosity comprises vintage footage that surfaced, showing several of the WMYA creations in motion at the Fair. Some of the more prominent, promoted and popular 'monsters' included fin-backed *Dimetrodon* which held a struggling reptilian *Varanops* in its toothy jaws, aforementioned *Brontosaurus* (e.g. - a genus by then popularized in two major monster movies, both 1925's The Lost World and RKO's *King Kong*), plated *Stegosaurus*, and a *Tyrannosaurus* whose mouth opened and closed, neck turned and diminutive arms raised and lowered. *Brontosaurus'* neck and tail swiveled, eyes blinked, abdomen 'breathed' and its mouth opened and 'chewed' food. There was also an enormous bat-winged Pterodactyl (now in the collection of Don Glut), as well as several prehistoric mammals, such as a saber-toothed tiger, Giant Ground Sloth and a Woolly Mammoth. A Kong-like giant Gorilla opened and closed its mouth, and turned its neck while arms moved up and down while it also 'displayed' distinct abdominal 'breathing.' While footage on this DVD is silent, patrons to the exhibit must have heard a perpetual cacophony of roaring, growling sounds and hair-raising jungle animal screams.

Another curiosity on this DVD is a meaty segment titled "Dancing Stegosaurus: (Man or Machine?)" that clearly shows a man in a plated *Stegosaurus* suit cavorting around a stage, rising into tripodal position (balanced on its tail) while performing in front of two cave people that appear smaller than life-size due to trick photography enhancements. This may be the earliest example of filmed 'suitmation' involving a "giant" dino-monster and trick photography. Note that this footage and suit action must predate the *Tyrannosaurus* suit briefly seen on camera in 1940's *One Million B. C.*. "Dancing Stegosaurus" was also devised considerably before any of Toho's suitmation monsters, such as 1954's Godzilla. According to Jeff Quinn who compiled this delectable DVD, "… it is definitely a Messmore & Damon film. … The last second or so … reveals the ending leader… 3178 Messmore... I would venture to guess this (short) film was used around the time of the 1933 World's Fair as a promo shown as a filler in movie theaters being it is a silent film."

During December 1932, paleontologists Roy Chapman Andrews, who along with Barnum Brown and Walter Granger assisted with their scientific knowledge toward reconstruction of the mechanical brutes, expressed how "very impressed" he was with "…M&D's craftsmanship and state-of-the-art mechanisms utilized." "It almost makes fossil animals live and I look forward to the time when our scientific museums will have exhibits of this sort."

While displays were rather admixed, or congregated within close quarter confines of the planetary globe, each individual scene offered glimpses into successive stages of Life's history, ranging from the Late Paleozoic Era to the Late Pleistocene. Additionally, there were six separate dioramas portraying early stages in the evolutionary history of man (i.e. from time of the Giant Gorilla's Ice Age through Cro-Magnon Man) situated along the "main arena on the globe's outer wall." There was even a "living and speaking" Piltdown Man on show! There were additional prehistoric displays at the Chicago World's Fair too which we'll come to, but Messmore and Damon's creations, reflecting a quasi-general, life-through-time tour sequence were most ambitious. Interestingly, in a vintage magazine advertisement for Gimbels Toy World, titled "Interesting Information About the World A Million Years Ago," when they exhibited there on the 6th floor during the holiday season in 1934 through March 1935, a 'bullet' in the ad states, "The creators of this exhibit, Messmore & Damon, made the 'King Kong' for RKO, and have made all kinds of animated things for the movies." Would this Kong reference allude to the large prop used for filmed facial close-ups?

In a gift shop, patrons—such as my grandparents, purchased two small cast iron prehistoric animal "toys" (i.e. a *Brontosaurus* or Saber-tooth tiger). A third, iron-cast metal figure – *Triceratops*, was also sold at the Fair, but today collectors are aware of only one existing. These 3 iron metal figures are commemorative but were not specifically associated with Messmore & Damon. Later, during the 1939/40 New York World's Fair, when several of Messmore & Damon's prehistoric animals triumphantly returned, the WMYA toy line had expanded into two "Creatures of the Past" sets of lead-cast figures – a total of 19 creatures, including cavemen, which were of course miniature sculptures of some of the larger recreations on display. Attending both Fairs, my dad acquired both iron-cast and lead dino-monsters; it is possible that the lead-cast figures were also available during (as well as after) the Chicago Fair as well. According to Mike Fredericks, editor of *Prehistoric Times* magazine writing in 2000, "Very few of these toys are known to have survived the past 65 years and are possibly the most desirable prehistoric animal figures for collectors today." [3] To my knowledge and possibly for copyright reasons, a giant prehistoric gorilla 'King Kong' toy was not produced as part of either set.

During 1934, the theme of the display became "Down the Lost River," when patrons floated in boats along Chicago's nearby Northerly Island past the prehistoric animal displays, perhaps a peculiar 'real' precursor to Czech Karel Zeman's 1955 film *Journey to the Beginning of Time*. As recounted by Francis B. Messmore (George's son) in a 1997 *Dinosaur World* article, in 1934, the World A Million Years Ago exhibit moved to another location, although now isolated from the globe housing. Using a "time-honored" river of time metaphor, the "… new show was called Down the Lost River to the World A Million Years Ago." [4] Now visitors were carried in boats across Northerly Island's lagoon to see the recreations. Thanks to visionary artists like Benjamin Waterhouse Hawkins and later Messmore and Damon, the theme of illustrating Life's vertebrate history in the form of *life-sized* statues arranged in geologically ordered settings has persisted until the present day.

After the Fair ended in 1934, following their Chicagoland sojourn, The World A Million Years Ago mad robotic creations escaped from Chicagoland, and were unleashed to the world! The WMYA was seen at the Detroit and Michigan Exposition in 1936. The dino-monsters also traveled to Montreal, returned briefly to Chicago, headed to Broadway's Warner Brothers Theater, then toured the nation and other countries as well until the 1970s when they were retired in various locations. The exhibition traveled to Luna Park in Paris in 1935, moved to Cleveland for the following summer season in 1936. By now, however, the monstrous "troupe" was being split up.

In 1937 "half of the show" was off to Dallas while several eventually moved on to New York for the 1939/40 World's Fair, therein included as part of General Motors' "Futurama" exhibit (although Messmore and Damon's display was whimsically re-named "Pastorama"). Meanwhile, many of the World A Million Years Ago monsters were witnessed on San Francisco's Treasure Island. For the New York World's Fair, the nature of the prehistorical display changed significantly, in that dinosaurs were incorporated into a "Rocketship to Venus" 'ride.' After completing an imaginary space journey to Venus, patrons toured through weird, alien landscape where the WMYA *Brontosaurus*, *Triceratops*, *Tyrannosaurus* a few other prehistoria could be witnessed. This made "sense" then because, according to the nebular hypothesis, astronomers believed that Venus was geologically younger than the Earth by millions of years, and thus could still exist in the primeval stages of ecological development. Several dino-monsters were exhibited in Baltimore during 1941, and also in department stores in Winnipeg, Toronto, Pittsburgh and St. Louis (the latter in 1936). In 1937, George's son, the young Francis B. Messmore, who passed away in late 1997, wrote a term paper on details of the electrical apparatus powering the *Brontosaurus* "robot" for his physics class at Fordham University (receiving high accolades). During World War II, WMYA's exhibition options declined (or according to Francis Messmore, the war "cramped the style of WMYA"). However, during a New York City parade intended to support the war effort, the WMYA *Brontosaurus* was bedecked with Nazi accoutrements, such that it was added to the lead parade float labeled, "Hitler the Axis War Monster."

By 1951, WMYA "dinosaurs" were being loaned to television producers. In his *Jurassic Classics* (2001), Don Glut recounts seeing "... alien dinosaurs, actually no more than stationary models produced by Messmore and Damon to promote their 'World A Million Years Ago' attraction ... in 1951 episodes of the television science-fiction series *Tom Corbett, Space Cadet.*" In one episode, after landing the spaceship *Polaris* on a world that is still in its "Mesozoic stage" of planetary development, both the WMYA *Brontosaurus* and a fin-backed *Dimetrodon* were used as 'props' "zapped" motionless by the spacemen with "parallo-ray" guns. Glut noted, "The fact that the creatures do not move is helpful..."

Now, here's the thing—while Donald F. Glut happens to own several of the smaller displays that I've seen firsthand, including a pterodactyl and cavebear, which when plugged into a wall socket still rotate their heads, I MAY have also seen several of the models much earlier when they were still touring as the WMYA. This MIGHT have been in 1972, when the models came out of storage and went back on exhibition in Chicago's Old Town sector. I do recall visiting Old Town, walking around with some college buddies back then there one night, and have a dim recollection how I was amazed by the fact that I had encountered at least a few of the old WMYA creatures tucked away in an establishment, which I of course I would have instantly recognized them from my father's descriptions. Nobody else I was with seemed to care and we didn't have much money to spend. I peered through an open doorway, but regrettably we didn't go inside. But you know ... I do seem to also vaguely recall that that several of the WMYA dino-monsters were exhibited in conjunction with another—a torture chamber display that was also created by Messmore and Damon, also 'in the house.' Merely a false memory? As recounted by Francis Messmore decades later in 1997 describing the latter largely forgotten display, "... scenes showing various tortures .... Figures were life-size and there were sound effects accompanying them."

The Old Town dino-monsters went on to Japan, displayed at an amusement park called "Nagashima Onsen" in Nagoya for the full 1973 season. According to Francis Messmore, these were sold there, never returning to the states. Meanwhile, another part of the original 1933/34 display toured that same season at Niagara Falls' Skylon, from there moving on to Toronto's Canadian Exposition. The rest of the mechanical monsters had been stored for a lengthy hiatus in Saddle River, New Jersey, but by the summer of 1973, Francis Messmore—then, head of the company—was taking them out of mothballs, and creating a King Kong "robot" as well to join them on their Lake George 'stomp' in New York, where they had further life. According to journalist Michael T. Kaufman (July 26, 1973), "Just 11 blocks from the Empire State Building, where King Kong was shot down for the love of Fay Wray, ... the great ape is being reborn. So far, the vital parts of his 10-foot high torso have been cast in Fiber-glass, and within the next few weeks he will be outfitted with a growling tape recorder and with a mechanism that will enable him to move his synthetic shag-covered arms, his mouth, his neck and his eyes. And when he is finished ... he will scare the tourists at an exhibition that includes mechanized dinosaurs made by the firm more than 40 years ago."

The other, aforementioned 9-foot tall "Prehistoric Gorilla" had been exhibited during the Chicago World's Fair in 1933 too, representing "Gorilla life in the Pleistocene Period," and I am uncertain if Messmore's 1973 creation was a reboot of that original gorilla, versus an entirely new monster, given that its (new) body was made of fiberglass. (Incidentally, there was even a very genuine fossil "Kong" which lived in what is now Southeast Asia during the Pleistocene and once grew to 10-feet tall. This genus, named *Gigantopithecus,* was discovered in 1935 – too late for inspiring RKO's "King Kong" or exhibits at the Chicago World's Fair. But then these, and presumably this latest 'Kong' version as well, were eventually sold to the Amusement Park Foundation in Indianapolis, Indiana (for posterity). Wouldn't it be cool to find the rest of the dino-monsters, those not owned by Don Glut, still 'living' somewhere in a 'lost' storeroom?

Dinosaur-related memorabilia from the Fair are rare and valuable, but turn up every so often. One finds, for example, postcards, and a colorful, 19-page booklet titled "The World A Million Years Ago" by Leon Morgan, vividly illustrated by H. G. Arbo. Morgan and Arbo outline Life's history through time from the beginning to the evolution of man, emphasizing displays patrons would witness within the exhibition globe. However, perhaps the most sought-after collectibles are those rare lead-cast pot metal (i.e. made of "Medamet") figures featuring several of the most popular prehistoric animals, such as the *Tyrannosaurus*, three-horned *Triceratops*, plated *Stegosaurus*, etc. Before the days of plastic toys, the metal figures would prove especially intriguing for youngsters yearning to stage imaginative journeys to prehistoric 'lost worlds' on their bedroom floors. Although fortunately several of my dad's figures have survived, collector Larry Blincoe stated in 2004 that due to their fragility, "collecting these figures is kind of an ephemeral experience; you know you may be the last one to own them, and they are probably aging at about the same pace as you." Messmore and Damon's Medieval Torture chamber display was also promoted with an instructional booklet penned by Leon Morgan. Of course, given their creativity and abilities, Messmore and Damon did many other, numerous types of mechanized displays, but the World A Million Years Ago show really put them on the map.

Perhaps the most generally "visible" and memorable dinosaurs, however, were those manufactured by the Sinclair Refining Company, all six of which appeared out of doors as a free exhibition, mounted on an artificial rocky dino-diorama. These were built by P. G. Alen, under advisement of several prominent American Museum of Natural History scientists, including eminent paleontologist—Barnum Brown. P. G. Alen consulted with Brown who offered critique of his small-scale clay sculptures, which were based on Charles R. Knight's world-famous Chicago Natural History Museum mural restorations. According to Don Glut, a keen feature of the exhibit was the Grotto, stocked with numerous displays, which visitors passed through only to "emerge in a facsimile of the Mesozoic era." P. G. Alen's imposing 21-foot tall *Brontosaurus* (with 31-foot long tail and 20-foot long neck) stood above the Grotto's entrance, greeting visitors with its gently swaying neck as they entered the Sinclair exhibit. Glut states further, "Not only did its ... neck and ... tail twist and curl to the spectators' delight, but its sides pulsated realistically, simulating respiration."

Within the Grotto itself, patrons could enjoy "a number of striking photographic murals, ten feet long and four feet, six-inches high, depicting more re-creations of extinct fauna." Furthermore, Glut states, "P. G. Alen determined to have them virtually come alive, powered by electric motors implanted in their innards." Besides, *Brontosaurus*, also on display were five others, including plated *Stegosaurus* and the showcase diorama, a thrilling life-sized representation of THE most famous dinosaur painting of all-time: Charles R. Knight's restoration of *Tyrannosaurus* engaging three-horned *Triceratops* in a battle to the death! (In fact, Knight's painting was so famous and still is, that it was also re-created in miniature for the Fair, although this time in a diorama contributed by Century Diorama Studios.) Sinclair also published a booklet coinciding with the Fair, "The Sinclair Dinosaur Book," written by Barnum Brown and featuring artwork by James E. Allen.

Why this curious association between gasoline and dinosaurs? For the baby boomers (and older) among us, the connection may still seem entirely natural, thanks to Sinclair. Basically, Sinclair executives wanted a great 'mind worm-ish' type of "mascot" to promote their petroleum products. Their advertising campaign proved successful. ("Dino" the brontosaur symbol officially became their mascot.) One representative stated, "... the oldest crudes make the finest lubricants. To present this fact in an impressive manner, we constructed the Sinclair Dinosaur Exhibit. Nothing we could use would graphically portray tremendous age as well as do these dinosaurs."

But as noticeable and effective as the Sinclair display was in associating prehistoric monsters with their petroleum products, Francis Messmore was not as impressed, that is, relative to what his father and Joseph Damon had achieved under their dome. In 1997 Francis Messmore wrote, "Next to the 'WMYA' was Sinclair Oil's exhibit of dinosaurs. Their ad slogan was 'Mellowed 100 Million Years.' They showed a *Brontosaurus*, *Tyrannosaurus*, and a *Triceratops* among other pieces which were made of concrete for outdoor use. The animation of these pieces was minuscule. This was the competition that the 'WMYA' was up against, notwithstanding the fact that the Sinclair exhibit was free and the 'WMYA' was a paid attraction. WMYA grossed over $1 million ... charging 25 cents for adults and 10 cents for children." And the whereabouts of the original 1933/34 Sinclair dino-monsters? Unlike the WMYA dino-monsters which repeatedly invaded numerous cities, Fairs, department stores and the earliest shopping malls over ensuing decades before their ultimate demise, the Sinclair prehistoric monsters evidently were scrapped, following a 1936 appearance at the Texas Centennial exhibition. Glut claimed "... they survive now only in memory and memorabilia." Today, collectors may come across any of several "Picture News" leaflets distributed by the Sinclair Refining Company highlighting the fun, frivolity and general awesomeness of their imposing dino-monster display, or any of three vintage Sinclair dinosaur collecting sticker-stamp albums.

Another dino-inspired exhibit was the "Clock of Ages," placed in the Hall of Science. As stated in the Official Guide-book of the Fair (p. 38), "The science of geology is epitomized by a giant "Clock of the Ages" which ticks off the two billion years or more of the earth's history on a conventional clock dial. Geological pictures (i.e. including prehistoric animals) appear on a screen in the center of the clock face, and they are described by a synchronized phonographic record." Each revolution of the clock dial corresponded to 100,000 years of elapsed geological time. Century Dioramas Studio miniature dioramas also contributed three miniature dinosaur dioramas, each based upon a Charles Knight dinosaur mural—one of which was the ever-popular *Tyrannosaurus* vs. Triceratops battle, that has influenced many scenes of battling dino-monsters in movies such as Toho's *Gigantis the Fire Monster* and *The Last Dinosaur*.

George H. Messmore died on Feb. 15 1963 at the age of 76; Damon retired from the industry in 1940. Their business was still thriving through the mid-1950s, after moving to Park Avenue. Their one-time competitor, Earl Carroll, died in 1948. Francis Messmore passed on in 1997.

Time has marched on since the dimly recalled days of the 1933 Chicago World's Fair. Yet since then Chicagoland has suffered repeated invasions of impressive dinosaurs and those proverbial 'other' prehistoric animals. We've since captured a giant reconstructed *Brachiosaurus* and "Sue" our mascot *T. rex* at the Field Museum, cringed at 1980s mechanical recreations built by Dinamation technicians in many a shopping mall, and witnessed numerous other traveling dino-related exhibitions at the Field Museum and other whereabouts and nearby institutions. Those latest Brookfield Zoo denizens, including many of the latest known varieties of scary-looking feathered dinosaurs reflect a host of scientific ideas which were not even a gleam then in any 1930s scientist's eye! The new computerized dino-monsters are exhibited in the most recently accepted anatomically correct poses, yet are really no better than what Messmore and Damon accomplished and (along with Sinclair) offered to an adoring public years ago. But today at Brookfield, there are no 'other' mechanistic prehistoric mammals on sight, or a Kong. At least Messmore and Damon gave us a 'Kong' to dwell upon.

Notes: (1.) 'Mokele mbembe' is a term Roy P. Mackal assigned during the 1980s to a "Brontosaur," allegedly still living in the Congo which he searched for in vain. (2.) Allen G. Debus, "A Dinosaur Hunt in Chicagoland in the 1930s," in *Dinosaur World* vol. 1, no.1, Feb. 1997, pp.31-33. The cited article was reprinted in more permanent form in an essay collection titled "Dinosaur Memories," by Allen A. Debus, 2002, pp.549-554. (3.) Mike Fredericks in *Prehistoric Times* magazine, no. 45, 2001, pp. 48-49. (4.) Francis B. Messmore, Dinosaur World magazine, no. 3, October 1997, pp.27-34. Also, a portion of this article was excerpted from Chapter Two of Allen's *Dinosaur Sculpting: A Complete Guide, 2nd ed.*, McFarland Publishing, 2013. Other references include the following: Donald F. Glut, *The Dinosaur Scrapbook*, Chapter Two, 1980; Jeffrey Quinn's article in Prehistoric Times magazine no. 99, Fall, 2011, pp.54-56.)

Chapter Seven—*My 'Gorgosaurus'*

Few would remember.

Back during the late 1950s the world's second-most famous tyrannosaurid skeletal mount was first displayed in Chicago's Field Museum. This was the "Dinosaurs: Predator and Prey - *Gorgosaurus-Lambeosaurus*" display situated in the Museum's main hall - where 'Sue' the *T. rex* now reigns instead. I loved "Predator & Prey"—it was the best dinosaur exhibit my young eyes had ever feasted upon... but then, as of the early 1990s, it was gone. Our - excuse me, *my* - *Gorgosaurus* was once the first and only free-standing dinosaur skeleton in the world, that is, where visible vertical supports had been dispensed with. Sensational, but since vanished!

No, don't get me wrong. The specimens remain on exhibition, upstairs in the main 'dinosaur hall' where new remarkable displays chart the Earth's pageant of life. But the pose of our since dethroned 'Gorgy' is considerably different than before ... and herein there's an important story to relate.

During the 1950s and 1960s when Rexes truly were king, *Tyrannosaurus* ruled over Prehistory in its illustrious 'Lordly Rex' era. Like the American Museum's specimen, Chicago's 'Rex' also stood upright triumphantly, lord of all it surveyed in prehistoric nature, or at least in the Stanley Field Hall. But as we were to learn shortly, this upright pose—so masterfully constructed—was quite erroneous and illusional.

The Field Museum's beautiful 'Gorgy' specimen was collected by Barnum Brown in 1914, and by 1956 was erected by Field Museum staff. Our centerpiece fossil in the Chicago Natural History Museum's (then as it was named, but now the Field Museum of Natural History) was of course, *Gorgosaurus libratus*. Technically, our 30-foot long *Gorgosaurus* wasn't a *T. rex*: they didn't even live during the same time (*Gorgosaurus* lived a few million years earlier during an earlier stage, the Campanian, of the Cretaceous). However, it highly resembled *T. rex*, and for me that mattered.

Gorgosaurus stood upright, fully erect as the museum's 'mascot'—the Stanley Field Hall's main attraction.

As Rainer Zangerl related in a 1956 souvenir booklet (*Dinosaurs, Predator and Prey: The Gorgosaurus and Lambeosaurus Exhibit in Chicago Natural History Museum.* 1961), the composition, perhaps then the world's *second* most impressive tyrannosaurid skeletal mount, intended to show *Gorgosaurus* "... as it has just come upon the carcass of another animal (*Lambeosaurus*) that had recently died. While the predator was looking it over, something disturbed him: he has reared up, startled."

'Gorgy's' magnificent pose formerly rather resembled Rudolph Zallinger's masterful "Age of Reptiles" *T. rex* rendition at Yale University's Peabody Museum (as well as contemporary life-sized restorations sculpted by fellow paleoartists, Louis Paul Jonas and Elbert H. Porter.) With its prominent gastralia (belly ribs), our Gorgy even looked pot-bellied much like Zallinger's Rex restoration. Those who were seeking evidence of the symbolic and powerful imagery of a Rex battling with its eternal adversary *Triceratops* could venture upstairs to see a disembodied *Triceratops* skull (as well as Knight's famous Rex vs. Tops mural where it all mystically 'came together').

Through the 1910s, speculations as to the natural history of the peculiar morphological form represented by tyrannosaurids (gigantic bipeds with shortened arms and equipped with massive toothy heads and hind limbs) abounded. In 1917, Lawrence Lambe pondered simple questions - in what stance would tyrannosaurs most comfortably rest and were they scavengers? Such contemplations led to further related matters such as their posture, maximum running speed and their probable physiologies (e.g. were tyrannosaurids warm- or cold-blooded animals?)—questions which in one form or another have been debated ever since, *especially* during the current phase of tyrannosaurid lore and love which I have dubbed the era of 'Renaissance Rex.'

Lambe, convinced that animals such as 30-foot long *Gorgosaurus* were cold-blooded reptilians, championed the idea that such meat-eaters scavenged food rather than actively preying on huge contemporary herbivores. To him, Gorgosaur teeth seemed relatively free of wear, a result which could only signal its presumed scavenging lifestyle. In other words, Lambe would not have accepted the 'authenticity' of a 'Knightian' *Tyrannosaurus - Triceratops* battle. Lambe believed, "... therefore, that Gorgosaurus confined itself to feeding upon carcasses of animals that had not been freshly killed, that it was not as an intrepid hunter but as a scavenger that it played its useful part in nature, and no doubt its services were fully required when we consider the immense numbers of trachodonts, ceratopsians, stegosaurs, and other dinosaurs and reptiles that lived and died in this particular time of the Cretaceous period." (Lambe quoted in Adrian J. Desmond's *The Hot-Blooded Dinosaurs*, 1975, pp. 90-91)

Lambe attributed the gastralia and broadened pubic bones found in *Gorgosaurus* as special adaptations allowing gorgosaurids (and presumably all tyrannosaurids) to snooze on their bloated bellies through most of the day. Definitely not an inspiring picture for those of us who prefer imagining roaring Rexes charging across the prehistoric plains in search of danger, mayhem and Nature red in tooth & claw bloody combat with others of its kind over the spoils of war with packs of sickle-clawed raptors, or with three-horns and club-wielding ankylosaurs! However, Lambe's theory wasn't so popular even among his peers, as William D. Matthew oversaw the mounting of the American Museum's own gorgosaur skeleton in an active running pose, while instead Christman illustrated the "terrible lizard" charging (without a credit card) after a dashing duckbill dinner.

And so, despite Lambe's erudite speculations, for years, *T. rex* was often conventionally figured in a powerful, relentlessly stalking pose. This was because *Gorgosaurus* was - skeletally - considered a more gracile form. Therefore, paleontologists mounted its bones at the American Museum in a running pose, inspiring Christman's 1921 life restoration of the carnivore dashing after a pair of *Saurolophuses*. By 1933, with *King Kong's* active fighting Rex in public consciousness, Lambe's boring, slovenly, scavenging (and definitely uncool) tyrannosaurids were left snoozing by the wayside, that is, until 1970 when such questions resurfaced with a vengeance. Eventually, understanding *Gorgosaurus* led to reconsideration of *Tyrannosaurus rex's* natural history.

In spite of how the Field Museum's *Gorgosaurus* was mounted - emulating the American Museum's skeletal mount (AMNH-5027) as it stood commandingly for 80 years during the 20th century, Lambe had envisioned a different posture for the former. As stated by Adrian J. Desmond in *The Hot-Blooded Dinosaurs: A Revolution In Palaeontology* (1975), in 1917, "Lambe depicted this beast walking with a definite stoop rather than adopting the imposing upright stance always accorded to *Tyrannosaurus*. He also found in the leg joints signs that it had been bowlegged in life... *Gorgosaurus*' stooping walk with tail raised of the ground as a counterbalance must have been exhausting..." (p. 78)

But, to my dino-loving eyes, in October 1959, appeared a magnificent visage of another *Gorgosaurus* in a monochrome Plate (no.35) painted by paleoartist Zdenek Burian. This eye-riveting 1966 restoration published in Josef Augusta's *Prehistoric Animals* (1956), dramatically posed a muscular, ferocious-looking *Gorgosaurus* stooping over to menace an armored "Scolosaurus." At the time, and for the next two decades Burian's *Gorgosaurus* seemed the most accurate portrayal of my then favorite 'real' dinosaur. Of this hypothetical confrontation, Augusta opined:

> "...gorgosaurians ... certainly attacked, and mostly with success. Such a prey might ... be the genus Scolosaurus ... Gorgosaurus, with its powerful clawed legs, succeeded in turning Scolosaurus on its back ... then it could plunge its terrible toothed mouth into Scolosaurus' soft abdomen, tear it open, and inflict ... heavy wounds..."

Why was *Gorgosaurus* my favorite dinosaur so long ago? The display seemed to be a well-kept secret, one which only I seemed to know about: the few kids I knew then (very few indeed!) who liked dinosaurs were all into *Tyrannosaurus* or *Triceratops*. But strangely there was little knowledge among my peers about the Field Museum's fantastic *Gorgosaurus*—a name, incidentally, riffing off that of another cool, yet fictional dino-monster, "Gorgo." (Later in life my chief dinosaurian interests shifted toward *Stegosaurus*.)

Later in 1970, British Museum paleontologist Barney Newman reinterpreted the contemporary view of the more evolutionarily derived *T. rex*, rather aligning his thoughts with Lambe's views on *Gorgosaurus'* posture. Newman thought *T. rex* walked with a stoop, like *Gorgosaurus*, and that it used its relatively tiny forearms to push itself from the ground in raising from a reclined resting position. Newman also quite correctly chopped twelve feet of tail from *T. rex's* behind, most of which had been added conjecturally by Henry F. Osborn over half a century before, thereby providing proper balance to the now more horizontally aligned animal. According to Newman, *T. rex* no longer dragged, but waved its abbreviated tail triumphantly behind, while in front the head was also held horizontally aloft capping a flexible 'swan neck.' In fact, the British Museum's *T. rex* skeletal mount reflected Newman's then radical views.

But now, given Newman's musculo-skeletal realignment of the hind limbs, *T. rex* appeared pigeon - toed, perhaps a metaphorical sign of the times that were then dawning, as ever since the early 1970s birds were increasingly viewed as legitimate dinosaur descendants. As Glut outlined in his *Dinosaurs: The Encyclopedia* (1997), p.951, "With an *Albertosaurus* for comparison... Newman ... stated that the hind legs could not have borne the stresses in walking if the body were oriented so that the tail dragged, and speculated that the weight of the walking animal must have been carried by one leg at a time, the gait most likely sinuous, producing a pigeon-toed waddling as in birds." No longer a 'Gorgo-like' tail-dragger, or an anthropomorphized Godzilla-esque ground stomping creature, *T. rex's* science was catching up to *Gorgosaurus'*, taking on novel proportions that were unfamiliar to most!

If not for *Gorgosaurus*, fairly complete specimens of which had been found by Barnum Brown, Charles H. Sternberg, and Lawrence Lambe, by the 1930s paleontologists wouldn't have been able to make educated guesses as to how the forearms of *T. rex* appeared—before they had been found, or how many fingers *T. rex* had on each hand. In 2006, Donald F. Glut summarized this accordingly in context of a potentially new 'Rex' specimen: "*Tyrannosaurus*, as well as other tyrannosaurids, are usually described as possessing but two fingers on each manus (a comparatively slightly vestigial third finger had already been known in the older and smaller tyrannosaurid *Gorgosaurus libratus*." (Glut, *Dinosaurs: The Encyclopedia - Supplement 4*, 2006, p.551.)

Years later *Gorgosaurus* (described in 1914), became synonymized and/or technically conflated with another generic name, *Albertosaurus*, defined by Osborn in the same 1905 paper where he described *T. rex*. However, as of 2006, the Field Museum's marvelous 'Gorgosaurus' specimen was referred to tyrannosaurid genus, *Daspletosaurus* (a genus erected in 1970). Although paleontologist splitters and lumpers have flip-flopped on these names over the past three decades, for our purposes here, I'll always think of the Field Museum's mounted specimen as '*Gorgosaurus*' instead (even though as just stated you'll read the name *Daspletosaurus* instead on museum signage). Later, Thomas D. Carr offered rationale fortifying for why the Field Museum's *Gorgosaurus* was referred to *Daspleosaurus*.[1]

By 1994, Field Museum artists *horizontally* aligned Gorgy's vertebral anatomy, posed as if eating the *Lambeosaurus* corpse. This reflected current 'Renaissance Rex' science, and rather resembled posture and theme conveyed in the Milwaukee Public Museum's haunting, life-size *T. rex-Triceratops* diorama restoration (first displayed in 1983). Now *our* Field Museum's *Triceratops* cast skeleton fends off our feasting *Daspletosaurus* (as if conveying the message to the carnivore, 'don't even think about it!'). The presence of these dinosaurs accentuates Knight's world famous 'Rex vs. Tops' mural hanging on an adjacent wall.

Yes—it's still a terrific display, but somehow not as captivating as before, because let's face it; Gorgy now plays second fiddle to Sue.

By 1990 during the fleeting final days of Gorgy's Chicago reign, as I avidly honed prehistoric animal sculpting skills, a miniature sculpture of 'Gorgy & its intended Prey' resulted. This sculptural design was far more dramatically posed than in Maidi Wiebe's original miniature sculpture, which for many years had been formerly displayed in the Stanley Field Hall adjacent to the skeletons. However, the absolute best restoration of a similarly posed 'Gorgy & Prey' scenario is Michael Skrepnick's late 1990s painted rendition—formerly in the John Lanzendorf collection. Take a look sometime at this breathtaking scene—I mean, Wow![2]

Unquestionably, without 'Gorgy,' 'Rex' would be a lesser known entity today.[3]

Oops - I meant to say, *Daspletosaurus*.

Notes: (1.) Thomas Carr wrote the following publications - "Craniofacial ontogeny in Tyrannosauridae (Dinosauria, Coelosauria)," *Journal of Vertebrate Paleontology*, Vol. 19 (3), Sept, 1990, pp. 497-520, and Carr, Thomas D. "FMNH PR308 - Part I: or analyzing an Enigmatic Tyrannosaurid Specimen," *Dinosaur World*, no. 6, Spring/Winter 1999, pp.16-18, and Carr, Thomas D. "FMNH PR308 Part II: The Cross-Dressing *Daspletosaurus*," *Dinosaur World*, no. 7, Winter 1999/2000, pp.21-24. (2.) Skrepnick's art was published as a color Plate two pages before page 747 in the *Encyclopedia of Dinosaurs* (Academic Press, 1997), edited by Philip J. Currie and Kevin Padian. (3.) For excellent summaries of the classifications and reclassifications of *Gorgosaurus Daspletosaurus*, and *Albertosaurus* under individual entries for each dinosaur name in Donald F. Glut's *Dinosaurs: The Encyclopedia,* Jefferson, NC: McFarland Publishers & Company, Inc., 1997; Glut, Donald F. *Dinosaurs: The Encyclopedia - Supplement 1*. Jefferson, NC: McFarland Publishers & Company, Inc., 1999; Glut, Donald F. *Dinosaurs: The Encyclopedia - Supplement 2.* Jefferson, NC: McFarland Publishers & Company, Inc., 2002; Glut, Donald F. *Dinosaurs: The Encyclopedia - Supplement 3*. Jefferson, NC: McFarland Publishers & Company, Inc., 2003, and Glut, Donald F. *Dinosaurs: The Encyclopedia - Supplement 4.* Jefferson, NC: McFarland Publishers & Company, Inc., 2006.

THREE

THE HORNED DINOSAUR.

The *Agathaumas sphenocerus* (Cope) is based on the reconstruction of a possibly identical and prior restoration, *Triceratops prorsus* (Marsh). This elephantine Laramie Cretaceous dinosaur, twelve to fourteen feet in length, was herbivorous and harmless, but so well protected as to be free from molestation.

(Top) Charles R. Knight's 1897 Agathaumas originally appearing in Century Magazine. (Bottom) A British Museum postcard issued in the late 1950s showing Neave Parker's paleoart. See Chapters Ten, Twelve (and Twenty-two) for more. (postcard - A. Debus collection)

Paleoimagery versus Paleoart

Paleoimagery is the ubiquitous means by which most dinophiles—often young of age or of heart—become acquainted with former denizens of our primeval planet. Thus, it is peculiar that minimal attention was afforded the most brilliant of paleoartists (themselves), with their artwork taken much for granted, until a time when the burgeoning dinosaur renaissance was still in early stages during the late 1970s. The five chapters herein are loosely linked by their reliance upon, or fostering of, forms of paleoimagery and associated artists—a disparate array. Although several books addressing forms of paleoimagery have been published, in my opinion (as of late summer 2017) the 'true' volume on this topic has yet to be written. Obviously, such a generalized topic addressing a concept such as paleoimagery cannot be constrained within so few chapters, and therefore must become a recurrent theme throughout this book. Consider this segment merely a 'sampler.'

As a youngster, I was especially fond of absorbing paleo-landscapes portrayed by Charles Knight, Neave Parker, and Zdenek Burian, which I found in books and museum pamphlets, or in the case of the Field Museum—displayed on hallowed walls of Hall 38. I often tried to imagine myself stepping *through* the canvases into 'actual' primeval times to experience the strange and majestic animals depicted therein! As mentioned previously, by the late 1970s, I began a project in earnest to research the classic paleoartists of the 19th and 20th centuries, to see what I could learn about them. Over a decade later, I had come to be regarded as something of an 'authority' on (at least some aspects) of paleoart history. I recall writing Mike Fredericks during the early days of his publishing *Prehistoric Times* (and before we had the internet to rely on) that he should consider including articles on paleoart and paleoartists, thus re-emphasizing certain topical areas of his magazine. Material presented here as Chapter Ten is one of the early such articles to appear in *Prehistoric Times*.

In 2003 I lectured at Rockford, Illinois' Burpee Museum on the 'history of paleoart & paleoimagery,' and then again in 2004 on how to sculpt prehistoric animals. A couple years after, I presented another slide show on paleoimagery, this time encompassing a bit of 'Godzimagery' for an audience at Chicago's G-Fest. For, as I've suggested elsewhere, how pivotal is imagery of real paleo-monsters known to science in suspending disbelief in the matter of totally science fictional dino-monsters! Yes, paleoimgery and the allied history of paleoart were at the core of my avocational research interests back during the very late 1970s (& beyond).

Chapter Eight was perhaps the most widely spread article of mine appearing in this compilation. I remember what a thrill it was after the magazine issue came out in the spring of 1993. Not only did Don Glut's article, "Real Dinosaurs" also appear in the very same issue, but because one could buy the 'zine in every book and drugstore one could find (at least for a month or so), it was great fun watching wife Diane pointing to the magazine rack where the issue was displayed—demonstrating great marketing skills, beaming and exclaiming to store clerks and nearby patrons that "our article is in that magazine!"

Chapter Eleven concerning the "Question" of Benjamin Waterhouse Hawkins' spiritual nature was first printed in a 1995 issue of *Archosaurian Archive* edited by Gary Williams and Steve Harvey. While in this volume, I've endeavored to minimize overlap with my previous books, elements of this short piece were subsequently also incorporated into a longer article I co-wrote with Steve McCarthy on Hawkins for issue Vol. VI of *The Mosasaur: The Journal of the Delaware Valley Paleontological Society,* pp.105-115, May 1999 (not included in this collection). Of recent interest, is the revelation that pioneering paleoartist Benjamin Waterhouse Hawkins' preliminary small model sculpts for the 1854 Crystal Palace exhibition have been 'rediscovered' on display in a Cambridgeshire museum by British dino-aficionado Mike Howgate, as detailed in his interesting *Prehistoric Times* article (issue no.114, Summer 2015).

One significant Czech paleoartist of whom I've written previously in books and articles is Zdenek Burian. As I was winding down preliminaries concerning my *Dinosaurs Ever Evolving*, a series of important articles on Burian, written by John Lavas, commenced in the pages of *Prehistoric Times* (beginning with issue no. 116, Winter 2016). Burian's art was a 'constant' in the formative years of my youthful paleo-life [as I was then learning to appreciate all manner of paleoimagery], since October 1959 when my brother Rick and I first feasted our eyes upon Burian's paintings reproduced in Josef Augusta's *Prehistoric Animals* volume. My fascination with Burian continued on into adulthood, as I researched his life during the early 1980s. I wrote an early manuscript, "They Painted the Dinosaurs," one of the first in English including an outline of Burian's career. I encourage readers to seek Lavas's *PT* series on this important pioneering paleoartist, who excelled in 'monochromatic' depictions of prehistory (which have a nearly 'photographic' quality), as well as rather impressionistic paintings—in vivid color—showing life 'witnessed' through portals of geological time.,

In looking through my old correspondence and more obscure samples of Neave Parker's art—including a nice restoration of tiny Triassic mammals, *Morganucodon,* printed in Adrian J. Desmond's *The Hot-Blooded Dinosaurs* (1975)—I'm reminded that my ambition to write about the classic paleoartists sprang from a positive experience involving an individual in Poland who supplied me with information about a native artist …. one whom I was seeking further information about for a prior article of mine titled "Paleophilately: Stamps, Stones and Stories" (printed in the *Earth Science Club of Northern Illinois Newsletter*, Oct. 1981). So, by the fall of 2012, although it seemed everything that could possibly be known about Parker had already been published, Mike Fredericks of *PT* urged me to send him a new article on Neave Parker: material represented here as Chapter Eleven was the result.

> These chapters originally appeared as articles in the following publications:
> Chapter Eight—*Dinosaur: The Collectible Edition. Dinosaur Movies—The Complete History* (Starlog Communications, 1993, pp.81-86).

Chapter Nine—*Prehistoric Times* nos. 122-123, Summer & Fall 2017. (In press as of this writing).
Chapter Ten—*Prehistoric Times* no.6, May/June 1994, pp.19-21.
Chapter Eleven—*Archosaurian Archive* no.3, Oct/Nov 1995, p.22.
Chapter Twelve—*Prehistoric Times* no.105, Spring 2013, pp.42-44, 46.

Chapter Eight—*Prehistoric Scenes*

Throughout paleontological history, it has been the painters and sculptors who shaped our visions of dinosaurs. In a sense it was man who created the dinosaur. While paleontologists may have excavated and reassembled the bones of those long-dead creatures, it has been the artists who restored them, in the flesh, for all to see. Who are the people who have made these 'journeys' into the remote past, and what contributions have they made to contemporary science.

"Artists are the eyes of the paleontologists, and paintings are the window through which non-specialists can see the dinosaurian world" wrote vertebrate paleontologist Dale Russell. In a sense, we can look back into time's corridors through virtual 'windows' created by paleoartists. But because their restorations are often so effectively done, one easily forgets that those who crafted these images of dinosaurs and other Mesozoic animals were not actually on the scene to paint them. And yet, were it not for the artists who team with professional paleontologists, it would be difficult for most of us to picture how these ancient dragons appeared in life.

In 1834, desperate for additional income to clear his growing debt, physician and famed fossil hunter Gideon Mantell recruited artist John Martin to create a scene from prehistory, involving the *Iguanodon*, a dinosaur described by Mantell in 1825. The result was "The Country of the Iguanodon," a mezzotint showing three great reptiles struggling in a primeval setting. It was used as the frontispiece for Mantell's *Wonders of Geology*, published in 1838.

Both author and artist straddled theological boundaries with their presentations of the deceased creatures. Mantell referred to the Creator as the cause of apparent evolutionary change, while maintaining that Scripture bore no implications for science. Martin was a painter of apocalyptical historical events who won particular renown in France and England, specializing in such biblical scenes as the Great Deluge and "The Fall of Babylon." "The Country of the Iguanodon" symbolically united historical biblical tradition with the more scientifically founded genre of natural history illustration.

Despite the success of his book, poor Mantell eventually had to sell his fossil collection. Then, to make matters worse, his marriage became another extinction in his life. However, the incorporation of Martin's nightmarish scene into Mantell's popular book illustrates how different paleontology was in the early 19th century, compared to where the science is today. Prior to publication of Charles Darwin's *The Origin of Species* (1st ed. 1859), it was difficult for scientists to relinquish theological implications of dinosaurs.

Benjamin Waterhouse Hawkins was the most influential dinosaur artist of the 19th century. His prominence spans the Darwinian evolutionary debates and the period during which the first remains of relatively complete dinosaurs were discovered. Because fossils studied until then were incomplete, Hawkins' recreations of *Iguanodon, Megalosaurus,* and *Hylaeosaurus,* completed in 1854 for the Crystal Palace grounds at Sydenham, were realized to be inaccurate representations over a century ago.

These sculptures were completed under guidance of Richard Owen, an eminent British paleontologist who sometimes got on Hawkins' nerves. For instance: While Owen was seated in the belly of the nearly completed *Iguanodon* model during a festive New Year's Eve affair, he picked a most unappetizing moment to critique Hawkins' creations. When informed by Owen that his *Iguanodon* had too many toes, Hawkins steadfastly replied that if they were corns, he would have removed them. But since they were toes, they would have to stay!

Owen enlisted the dinosaurs in his crusade to defend paleontology from evolutionists, who claimed species transmuted naturally and progressively into more advanced forms. Interested spectators may not have understood all the implications of Hawkins' dinosaurs for the science of paleontology, but remarkable reconstructions and eerie primeval setting at Sydenham certainly proved captivating to many visitors.

After it was recognized that dinosaurs such as "Laelaps," and *Hadrosaurus* (both discovered in North America) may have been bipedal instead of quadrupedal, Hawkins had another try at a "Palaeozoic Museum," this one in New York. It's possible the antireligious theme of this museum proved objectionable to local politicians, and this factor may have partially contributed to the destruction of Hawkins' unfinished models and casts.

Hawkins retains credit for mounting the first relatively complete dinosaur skeleton, a *Hadrosaurus* for the Philadelphia Academy of Sciences in 1868. It *was* a fairly complete specimen as dinosaur fossils generally go—but was it all dinosaur? Unable to obtain an actual skull, and mindful of the anatomical similarities between the living iguana and the known portion of *Hadrosaurus'* remains, cast an over-sized iguana skull with jaws sufficiently large to accommodate the teeth. Hawkins also scaled a human collarbone to fit analogous missing parts.

In the 1860s, several editions of Louis Figuier's *Earth Before the Deluge* were published. Figuier targeted artist Edouard Riou to complete prehistoric scenes. There were only two scenes depicting dinosaurs, both resembling the models Hawkins had sculpted a decade earlier. However, many of Riou's Mesozoic animals are depicted in the Darwin-inspired "nature red in tooth and claw" struggle for survival.

Although blind in his right eye, Charles R. Knight was blessed with an extraordinary ability to project himself and his artistic sense into ages past. Expressing kinship with the ancient European Cro-Magnon artists, Knight believed that within the soul of man existed a primitive artistic urge to study the form of animals.

Early in his career, Knight conspired with famous paleontologist Edward Drinker Cope, and during the late 1890s, they produced the 19th century's most beautiful dinosaur paintings. These works depicted the leaping carnivore "Laelaps" and the *Agathaumas*. Because there were hardly any fossils attributed to these genera, Knight's restorations probed deeply into Cope's imaginative sense. In particular, *Agathaumas'* beautiful crest of horns is utter speculation, for Cope had discovered no skull. And yet, could any fossil compare with the stately creature Knight divined?

Particularly through his work at the American Museum of Natural History and the Field Museum in Chicago, the temperamental Knight transformed the science of fossil vertebrate restoration. Knight didn't specialize in dinosaurs, although much of his notoriety stemmed from brilliant portrayals of animals in fantastic primordial settings. Knight accepted Darwinian evolution, and even made clever observations concerning skeletal similarities between dinosaurs, birds and the extinct bird, *Archaeopteryx*. Reproductions of his illustrations, paintings and models have graced the pages of many scientific books and popular magazines. Knight's depictions served as basis for movie dinosaurs (e.g. those in 1925's *The Lost World* and 1933's *King Kong*), as well as the mechanical Sinclair dinosaurs exhibited at Chicago's Century of Progress exhibition.

Did he somehow perform a magic trick? Some of Knight's scenes appear so realistic that it seems possible to step through the picture frame and into the world of yesteryear. One is often left with a yearning for the time beyond the canvas. Indeed, the spirit and legacy of Charles R. Knight live on today.

For its time, one of the most exact three-dimensional representations of a dinosaur was created by Erwin S. Christman for the American Museum of Natural History. This was a miniature sculpture of the Jurassic sauropod, *Camarasaurus*. Christman, a playful but virtuous character, began his professional career at the tender age of 15 and died in his prime at 36. During his 21-year career, the artist spent many hours studying and drawing fossil bones, steadily increasing his familiarity with dinosaur features to the point where he could build small-scale replicas of the skeletons. Such painstaking efforts greatly facilitated paleontologists' study of prehistoric animals.

Christman then constructed a flexible miniature of the skeleton, which was used to select a probable life pose. Christman determined where musculature adhered to the skeleton and added a 'fleshy' exterior. Such efforts vastly improved the appearance of mounted skeletons, which had previously been assembled on metal rods—sometimes without great regard for how these supports should be curved and angled.

Consensus opinion concerning sauropod natural history has changed considerably since the early 20th century. However, many museum artists still rely on techniques introduced by Christman in his early study of *Camarasaurus*.

Zdenek Burian's association with paleontology began in 1935, when he completed an oil painting depicting two species of Paleozoic amphibians for the paleontologist Josef Augusta. Books published by Augusta, which Burian illustrated, have become treasured dinosaur memorabilia. Claiming to have been influenced by Knight's expert renderings, Burian became the most prolific and popular European paleoartist. Of his (estimated) 1,023 oil paintings, 381 are on prehistoric themes. Burian also completed another 114 watercolors on the subject of prehistory. During his life, the artist finished over 15,000 illustrations. Sixty of his paintings are on display at a Burian Museum, maintained by the Czech government.

Beyond the artist's vision, psychical energies may also be of value in discerning the world of yesterday, as in the eerie instance of Neave Parker. Parker teamed with British paleontologist William E. Swinton during the 1950s to produce some of the cold-blooded dinosaur era's greatest restorations. Parker's photographic sense of the dim past innovatively captured dinosaur fauns during their heyday in different regional settings.

Parker wrote and illustrated children's books prior to his association with Swinton and mammologist Maurice Burton. However, those who knew Parker spoke of his uncanny ability to 'see' the ghosts of dogs, animals of which he was very fond of (presumably when they were alive). Apparently, Parker who had mastered the art of restoring deceased animals, would also on occasion describe with alarming accuracy the appearance of dead dogs to their bewildered former owners. Despite this idiosyncrasy, Parker was very highly regarded by his associates at the British Museum. His restorations most vividly represent formerly held opinions of how such creatures appeared and lived.

Rudolph F. Zallinger created the world's largest mural containing dinosaurs for Yale University. In 1942, he was asked by the director of Yale's Peabody Museum to paint a wall in the museum's Great Hall. Five years later, the vacant space was magnificently adorned with a mural known as "The Age of Reptiles." This impressive creation, which features majestic dinosaurs in the Olympian setting, garnered Zallinger the 1949 Pulitzer Prize for painting.

This "panorama of time" is 16 feet tall, spanning over 300 million years of represented geological time along its 110-foot length, and nearly 75 feet of this mural is devoted to dinosaurs. Although the painting's vivid hues immediately draw the eye, Zallinger had no idea what colors dinosaurs sported. The good-natured artist relied, most successfully, on intuition, and has often chuckled when paleontologists mistook them for the creatures' true colorations.

Robert T. Bakker's early fascination with dinosaurs was fired upon viewing a 1953 issue of *Life* magazine containing a foldout of the "Age of Reptiles" mural. Known chiefly for his heretical views on the science of dinosaurs, Bakker (along with John Ostrom and Peter Galton) ushered in a renaissance of understanding concerning warm-blooded dinosaur physiology, evolution and movement. Over two decades later, some of these ideas remain controversial. But despite this, the flamboyant Bakker may have made his greatest mark in the artistic instead of the scientific arena.

In his first article, Bakker defied convention by publishing restorations of a pair of galloping, horned chasmosaurs, a trotting *Protoceratops*, and a sleek giraffe-like *Barosaurus,* and then challenged dogma that dinosaurs were inefficient, lumbering beasts fit only for eventual extinction. In Bakker's view, supported thereafter in many spectacular illustrations, dinosaurs were "superior" to other contemporary organisms. His fast-running *Deinonychus* illustration, published in

Ostrom's 1969 monograph on that animal, symbolically ended the reign of cold-blooded dinosaur restorations.

Today, during the celebrated dinosaur craze, a new generation of talented artists, including sculptors David Thomas and Stephen A. Czerkas, along with illustrators Mark Hallett, John Sibbick, Douglas Henderson and Gregory S. Paul, have produced magnificent renderings of living dinosaurs for museum displays and popular books. Such individuals bear great responsibility of how dinosaurs appeared in life. And yet, someday, even their visions may be altered in the course of new scientific discovery or reinterpretation.

How accurate are modern restorations? Think of it this way. Fifty million years from now, how would a future intelligent race restore *our* fossils? Would we be adorned with antlers, plumage or tiger-striped skin? Our perception of 'accuracy'—in terms of color, body posturing and proportion, skin details and other speculative fleshy or feathery accoutrements—is conveyed through the artist's vision, but is also founded on factual knowledge accepted by a consensus of dinosaur paleontologists. And yet, many of the current restorations must certainly be closer to reality than in Victorian times, given that so much more fossil material has been studied during the past 150 years.

Until time machines capable of traveling into the remote past are invented, or until biotechnologists create genetically engineered dinosaurs, our images of them will always contain a fanciful element. Man's dinosaurs shall continually evolve.

(Co-authored with Diane Debus, 1993.)

Chapter Nine—*New-Millennial Paleoimagery*

So it's time to consider several newly minted terms like John Lavas' 2017 "paleo fiction art" (i.e. made in reference to Zdenek Burian), and ponder whether paleo-people will make *T. rex's* lip-s – "*tick.*" (True dino-philes might see what I did there. Maybe?) What's new: absolutely so much! But you need perspective. In this 2-part installment, let's explore how some commonly used PT conventions and terms link to the practice of paleo-image making.

When I first began researching a 'who's who of paleoartists' back in 1980, there weren't as many individuals to delve into. And when I completed the manuscript to a book, *Paleoimagery: The Evolution of Dinosaurs in Art* (2002), in early 2001, nearly 18 years ago, the number of described dinosaur genera was still modest by comparison to today's annually swelling fossil ranks. Thanks, especially, to *Prehistoric Times* magazine and developments in vertebrate paleontology, many 'new' artists (and scientists) have come to the fore. Although paleoart is a 'survival of the fittest' field, practicing is a labor of love. Paleoartist Don LoRusso once said, "You do it for the love of the art and the personal satisfaction of the finished piece." (*Prehistoric Times* # 111, p.53) During new-millennial years so far, the field attracted more talented paleo-image makers than ever before, producing an eye-popping, bewildering array of images and sculptures along the way. Dino-science and prehistoric animal art are thriving like never before (an escalation in popularity burgeoning since the 1980s), in books, magazines, museum exhibitions, toys, dino-documentaries and many other forms, some of which would have been inconceivable decades ago.

It seems we've only just emerged into "paleontology's artistic renaissance." Yes, we're way beyond the pioneering 'era' of Benjamin Waterhouse Hawkins, Charles Knight, Rudolph Zallinger, Neave Parker and Zdenek Burian, and well into a 'renaissance' phase heralded by Robert Bakker in 1969 with his sprinting deinonychosaur drawing. Yet fresh studies of their magnificent works continue to enthrall us. Over a century ago, would-be aspiring paleoartists had Frederik A. Lucas' *Animals of the Past* (1901) to reflect upon. Today, we have Gregory S. Paul's magisterial

The Princeton Field Guide to Dinosaurs, 2nd ed. (2016), perhaps the paleobiological "CRC" (*Handbook of Chemistry and Physics*) equivalent for guiding paleoart restoration and reconstruction of dinosaurs. Imagine how startled we'd have been if *his* book had hurtled out of a space-time warp from the cosmic void, six decades ago—back during our limited mid-20th century state of knowledge!

The new millennium has only just begun, but already we've witnessed significant strides in our depiction of prehistoric life, dinosaurs particularly. (And now paleontology shifts and rumbles so rapidly. In March 2017, we learned of a revised dinosaur phylogeny and re-diagnosis linking clades: "theropods" with "ornithischia" into "ornithoscelida," and "sauropodomorphs" with "herrerasauridae.") On many levels we've gained vast new knowledge concerning how these extinct forms should be considered and depicted. So here I will comment on the current status of paleoimagery, focusing more on the 'now' instead of 'then.'

"Paleo-artist," now a familiar term coined by geologist Alexander D. Winchell in 1870, became more commonly used by dinosaur enthusiasts during the 1990s and beyond. A related term, "paleoimagery," carries broader connotations. Paleontology is a highly visual science, and our enchantment with such imagery has fueled public fascination with prehistoric life (particularly vertebrate forms), notably in television and movies. We may even discern a basic twofold overarching, historical outline in "paleoart"- that revered, scientifically oriented practice of visually and realistically reconstructing extinct life forms, pertaining to (1.) a traditional life-through-geological-time edifying theme, then (2.) transforming into visual imagery conforming to a "dinosaur renaissance" phase, which is ongoing, branching at accelerated pace. In his article, "Dinosaur Renaissance: A Brief History of Paleoart," Steve White adds, "Paleoart is a rather unique scientific discipline in that many scientists are pretty solid artists, while a number of artists have acquired serious scholarly reputations." (*Dinosaur Art: The World's Greatest Paleoart*, 2012).

Paleoimagery's historical development is more complicated, diversified and variegated than what we see in the (narrower) paleoart realm. Prehistoric animal (and fantasy and science fictional dino-monster) *paleoimagery* has always been introduced via many levels of media and sophistication, for numerous purposes, while intended for a broad demographic. Yes, those plastic dinosaur toys, dino-monster films, or even your old Aurora dino-model kits, for example, would qualify as 'paleoimagery.' With the care, consultation and precision invested in preparation of small figures these days, dino-toys, in fact can sometimes approximate museum-ready paleoart. Admittedly, in my 2002 book, I emphasized the more scientifically founded 'paleoart' end of the paleo-image spectrum.

Art historian Jane P. Davidson, who introduced another (titular) term in her 2008 book *A History of Paleontology Illustration*, mentioned that when preparing a college course concerning this field (novel for its time; no such courses existed in the 1990s), a student commented that "…she was astonished by how much 'stuff was there.'" (p.xiii) Yes, truly there was then, and continues to be, so "much stuff" to assimilate, making the task (a fun task nonetheless) difficult. This is in utter contrast to another academic W.J.T. Mitchell's misguided forecast (made shortly after announcement of the landmark 1998 Chinese feathered dinosaur discoveries), that dinosaurs will become blasé' and possibly boring in years following publication of his 1998 *The Last Dinosaur Book*. "How much mystique will remain when the creatures formerly known as dinosaurs take on their proper name of 'early birds,' and dinosaurology disappears to be replaced by paleo-ornithology?" (p.25) Simply judging from the outpouring of new books reviewed in each new *Prehistoric Times* issue, nearly 20 years later, it seems the "last" dinosaur book has yet to be published! Most dinophiles evidently have embraced the 'feather revolution' after all (perhaps with exception of paleoartist Raul Martin who seems less enamored, per remarks in 2012's *Dinosaur Art*, p. 187).

Permit this odd analogy that oldsters (like me) may appreciate. Indeed, isn't it rather like the numbers of television cable stations and online networks which have arisen and multiplied over time since the mid-1960s – (e.g. unrelated to in context, but since the dinosaur renaissance dawn)? There are almost too many 'shows' available to watch now, as opposed to the late 60s when you only had ABC, NBC and CBS (or WGN if you lived in Chicago). And so it goes with the swelling ranks of "dinophiles," paleo-image creators and favored dino-creatures, with new genera described every month! Today, just as there are so many more 'channels' than ever before to choose from or means for viewing live or pre-recorded programs and VLOGs, analogously, we seemingly have an embarrassment of riches in cases of newly published dinosaur genera, and paleoartists who frame their primeval visages.

So, what is Davidson's "paleontology illustration," and how might it differ from, or ally with other commonly encountered terms like "restoration" and "reconstruction"? Davidson associates her 2008 term with an artistic striving for *realism* in depicted art showing fossils and prehistoric life restorations. Think, perhaps, of her "paleontology illustration" as the visual presentation of paleo-evidence, ideas, interpretation and realistic documentation of scientific visual data (including metadata) in various art media and forms (including lithographic engraving, photography, etc.). Davidson, doesn't fully address theoretical or rhetorical notions surrounding many aspects of paleoart, however, as her book is presented as more or less a survey or historical outline of 'stuff' that's been published, or out there.

Davidson tends to draw the line (more or less like paleontologists did two centuries ago) at more fanciful representations, and questions status of other media such as sculptures, and internet art - due to its "ephemeral," digitized online nature. (pp.180-181) Note that this isn't a criticism of the use of original restorations facilitated via use of computer digitization techniques (e.g. "digital paintings"). In the matter of internet postings, Davidson opines "... unless the art itself is digital and created solely for that website, then we may suggest that this is not paleontology illustration, but only a way to see it for a period of time." Yet there are highly popular Facebook pages devoted to "paleoart" and "paleoartists" where a continual, fresh stream of images may be seen daily. Many of these members might not accept that, when posted, their works are not "paleontology illustration." And while apparently accepting that table-top (indoor), albeit 'scientifically accurate' sculptures can fall within purview of paleontology illustration, she further states "... large-scale outdoor sculptures of dinosaurs ... may stretch the definition of scientific illustration somewhat, but there is no doubt that they reinforce the contention that sculptures of fossil restorations have become increasingly popular."

Whether they're large or small, situated indoors or outdoors (scientifically rendered, or spirited do-it-yourself, home grown projects), such sculptures clearly are forms of paleoimagery. Readers also realize that whereas large (outdoor) paleo-sculptures have been familiar for well over a century since the Victorian age, it is instead the smaller variety of sculptures that have only in recent years become so eminently popular, as fortified in Robert Telleria's 2012 book, *The Visual Guide to Scale Model Dinosaurs*, and Telleria's and Joe DeMarco's CD, *Dinosauriana: The Essential Guide to Collectible, Figural Toy and Model Dinosaurs*. Due to its prevalence, one may even refer to a prehistoric animal sculptural art "industry," perhaps instigated, to at least minor degree, by a certain 1995 book, *Dinosaur Sculpting*. But, despite W.J.T. Mitchell's 1998 claim, other forms of sculptural iconography championed in his *LDB*, to me, don't qualify as "paleoart" or even paleontology illustration; see his pp.264-275. A simple test may be applied. If the thing you're looking at doesn't remotely resemble something 'prehistoric' (yes, conceivably even that handy dinosaur-shaped cookie cutter would do!), then it's truly difficult to categorize it as a kind of paleoimagery. Of course, everything is subjective, and such considerations may at times seem overly philosophical. Consider—when might a CAT scan of, say, a coprolite be regarded as paleoart? Humor aside though, paleoartist Tyler Keillor discusses how critical some CAT scans can be in "Digital Paleontology," toward accurately restoring dinosaur crania. (See *Prehistoric Times* # 112, pp. 28-29.)

Sometimes, the fabric of scientific realism (which, historically, may seem like a moving target as paleontology evolves with new discoveries) may be extended, acquiring speculative or hypothetical tones, so often introduced and welcomed today as opposed to 200 years ago. LoRusso stated in *Prehistoric Times* # 111 (p.53), "…I try to follow the science as far as it will take me, then the artistic license comes in. What will make this look realistic and lifelike as well as plausible." Let's not forget that paleoimagery (or sometimes even paleoart) often is deliberately speculative. In this vein, Luis Rey's precocious fancifully colored, feathery dinosaurs, or even filmic, often fateful, metaphorical dino-monsters such as Ray Harryhausen's 'Rhedosaur,' *Jurassic Park's* "raptors" or Toho's mutated dino-dai*kaiju* monster, Godzilla, come to mind.

According to Davidson, BBC's *Walking With Dinosaurs* (1999) strained the limits of what might be regarded as "paleontology illustration." Yes, here we straddle the boundary where paleoimagery seems more "guided by fantasy," falling out of favor with concepts of 'traditional' paleoart. (Davidson, p.156) Can a cgi-operated dinosaur paleo-image teach us more about the science of dinosaurs than a highly detailed coprolite CAT scan? It depends on research objective. At the very least, those lively, sometimes anthropomorphized "cgi" restorations do force us to think. And then there's aforementioned "paleo-fiction art": that accompanying or used to promote science fiction.

A curious case in point might be Dougal Dixon's illustrated "new dinosaurs" of 1988 (granted further evolutionary time because the asteroid never impacted 66 mya). How much less then is G. S. Paul's fantasy extrapolation, a 1989 illustration, "Dinosaurs of the Cenozoic on the Great Plains of the Neogene" (pp. 36-37 in *Dinosaur Art: The World's Greatest Paleoart,* 2012)? Both are exercises in extrapolating hypothetical evolutionary pathways: both boldly reflect a similar 'what if' universe—representations of Gouldian contingency!

For modern paleontologists and artists intrigued by prehistoric life and landscapes, freedom to speculate is a key lure into the profession. For example, in restoring life appearances of embryonic pterosaurs (*Fossil News*, Summer 2016, p.42), paleoartist Gareth Monger invented another descriptive term, "speculative paleoillustration," for his article titled "Speculative Paleoart." Much may be inferred from exceptionally preserved fossils—conveyed via paleoart--such as soft organ, internal anatomy (e.g. pneumatic respiratory structures). To many, perhaps, Monger's term may seem redundant because paleoart is often inherently "speculative." For further interesting perspective, see Todd Marshall's entry in the aforementioned *Dinosaur Art* (pp.92-93, 97, 100).

Meanwhile, by the early 1990s, paleontologist Stephen J. Gould, a staunch admirer of Charles R. Knight's artistry, had become a vocal supporter of 'paleontological iconography,' echoed through several of his popular essays on how (abstract) evolutionary 'contingency' might be effectively portrayed. Then there's "paleo-fanart," which I encountered on a Facebook posting in early 2017. The author of this original term had dismissed an artist's furry "grizzly *T. rex*" that, he stated, couldn't qualify as true "paleoart." I suggest that instead this may have been an example of speculative paleoart, depending on the factors underlying its creation. But the question lingers; have we carved out a new paleoimagery niche for "paleo-fanart," apart from paleoart? Are we not all in some capacity … "fans," or at least paleo-aficionados? Do we now need terms to describe ourselves too, besides "dinosaurologist" which applies to some?

Paleoimagery is the ubiquitous means by which most dinophiles – often young of age or of heart –become acquainted with former denizens of our primeval planet. Thus, it is peculiar that not much attention was afforded former paleoartists (themselves), with their artwork often taken for granted, until a time when the burgeoning dinosaur renaissance was still in early stages.

Decades later, in mid-February 2017, concerned members of the Facebook "Paleoartists" group wanted to qualify posted content of certain kinds of visuals (i.e. paleoimages that aren't regarded by a consensus as 'true' scientifically founded paleoart). So, based on a set of rules, an interesting suggestion was made to weed out, for example, fantasy dinosaurs, artwork depicting 'real dinosaurs,' yet placed /cast in fantastic, unlikely settings, varieties of dino-monsters (perhaps of sci-fi/horror derivation), restorations that otherwise seemed too conjectural. I understand the rationale, but it's a bit of a slippery slope to make (fine) judgements of this nature. It's a tough 'call.' Ultimately who must sanction how much detailed research and contemplation is needed before the image rises from the more common ranks of paleoimagery to a more exalted paleoart form?

Where does one draw the 'line'? Regardless, don't many of us enjoy all the diversified categories? What qualifications are necessary to make such calls?

Apart from accurate depiction of fossil specimens in a drawing, engraving or photo, terms such as "restoration" and "reconstruction" are often loosely intertwined. However, to make proper distinction, generally, I've found it useful in referring to vertebrate paleontologist William E. Swinton's mid-20th century definitions. Swinton reserved "reconstruction" for organizing or reconfiguring the skeleton of an out of the matrix prehistoric vertebrate either on paper or as a skeletal mount, and "restoration" used in reference to an artistic, interpretive impression of the animal as it may have appeared as a living organism. Unsurprisingly, paleoart restorations may introduce more speculation than skeletal reconstructions. In both cases, however, at least some degree of artistic license is unavoidable. I've noted that G. S. Paul reserves a term – "full-life restoration"—for specimens deemed suitably complete for paleoart. Conventionally, drawn reconstructions are now "shaded" to show probable body outlines.

Did I coin the term, "paleoimagery"? My use of it (i.e. perhaps first used in print – 2002) stemmed from a similar term, "Dinosaur Imagery" which happened to be the title of a 2000 book showcasing several items in John Lanzendorf's private paleoart collection, setting a new-millennial wave for collectors of original paintings and bronze figures, as well as aficionados of books featuring reproductions of such art. Due to the impressive nature of his (former) high quality dinosaur collection, and in thanks for supporting paleoartists, Lanzendorf's name was allied to the Society of Vertebrate Paleontology's "Lanzy," (i.e. the Lanzendorf PaleoArt Prize) awarded to paleoartists, beginning in October 2000.[1] But, expectedly, many have personal favorites, eye-catching examples such as paleoartist Danielle Dufault conveyed in a Dec. 27, 2016 online article, "Mud dragons, Tully Monsters, and Toothed Whales: The Best Paleoart of 2016."

Two additional, adjectival terms are "museum quality" and "scientifically accurate," both usually associated with paleoart advertising. If works are exhibited in a museum, by definition one must agree they're of sufficient museum 'quality.' My impression is that the general level in sculptural figure quality has escalated considerably during the dinosaur renaissance, even in the case of toys, some of which certainly are 'museum quality.' And being 'scientifically accurate,' usually suggesting a scientist may have directed creation of the art or made it her/himself, remains a moving target. As dinosaur science moves on, "progresses" some might say, historically older paleo-images (restorations and reconstructions) remain static, memorialized in time – bearing testimony to those who imagined (accurately) such specimens long ago. As one recent example, in 2016 paleoartist Robert Nicholls sculpted a life-sized *Psittacosaurus* sculpture earning high praise, deemed as the "most accurate depiction of a dinosaur ever created," for reasons explained in *The Guardian (*Sept. 14, 2016) and mentioned further below.

The meaning of such artworks, especially paleoart, is often conveyed via "imagetext," which may also suggest hypothetical or rhetorical ideas, visually and through associated cues or written 'captions.' Interpretive meaning of paleo-images and associated iconography is essential to the viewing of paleoimagery. When we look at dinosaur imagery, it connotes an idea, conjuring something in the mind that its creator(s) perhaps wished us to see, comprehend, though always subject to interpretation and distortion. Otherwise, the artwork and images would be of lesser interest to viewers, instilling far less frisson. W. J. T. Mitchell applies the term "imagetext" to the concept, "… a combination of verbal and visual signs … when we see a dinosaur … we are seeing a constructed image with an assigned name and description … an artifact, a visual-verbal-tactile construction based on its remains and an array of prototypes we use to make sense of those remains." (p.52). Upon this framework, in 2004 Kathryn Northcut intensively studied the manner in which dinosaur restorations and reconstructions, viewed in context as paleoart, are "rhetorical" in nature.

Is the term "dinosaur renaissance" an outmoded term by now? No, it remains relevant, although evolving (as outlined in Chapter 36 of *Paleoimagery*). Two decades ago I inconclusively attempted to discover who exactly coined this apt term. Its author may have been Robert Bakker – the prime suspect—or, alternatively, an editor then on staff at *Scientific American*, for the term first appears in print on the cover of their April 1975 issue, advertising a now famous article of same title by Bakker wherein he suggested that small dinosaurs like *Archaeopteryx* wore insulation. In my 2002 book I loosely structured the dinosaur renaissance into three distinguishing periods of scholarly activity, then culminating with the now persistent "feather revolution" paleophiles find themselves flocking to. But other terms have since arisen in recent publications, also acknowledging a feathered dinosaur craze, plus known bush-branching, bioevolutionary—saurischian/maniraptoriform—ties to dinosaur "paravians" (some having plausible flight capabilities) and modern birds.[2]

A competing, relevant term "dinosaur revolution" was applied to both a *Geotimes* 2006 article and a 2011 dino-documentary. Then in 2014, an author referred to the "2nd paleo-art revolution" (in *Prehistoric Times* # 111, p.27). "Revolution" appeared in the subtitle to Adrian J. Desmond's 1975 landmark book, *The Hot-Blooded Dinosaurs: A Revolution in Palaeontology*. So, I submit that "dinosaur renaissance" may, although in vogue, may be becoming rather vague, in need of further clarification. Have we moved from a "renaissance" phase into a "revolution"? What will the tides of history proclaim?

Then a new phrase capturing the spirit of our times coined by Jacqueline Ronson in referring to "paleontology's *artistic* renaissance" (my italics) appeared in the title of an online Dec. 16. 2016 article. The title alone triggered contemplation: for what reasons might today's paleoart be considered or implied to be that much 'better' perhaps than paleoart of yesteryear? Is it mainly technological approach, techniques and accessibility to amazing new fossils that have altered our picture significantly? Can we pin a date as to when paleontology's *artistic* renaissance began in throes or midst of a 'dinosaur renaissance'? It would seem fairly recent phenomena and discoveries fuel revivified or extrapolated paleoart inspiration. First, paleontology has increasingly moved from 'hard rock' to 'software.' We live in a highly computerized/digitized age, with access to amazing laboratory sophistication, unimaginable a century ago, wielding revelatory kinds of instrumental methods on (the best preserved) fossil specimens. So, relying on foundational principles of biology, chemistry, physics and engineering, aided by new technologies and analytical tools, amazingly refined assessments and results are permitted.

Also we've witnessed a sea change, cultivating revelation about integumentary appearances of dinosaurs that was mere speculation forty years ago when Robert Bakker and Adrian J. Desmond introduced us to an outlandish idea about "hot-blooded dinosaurs" and spoke of avant-garde creatures like *Sordes pilosus,* popularizing a once heretical notion since continually supported by accruing lines of evidence, which now more than barely hints at likely evolutionary lineage. As always, via the media, scientific knowledge filters down to the masses, often in forms of favored paleoimagery for many to ponder and enjoy.

Decades ago, Rudolph Zallinger could only speculate on true coloration of dinosaurs, based on those evinced in living reptiles, in his landmark "Age of Reptiles" Yale Peabody Museum mural, imaginatively introducing a striped plateosaur pair for the Triassic portion. Not long after the avian-dinosaur relationship was more properly established during the dinosaur renaissance did paleoartists like Luis Rey increasingly dare to show not only body coloration, but vivid feathers on their restored dinosaurians. As Rey stated in 2012's *Dinosaur Art* volume, "… science fiction with logic, not fantasy … The animals we depict are extinct and even if we follow strictly all the evidence …the external aspect will still be a conjecture, especially in color, ornaments, etc., and there you have the great opportunity to use your imaginative resources …"

And yet such (once) idle, hypothetical speculations may have been born out as fact, based on the most recent, post-2008 interpretations of "microbodies" found in fossilized feathers (and 2016 discovery of dino-feathers preserved in amber). Although a possibility remained that such microbodies were bacteria preserved on fossil feathers, by 2010, it seemed that key types, were identifiable as "melanosomes," responsible for pigmentation, indicative of color. Through detailed examination of melanosomes, therefore, one might infer color of ancient feathers, plumage in Late Jurassic/Early Cretaceous paravians, such as the *Anchiornis, Sinosauropteryx*, or most recently, *Eoconfuciusornis* .[3]

Paleontologist Jakob Vinther who published groundbreaking melanin studies (above) moved pigmentation intrigue further along with aid of paleoartist Robert Nicholls; together they produced that aforementioned "most accurate" dino-restoration in 2016, of an ornithischian dinosaur – *Psittacosaurus*—based on a "3-dimensional camouflaging" study. This was a foray into coloration patterning, not in "integumental appendages, such as feathers." The study, published in *Current Biology* (v.26), involving life-sized sculptures of the dinosaur remarkably proved beyond a 'shadow' of doubt that *Psittacosaurus'* body pigmentation was "countershaded with a light underbelly and tail, whereas the chest was more pigmented …" suggesting a preferred "… closed habitat such as a forest with a relatively dense canopy."

Research studies testing hypotheses originating within paleoart are more common now than in yesteryear. Nicholls' 2011 painting of a restored pair of *Carcharodontosaurus* hoisting an unlucky medium-sized sauropod in their jaws (a scene titled "Double Death," recently reproduced in *Dinosaur Art: The World's Greatest Paleoart* (2012), pp.160-161) inspirationally fueled another paper (*The Anatomical Record*, v.298, 2015) by Donald M. Henderson. He calculated with exacting bioengineering detail that indeed such a maneuver might be possible, provided the victim did not exceed 850 kilograms, 8.3 meters in length, and "load in the jaws would still be over the feet." Here, paleontology was driven by a paleoart 'sample,' rather than vice versa as is usually the case.

During October 28 to 29, 2016, the UK Dinosaur Society sponsored a special Paleoart Exhibition attended by paleoartists—the likes of (the aforementioned) Nicholls, John Sibbick, Luis Rey, pterosaurologist and paleoartist, Mark P. Witton, and Katrina Van Grouw. Well done! (There *was* such an extraordinary event once in California, the famed "Dinosaurs Past and Present" Symposium, in 1986, but to my knowledge nothing quite like that here since.) Increasingly, well-illustrated detailed 'techniques/interview' books are published featuring works by prominent paleoartists, such as the lavish *The Paleoart of Julius Csotoni: Dinosaurs, Sabre-Tooths and Beyond* (2014). And we continually spy new raw interpretations of favorite dinosaurs in *Prehistoric Times* or on the internet such as *Spinosaurus,* popularized via 2001's *Jurassic Park III*, yet whose form has since been substantially modified[4], long-armed *Deinocheirus* (now recognized as a giant, sail-backed ornithomimid), while upholding our affliction for the very largest of dinosaurs - record holder contenders such as *Dreadnoughtus*. New dinosaurs and interpretations thereof will just keep coming out of the bedrock into our living rooms!

Since the dawn of the dinosaur renaissance and particularly the new millennium, paleoart and paleoartists have become more integral to comprehending paleontological specimens and associated scientific hypotheses. In 2008, Jane P. Davidson perhaps espoused with conviction an extreme view when she commented that throughout the history of this science, "there is no paleontology without imagery" (p.183). I believe that my 2002 book (coauthored with Diane Debus) successfully captured the spirit of over 200 years of dinosaur paleoartistry (or as much as publisher limitations would then allow), capping developments up through the end of the 20th century, through the dawning of a later "quantum shift" (to borrow a Steve White term) in later 20th century dino-science and art. Yet, thanks to a bevy of new discoveries and a protean paradigm, as much as the practice of paleoimage-making has grown so vastly during the past 18 years, vibrantly fueled by a fascinating stream of new fossil discoveries, ideas, theories and methods, the captivating field of paleoimagery would seem about to soar far further than before!

Notes:(1.) **"Realm of the Dinosaur: The incredible dinosaur art collection of John Lanzendorf'**

It's always an adventure of sorts entering John Lanzendorf's Chicago apartment, which could really be referred to as the 'Realm of the Dinosaur.' Merely an apartment? No! Here it's wall-to-wall dinosaurs! His 'apartment' should, perhaps, more properly be referred to as a museum, an institution which should be rightfully mentioned in one of those popular dinosaur 'safari guides.' For here the mind can be boggled by the artistic renditions of numerous contemporary dinosaur artists, many of them well known, others rising stars. John's special fascination is with theropod dinosaurs, and so the cautious visitor must be particularly apprehensive about what lies in wait up the hallway, or around the next corner.

Since John is continually expanding his priceless collection, each successive visit offers new wonders. During our most recent 'tour,' John delightedly directed our attention to one of the most spectacular bronze sculptures we have ever encountered! This bronze statue depicts *Lambeosaurus* under attack by a pack of vicious dromaeosaurs, sculpted by Michael Trcic. We gazed in awe for many long moments at this magnificent work.

John has recently acquired a bronze sculpture (miniature version) of Stephen Czerkas' running *Deinonychus* trio. Many of you may have seen the original, full-sized sculpture exhibited in San Francisco's California Academy of Sciences. These, coupled with Stephen's *Carnotaurus* and life-sized bronze *Compsognathus* are by far the treasures of the 3-dimentional pieces in John's collection. And yet, there are many 'runners up,' including the Trcic *Daspletosaurus* and *Styracosaurus* pairing. Garfield Minot's life study of an *Allosaurus* head (it would not be possible for any sculptor to make it seem more alive) occupies a position of central prominence in the living room.

There are many other spectacular bronzes, some of which are shown in the accompanying photographs. Yet one of our special favorites is a one-of-a-kind Triassic scene by Maximo Salas. This finely detailed sculpture showing the *Plateosaurus* being attacked by a pair of feathery-looking *Coelophysis* is kept in a glass case due to the fragility of the polymer clay it is made of, to protect it from John's roving cats.

Moving on to the canvas portrayals, the casual visitor is confronted by a veritable 'wall of fame' of significant dino-art! Deserving special honorary mention, John always proudly leads us to the Michael Skrepnick "blue" wall where several original Skrepnick colorful paintings, showing dinosaurs living in "saurian splendor" under blue skies, are prominently displayed. Skrepnick's line drawings also adorn another wall, one of our favorites being a large carnivore peacefully resting under a tree.

Incredibly, within several feet of one another, visitors may gaze at original paintings by Gregory Paul, Donna Braginetz, Douglas Henderson, David Peters, and Mark Hallett. John also speaks very highly of artist friends Gary Staab, John Bindon, and Sean Murtha. An original Luis Rey painting of *Carnotaurus* is one of John's recent acquisitions. In fact, if you're wondering where the originals of many of those dinosaur paintings and illustrations you may have recently spied in magazine publications and popular books reside, well … John owns them!

We can't possibly name all of the talented artists represented in John's collection, but some of the photographs shown here may give you a feel for the breadth of his astounding collection. John also owns a vast dino-toy collection, yet the mind is far more riveted to the fine art on display.

One wonders how John's feisty mother, Ruth, can live there with John and those ravenous-looking Mesozoic carnivores! (Do the two 'talk dinosaur' into the wee hours of the morning?) John has many other pieces on order, but horrors, he is rapidly running out of available wall and floor space in which to mount new 'specimens.'! (Yale Peabody Museum curators need not worry because R. F. Zallinger's "Age of Reptiles" mural will no longer fit into John's apartment.

We've only provided a sneak preview of the full extent of John's collection. We haven't mentioned Brian Cooley's, Tony Merrithew's, Jon Neill's or Jim Gurney's artworks … so you'll just have to come out and see everything for yourself. John, let us say that we very much look forward to our next intriguing tour, back to that mystical "Realm of the Dinosaur." (Allen A. Debus & Gary Williams excerpted from *Prehistoric Times* no.23, March/April 1997, pp.40-41)

(2) Stephen L. Brusatte, "A Mesozoic Aviary," *Science*, v. 255, Feb. 24, 2017, pp.792-794; (3.) *Proceedings of the National Academy of Science*, v.113, 12/6/16; (4.) Nizar Ibrahim, Paul Sereno, et. al. in *Science*, v. 345, no. 6204, pp.1613-1616, 9/26/14.

Chapter Ten—*In Search of the Old Masters*

Why is it that sight of the Empire State Building conjures images of King Kong swatting Air Force biplanes, buzzing like stinging hornets? Here we are in New York City for our second visit. We have had a delightful evening of conversation with Rhoda Knight Kalt—granddaughter of the illustrious artist Charles R. Knight (1874-1953), whom she refers to as "Toppy." Knight of course was the influential artist most frequently lauded by aficionados of dinosaurs, fossil men and vertebrate paleontology during the past 100 years.

Rhoda is a charming, vivacious lady, with an inspirational personality. Lately, her life's mission revolves around the preservation and curating of her grandfather's work. The fact that Charles R. Knight was so prolific an artist makes this assignment unwieldy but, for her, a labor of love. Rhoda had written us some weeks before in appreciation for our words concerning her revered grandfather (published in *Dinosaur* magazine by Starlog, June 1993—also appearing as "Prehistoric Scenes" here as Chapter Eight). At her invitation we stopped in to discuss "Toppy" and see some of his original art.

There, occupying pride of place in her living room, was one of his favorite pieces. We are confronted by Nature's grandest living predator, the Bengal Tiger, glaring menacingly at the observer from a plane of canvas sporting three dimensional qualities. Tiger clutches a beautiful peacock under a massive paw.

How typical of Knight to place Nature's finest specimen on center stage (while in this case also providing a magnificent play on the cat versus bird-as-prey theme so often caricatured in the old Hanna-Barbara cartoons with Sylvester and Tweetypie). Beneath our feet, cavorted Rhoda's fluffy, feline friend, Elouise. Knight was fascinated with cats of all shapes and sizes. (Maybe he would have been interested to know that Allen is a 'Leo.')

After learning of our plans for the few days ahead, at 10:30 PM, Rhoda calls her friend Joyce Cloughly of the American Museum Exhibitions Department, who has been highly involved with the expensive restorations currently being rendered on some of the Knight prehistoric mammal murals. We learn that eight of Knight's famous 28 Field Museum murals are undergoing cleaning and restoration for display when the "Life Over Time" exhibit opens in 1994. Restoration is an expensive proposition, costing approximately $600,000 for the work performed on only six of the American Museum's prehistoric mammals scenes. Therefore, cost of restoring all the paintings at the Field Museum is certain to be extraordinary.

Plans are made for us to meet Joyce tomorrow. We reach our hotel room at 11:30 PM. The sound of garbage trucks operating at street level fills the night until an untold hour. Truly this is the city that never sleeps. (No wonder Kong climbed the Empire State. He was probably searching for a quieter room.)

The old Colbert exhibits have been dismantled in the American Museum's dinosaur halls, but the striking new *Barosaurus* display—a mother rearing to defend her young from a charging allosaur is breathtaking. The tiny head of the adult barosaur stretches, dizzyingly, to lofty heights.

Joyce Cloughly walks us through the vertebrate paleo halls upstairs, currently inaccessible to the public. On the walls we spy several of Knight's murals, including "Autumn in New Jersey," featuring extinct giant beavers and the Pleistocene moose, *Cervalces*. Knight's murals are being restored at great cost, but even greater magnanimity.

Cretaceous dinosaurs have been crated, but odd bits of bone protruding through unsealed portions of boxes expose the form of familiar skeletons—"Paleoscincus," *Triceratops,* "Trachodon," and *Styracosaurus.* In another room, the "bully for") *Brontosaurus* poses with its tail, now waving behind triumphantly in the air. Across the aisle, two *T. rex* hind limbs are being reassembled. A miniature 'to-scale' skeleton indicates that a decidedly 'Bakkerian' stance has been selected.

After leaving these gigantic structures, mounted mastodons, and supporting ironwork for the *Baluchitherium* and *Edaphosaurus*, we head toward the new Hall of Human Biology and Evolution, where we observe a band of grade school *Homo sapiens* tittering as they pass by full scale sculptures of fossil men replete with fully exposed genitalia. Grinning teachers warn unconvincingly of punishments for improper behavior.

Outside, Central Park displays a beautiful collage of fall color. Warm afternoon breezes stir golden leaves on the pavement. One can almost sense ghostly images of Benjamin Waterhouse Hawkins' never completed "Palaeozoic Museum" antediluvians basking quietly in the sun through the trees at the edge of a pond. If it hadn't been for Tweed's 'handiwork' over a century ago, Hawkins' restorations would not seem quite as imaginary today. With the pastoral image of "Autumn in New Jersey" in mind, we drove along a scenic highway into Princeton.

They must like popcorn here in New Jersey. In our hotel room we have seen the same television popcorn commercial play at least eight times during a half hour episode of *Bewitched*. Then we are handed baskets of popcorn to munch on in a restaurant before dinner is served.

In the morning we find Guyot Hall nestled amidst a campus of spectacular religiously styled architecture. Guyot Hall is named after the theologically-minded Swiss geographer/geologist Arnold H. Guyot who summoned the celebrated artist, Benjamin Waterhouse Hawkins, to Princeton in the 1870s. Here, Hawkins painted 17 murals spanning Earth's geological history. Two of these, both featuring Prehistoric Man, have been missing (at least) since 1975. Only one of Hawkins' murals remains on display—that showing "Moas and Maoris of New Zealand." Knight's similar, small canvas 1920s Moa scene is in storage, but 27 of his paintings (mainly his smaller 'practice' versions for the Field Museum murals in Chicago, as well as a small version of "Autumn in New Jersey") are placed prominently on display adjacent to a mounted *Antrodemus*. (Allen had seen all of Hawkins' murals in Guyot Hall back in September 1972, when they were displayed up high within the fossil hall.)

Maureen McCormick of the Princeton Art Library has granted permission for us to photograph art work by Hawkins and Knight. With Maureen's assistance, the dusty frames emerge from a well-protected storage area, and we begin what may be the first ever *color* photo session of Hawkins' entire series of existing paintings. Many of the frames are damaged. Some of the paintings are seriously in need of restoration. In one case, "Tertiary Mammals of Europe," the canvas had been punctured.

Maureen explains that due to insufficient funding, only certain paintings can be restored. Lately, only two have qualified, and as you would imagine, these scenes are of dinosaur fauna—"Cretaceous Period in New Jersey" (1877) and another titled "Jurassic Period in England" (undated). This is a peculiar selection process. The paintings most liable to survive here are those proving most 'popular,' due to limited funding and current fads. But will Hawkins' other paintings survive another century? Without a benefactor having interests ranging beyond the dinosaurs, further deterioration can only be expected.

Another curiosity is that because Knight's paintings are considered to be more scientifically 'accurate,' interest in his series of paintings has superseded that of Hawkins. But as we know, scientific 'accuracy' is a moving target. Often too, there are conflicting opinions in any age, making it difficult to frame a 'snapshot' of current ideology. The "old masters" remain visually captivating, and that's what largely matters.

Time is running short, so with a passing nod to the mounted *Smilodon, Eryops, Cervalces*, several plaster dinosaur sculptures by Knight (including a miniature *Triceratops* with a broken nasal horn), and some paintings of extinct mammals by R. Bruce Horsfall, we are off to lunch and Palmer Square. In Palmer Square we admire Knight's life-sized bronze lion, reclining in its resplendent setting of autumn foliage. Then off to the Mudd and Firestone Libraries, where we examine some old file information concerning Knight, Hawkins and Guyot. So absorbing are the files that we lost track of time, linger too long, and suffer a parking ticket. Afterward, there is time for reflection.

We have purchased from Rhoda Kalt, two of Knight's original drawings. One of these is an undated illustration of the Irish Elk. The Irish Elk was not really an elk or even a moose, but a deer of enormous proportions. Its famous headgear, the antlers for which it claims its fame, extended up to 12 feet in length, tip to tip. Its scientific name, *Megaloceros giganteus*, connotes majesty and power. Extinct in Europe and China since 11,000 and 2,500 years ago, respectively, the Irish Elk/Deer is one of the few 'prehistoric animals' to have survived into historic times. The second purchased drawing—a reclining female lion with cubs.

We feel closer to this great animal, having Knight's preliminary sketch in hand, and having seen two of Knight's 1920s paintings of the Irish Elk, one at Princeton, the other his famous mural at the Field Museum in Chicago. Also, we have seen on this trip the skeleton of this species mounted in Guyot Hall, which Knight and Hawkins referred to for their restorations (as well as a photo in Hawkins' Honorary Doctorate file showing his full-sized plaster restoration erected for his frustrated "Palaeozoic Museum" effort.

Hawkins and Knight saw great beauty in Nature's past and present ecological systems. Would they cringe at the accelerating pace of human-caused extinctions? Most likely! The Irish Elk well symbolizes the plight (or autumn?) of our inherited post-glacial fauna.

In one of his brilliant essays, "In Touch With Walcott, "Stephen J. Gould explained how he was able to symbolically shake the hand of two of his scientific heroes—Charles Darwin and Charles Doolittle Walcott (of Burgess Shale fauna fame) through only a 'handful' of intermediaries. (See Gould's book, *Eight Little Piggies* (1993), Chapter 15). Similarly, through Rhoda, we have touched her grandfather, and, intriguingly, along the chain also embraced Henry Fairfield Osborn, Edwin Drinker Cope and Othniel C.

Marsh (all presumably only 3 steps removed in the chain), and another step away, depending on which version of an Osborn tale you believe, either Charles Darwin or Thomas Huxley. Hawkins died in 1889, but we have handled his original canvases.

To a certain extent, our experiences have been disquieting. A third exhibit of Allen's original dinosaur sculptures begins in November. It is time to wipe the cobwebs off of horns, fins, scales, and bony armor once again. But what sort of condition will these creations be in a century from now? Slick and shiny in someone's cherished collection of dinosaur memorabilia, or wholly forgotten, broken and covered in mummy dust somewhere? Only time will tell.

Overall, this was a productive, interesting trip, made possible through Diane's particularly well-organized planning. We even took in a brief tour of the Dead Sea Scrolls at the New York Public Library, and Allen's father's exhibition on the early History of Medicine and the historical figure, early medical practitioner/'alchemist' Paracelsus, at the National Library of Medicine in Bethesda, Maryland. We always manage to fit 'dinosaurs' (using a *much* broader sense of the term) into our travel itinerary, whenever and wherever we go, so stay tuned for our next "In search of." (End – Dec. 6, 1993—co-written with Diane Debus)

Chapter Eleven—*On the Question of Hawkins' Spiritual Nature*

It has been questioned whether sculptures created by renown 'dinosaur artist' Benjamin Waterhouse Hawkins were destroyed by the infamous 'Tweed Gang' because of their supposed anti-religious nature, or, conversely, possibly due to overriding financial concerns.[1] Hawkins, of course, had partially constructed a group of full-sized prehistoric animal figures intended for exhibition in the "Palaeozoic Museum" on New York's Central Park grounds, before their eventual destruction in 1871. I believe some further light may be shed, bearing at least tangentially, on this intriguing matter.

Two explanations are usually contrasted given for why Hawkins' models were destroyed. For example, in Rev. Henry Neville Hutchinson's book, *Creatures of Other Days* (1894), the 'anti-religious' theory is presented. A similar opinion was stated in *O. C. Marsh: Pioneer in Paleontology* (1940). Authors Schuchert and LeVene claim on page 385 that: "they were broken up and thrown into the park lagoons by order of a former mayor of the City, not because they were incorrect scientifically, but because he considered them inconsistent with the doctrines of revealed religion!" Such a view contrasts with an alternative—the conventionalized notion that Hawkins' work was causing financial stresses for the City of New York.[2]

In a recent publication by Richard C. Ryder, "Dusting Off America's First Dinosaur," (*American Heritage*, March 1988, pp.66-72), following a careful consideration of contemporary circumstances involving Hawkins and the Ring, Hawkins is portrayed as a political antagonist. "Certainly Tweed and his associates were more likely to respond forcefully to a direct political confrontation than to some remote anti evolutionist crusade." Although this sentence speaks for itself rather well, the reader is referred to this article for further background on Ryder's reasoning, with which I am inclined to concur.

When one remembers that while in England, Hawkins had collaborated closely with Sir Richard Owen, who in turn battled against tenets of Darwinian evolution (and its fiercest defender, Thomas H. Huxley), it isn't too far-fetched a conclusion that Hawkins himself may not have been an advocate of Darwinian evolution either. (Owen had his own ideas about how organisms 'evolved' from an "Archetype," but you are referred to a book by Nicolaas Rupke, *Richard Owen: Victorian Naturalist* (1994) for further details.)

It is relatively unknown to what extent Hawkins may have been a "God-fearing" soul, versus an atheistic believer in evolutionary principles. Usually, such (dichotomous) questions (like the one regarding reasons for why Hawkins' models were destroyed) are of a simplified 'either/or', and therefore extremist nature anyway, when in fact, the real-world situations are laden with complexity. In fact, the matter would be of little interest today, were it not for the fact that Hawkins' greatest artistic creations involved the field of paleontology.

Hawkins may not have been an evolutionist, at least in a Darwinian sense. Hawkins had been summoned to Princeton University by geologist Arnold Guyot (1807-1884). Guyot featured several of Hawkins' Princeton murals showing scenes from Earth's geological history, as engravings in his theologically oriented science book, *Creation, or the Biblical Cosmogony in the Light of Modern Science* (1884). Hawkins lectured to the public occasionally and in one of his presentations to an audience in the Town Hall in Pendleton, concerning "The Age of Dragons" (October 29, 1879), expressed an opinion concerning the anatomical absurdities of mythical dragons. It was concluded that creatures most resembling dragons were the extinct winged pterosaurs. Although Hawkins did briefly discuss *Archaeopteryx*, his lecture was apparently conspicuously silent on the subject of evolution.

In a letter to the President and Board of Trustees of the College of New Jersey dated June 24, 1876, Guyot recommended that Hawkins be awarded with an honorary doctorate degree. After extolling his great skill and genius as an artist of vertebrate animals, Guyot stated, "…and I am happy to say that I have full reason to know that Prof. Hawkins bears a high Christian character." From today's perspective, the prospect that anyone so highly involved with restoration of fauna of prehistoric animals, now known to have lived in periods so many millions of years remote from one another would possibly be instilled with theologial proclivities, might seem ironic. Yet, Hawkins was a product of his time, and scientists of his day would have fostered his (possible) 'creationist' beliefs.

Whereas I do sense that Hawkins' views on evolution must have been rather closely allied to Owen's, I cannot presently 'prove' that this is the case. Also, the fact that Hawkins may have not intended to inflict the City of New York with 'Darwinian' themes through his artistry had little impact in sparing his handiwork! Suffice it to say that what may have arguably been the greatest 'dinosaur' exhibit ever planned until that time never got off the ground, a sad result which the City of New York may never live down!

Notes: (1.) Mike Howgate, "Yet More on Crystal Palace, Waterhouse Hawkins and Central Park," published in his *Dinosaur Collectors Club Newsletter*, no.6, March 1994 (2.) (For instance, see Adrian J. Desmond's *The Hot-Blooded Dinosaurs: A Revolution in Palaeontology* (1975), p. 55, where it is stated, "Since the Ring foresaw no financial gain, it was decided to fill in the foundations.")

Chapter Twelve—*Going 'Postal' over Neave Parker*

No, I never met Neave Parker (1910-1961) but our 'relationship' goes back to half a century ago when I first witnessed his amazing paleoart on a set of then just issued British Museum postcards purchased by my father. From the fall of 1959 to summer 1960 we lived in London, when my dad was conducting research toward his doctorate. He also performed his solemn duty in ensuring us kids learned to appreciate dinosaurs and paleontology. I was only in kindergarten, but thanks to him, already a budding dinosaur book and toy lover. In October we'd been introduced to a new book by Josef Augusta brimming with Zdenek Burian's masterful paleoart titled *Prehistoric Animals*. There was a new set of plastic Timpo dinosaurs on the toy shelf. And of course, by then I was very familiar with Charles R. Knight's artistry, having seen his magnificent murals at the Chicago Natural History Museum where I also picked out plastic models of American Museum dinosaur skeleton replicas as well as Marx toy dinosaurs. Around that time, I would have seen RKO's *King Kong* televised, and during the summer of 1960 my grandparents would take me to see Irwin Allen's remake of *The Lost World*. Yes, most anachronistically, dinosaurs just seemed a natural part of the terribly modern world I was growing into.

But Neave Parker represented a special, vibrant piece of the paleo-puzzle! Yes, his name appeared on those old British Museum postcards, signed as artist. As an aside, for those of you less than 20 years old who may be reading this, you see – before the days of smart phones and the internet, Skype and rapid-fire telecommunications, people used to purchase small cardboard, well, approx. 4" by 6" cards that could be posted in the mail (i.e. "hard copies" sent as "snail mail"), or collected. These usually showed a picture or photograph on one side commemorating places visited during one's travels and on the flip side, space to write a small note or message of good tidings—with additional space allocated for proper addressing and application of a postage stamp prior to mailing. These articles of sale were known as "postcards." Geez – I'm starting to feel old writing this! During the early 1960s, I used to create small 'museums' on my bedroom floor, displaying my fossil collection, plastic miniature replica skeletons of *Tyrannosaurus, Stegosaurus* and *Brontosaurus* and used dinosaur postcards as "murals" framing the museum walls.

Beginning in 1981 – during my early post-collegiate, early personal 'dinosaur renaissance' years, I began writing a manuscript titled, "They Painted the Dinosaurs," addressing careers and accomplishments of those whom I then considered as the classic paleoartists of history. I had written to a number of institutions and individuals who were familiar with artists I was most interested in. Questions were answered and considerable supplementary information was supplied with which to continue my project. Paleoartists in question were those whom I'd first been exposed to by my father through books, magazine articles and museum paintings and even some dino-films during the late 1950s and early 1960s. By 1983 (partly inspired by Parker as we shall soon see), the title of the manuscript evolved into "Seeking Ghosts of Eras Past: The Conquest of the Dinosaur in Modern Times." Looking back through my copy of this tidy little manuscript, segments of which were eventually printed in Chicagoland area earth science club newsletters during the 1980s, this 'book' was a template toward my much more comprehensive *Paleoimagery: The Evolution of Dinosaurs in Art* (McFarland, 2000).

So in my manuscript, following a concise Introduction to this then unusual topic about dinosaur artists, followed by sections addressing Benjamin Waterhouse Hawkins, Charles R. Knight, Rudolph Zallinger, etc., I tackled Neave Parker's life and career. By then I was well armed with information supplied through several letters from paleontologist William E. Swinton, who fostered my efforts. Swinton, who while at the British Museum had directed Parker on technical paleoart details, also kindly referred me to zoologist Maurice Burton, of whom more shall be stated shortly. I was also made aware of Parker's work at the *Illustrated London News* and so had written Liz Moore there for further information including copies of his *ILN* dinosaur paleoart. Also, strangely enough, for many years a copy of a 1956 children's publication, *Uncle Remus* (Bruce and Gawthorn, Ltd.) had been sitting on our home book shelf for many years, which as I then recognized had also been magnificently illustrated by Parker.

Although further interesting details about Parker, including a photographic portrait of the man, are available in Chapter 17, titled "Neave Parker's Prehistoric World" (pp. 111 to 114, 255) in the aforementioned *Paleoimagery* volume an outline concerning Parker should be presented here. After failing competency as a banker, and following stints both as a surveyor and as a photographer in the Royal Air Force, Parker honed artistic talents. Dr. Burton, who was then affiliated with the *ILN* as honorary science editor enlisted Parker in the illustration of animals, printed in issues of the *ILN*. Burton also introduced Parker to Swinton, whose mentoring led to several original dinosaur scenes published in the *ILN* from 1954 to 1960, usually 28" by 17" in size. These vivid scenes, produced as "monochrome gouache and wash drawings," certainly captured imaginations of many young dino-aficionados of the time. They were especially noteworthy because until Parker's arrival, few artists had endeavored to paint idealized scenes showing "all" the major types of dinosaurs and animals known to a particular space and time, coexisting—such as prehistoric ecosystems of North America in the Late Cretaceous or a then unusual theme like "Chinese Dinosaurs."

Parker did have one unusual proclivity, however. He seems to have been rather psychic, especially when it came to "seeing" ghosts of dogs. On Jan. 13, 1982, Swinton, then affiliated with the University of Toronto, encouraged me to write Burton. Burton's Feb. 18, 1982 letter to me came with his now rather famous "Toots and his Dogs" essay, which I circulated to several individuals during the early 1980s. From my old correspondence file, I read how Burton stated, "On receipt of your (Jan. 26th) letter I thought perhaps the best way I could help would be to write an essay about Neave Parker the man as I knew him and I have pleasure in enclosing a copy of this which you are welcome to keep." I know no better way to share my knowledge of Parker and (as a special treat for those of you who enjoy SYFY's *Ghosthunters* program) his uncanny psychic abilities than to use this forum now to insert Maurice Burton's essay herein:

"Toots and his Dogs"

"It was a lucky accident that brought me into contact with Neave Parker, leading to some years of pleasant and fruitful association. I had been commissioned to write a book to be entitled *The Story of Mammals*. Neave Parker had been invited to do the illustrations, six hundred pictures in colour. The book was written and reached the page-proof stage and the illustrations were finished. Even the dust-jacket was completed. All that remained to be done was the running off of the pages, the binding and distributing, when for reasons now unclear to us the publishers abandoned the project. We were both paid but money is not everything. We both suffered a loss of prestige because had publication gone ahead it would have been the first non-technical book on the world's mammals ever to have been published. Several have appeared since but that would have been the first.

Neave Parker was of medium height and very rotund with a shock of graying hair, a bushy moustache and horn-rimmed spectacles, very like G. K. Chesterton in appearance, almost the double of Gilbert Harding who was then enjoying immense popularity as a broadcaster, in the early days of television. In fact, on several occasions he was mistaken for Harding.

He was a most agreeable companion, wise beyond his years yet of the earth earthy. He enjoyed good food and above all was fond of beer which he drank in quantities without ever losing his equilibrium. An excellent artist, one of the leading animal artists of his day, and a first-class photographer, he always went around festooned with cameras, which seemed to add to his portliness. A bachelor, he never missed his weekly visits to the local cinemas, and, indeed, suffered a stroke while in a cinema, from which he never recovered, a great loss to the world of art.

Parker's burly figure was in complete contrast to the nickname, Toots, by which he was known to his intimates. The origin of the name is unknown to me. It seemed absurd for a man of his capabilities yet he carried it without loss of dignity. It did, however, seem not wholly inappropriate because of the impish streak in him that often came to the surface.

There was such an occasion when he was telling me how he came to be an artist. With the characteristic twinkle in his eyes he told me how his father refused to let him go to art school, insisting that he should seek a safe job in a bank. 'I was there for a week,' Toots explained. 'Each day there was an error in the books and the whole staff had to stay behind until the error was found. It always ended with me. At the end of the week the manager invited me into his office and suggested, kindly but firmly, I should take up something else as a career. So I became a surveyor and later dropped this to go to art school.'

His hobby was pistol-shooting and at this task he excelled. At his home was a cabinet containing the silver bowls and cups he had won. He held at one time the British Open Championship. Although I never saw him at the range I was told that his shooting was remarkable, because he used an ordinary pistol such as could be 'bought over the counter, with no special grip or anything special about it.'

Toots was very fond of dogs and was active with the Canine Defense League. He also appeared to have the faculty for seeing dogs that were not there. I discovered this one day when visiting him at the club, in Hampton Hill, near where we both lived. He had gone out of the room for a few minutes and I heard one of the members say: 'Toots hasn't been seeing things lately.'

The following day I was lunching with him and took the opportunity to mention this and ask him what was meant. 'Oh, that,' he said, almost as if brushing it aside. 'A club member brought a visitor in one evening and as they were drinking at the bar I said to the visitor, "That's a nice dog you have." "Dog," said the visitor, "I haven't got a dog."'

Apparently, Neave Parker then described the dog. The visitor went pale and was obviously emotionally disturbed. 'That's my dog, all right,' he said to Toots, 'but he died some time ago.'

It seems this sort of thing had happened several times. Neave Parker then told me how, when he was in the Royal Air Force during World War II, in the photographic unit, he was sitting in the canteen one evening when a RAF sergeant came in and sat at the table opposite him. 'On impulse I reached out for a sheet of paper that was lying on the table and drew a man's face on it. Then I pushed the paper over to the sergeant and said, Does that mean anything to you? I thought the sergeant was going to faint. That's my father, he said, but he died last year. I got in touch with the Psychical Research Society who instigated a number of tests, but nothing came of it.'

I started to question him but he was obviously unwilling to talk about these matters. Moreover, if ever our conversation drifted onto he occult he would dry up and refuse to discuss it, as if afraid of the subject.

Soon after I had made his acquaintance and long before the fate of *The Story of Mammals* was known, I introduced Neave Parker to the editor of the *Illustrated London News*. I was honorary science editor at the time. Parker was promptly commissioned as special artist to that paper and for some years his double-page spreads of animals were a feature of that weekly journal.

We cooperated closely in this and I was always impressed how, with a minimum of words from me, he would reconstruct a picture of animals in their natural setting merely from their bones and skins. He was very easy to work with and seldom was there need to correct what he had drawn. It was as if he could read one's thoughts. He also drew pictures of extinct animals from their bones, under the direction of Dr. W. E. Swinton, whose experience of working with him was similar to mine.

Parker also drew pictures for Dr. Kenneth Oakley, who was, among other things responsible for the unmasking of the fraud of

Piltdown Man. Oakley was a man of great perception himself, and I remember his saying, of Neave Parker, who drew for him several reconstructions of early man: 'It is quite remarkable how he seems to grasp the details of his subject with only a minimum of help from me, as if he could see into the past.'

A great artist, cut down in his prime!"

(Maurice Burton to Allen Debus, Feb. 1982)

Parker was truly a prolific artist; much of his original work (paleo-related or not) remains at large, or perhaps unrecognized. For instance, one wonders whatever what happened to the 600 color drawings he had done for the aborted *Story of Mammals* volume. As Peter Crowther stated in an August 13, 1999 letter, "Parker was commissioned regularly by the *ILN* on a wide range of subjects, not only vertebrate palaeontology."

Besides his *ILN* work Parker also teamed with Swinton in the painting of several prehistoric animals some of which were printed in Swinton's publications (e.g. *The Dinosaurs,* New York, Wiley Interscience, 1970), or then issued on British Museum postcards – 19 in all, each of which must now be considered valued collectibles. Several of these restorations were reproduced, for example, in a popular book, Edwin H. Colbert's *Dinosaurs: Their Discovery and Their World* (1961).

The postcards (individually numbered sequentially as "G.72 for *Icthyosaurus*," etc.) as originally issued contained brief identifying, descriptive information concerning each prehistoric animal, printed to the left or bottom of the image. These were printed on cardboard very lightly shaded in tan and, as I vaguely recall, came in a nice little sealable packet. Later, during the 1960s the postcards were (re-)issued with identifying information printed on the back of each card, each sold singly. By then the adopted coloration became the more familiar grayish, monochrome tone as printed in books. Parker postcards in my collection featuring restorations include the following prehistoric animals: *Paracyclotosaurus, Triceratops, Tyrannosaurus, Cetiosaurus, Polacanthus, Macroplata, Pteranodon, Icthyosaurus, Acantopholis, Scelidosaurus, Megalosaurus, Stegosaurus, Apatosaurus, Pterodactylus, Scleromochlus, Protoceratops, Hypsilophodon, Ignanodon* and *Ornithosuchus.* I also have an (approx.) 15" by 20.5" poster of Parker's *Pterodactylus* hanging in my office.

Swinton wrote a BMNH booklet titled *Dinosaurs* (Pub. No. 542) incorporating ten of Parker's restorations. Erroll White, Keeper of Palaeontology, stated in the Preface to the first edition (1962) that, "… the vivid monochrome restorations shown in the frontispiece and plates, although they have for some time been issued as part of a very successful series of postcards, are here published in a book for the first time. They are from the skilled brush of Mr. Neave Parker, whose recent death has robbed scientific journalism of one of its most outstanding artists." (Incidentally, an earlier British Museum publication, also by Swinton - *Fossil Amphibians and Reptiles* (1st ed. 1954, 4th ed. 1965), instead relied principally upon Maurice Wilson's restorations, as well as D. E. Woodall's line-drawings.) It seems as if Parker's original restorations were sold during the early 1970s, because Swinton recollected they were offered for sale, advertised for 80 to 90 pounds – which today would be a bargain!

However, while the whereabouts of the 19 original pictures that were later reprinted as BMNH postcards today is unknown to me, in 1989, an additional 17 restorations were purchased by the Ulster Museum in Belfast from the *ILN* at auction. During the summer of 1993 the Ulster Museum promoted a fascinating special exhibition titled "The Prehistoric World of Neave Parker," featuring the *ILN* originals. (An 18th *ILN* painting was retrieved in 1997.) Considering Parker's eerie predilections as recounted by Burton, appropriately, one of the paintings displayed was a study of a dozen early domesticated dogs "based upon bone-finds and sculpture, uncovered during archaeological excavations of tombs in Egypt and Mesopotamia," originally appearing in a June 23, 1956 *ILN*.

The *ILN* artwork itself required considerable conservation. As described by Peter Crowther, "… the high-quality artist's boards used by Parker had been fixed to poor millboard. While this kept the artwork flat for the purposes of reproduction, the millboard had a long-term damaging effect on the paper. Treatment involved the removal of the acidic backing millboard, and washing and de-acidifying the artist's board to remove acids and discoloration." We must humbly thank Peter Crowther (Keeper of Geology of the National Museums and Galleries of Northern Ireland) for perceiving the value of these items as well as his time, patience and care in curating this precious material.

In 1999, Peter Crowther sent me Xeroxes of the Neave Parker artwork they'd obtained from the *ILN*, which had been exhibited a decade before. Included in the collection were a dozen splendid restorations I'd never seen before (one of which – a life through time historical geology study—was not displayed during the 1993 Ulster Museum Neave Parker exhibition). Five of these featured large Cenozoic mammals, as in the depiction of "Trafalgar Square 100,000 Years Ago" (*ILN* June 14, 1958) and "London 50 Million Years Ago" (*ILN* Jan. 14, 1956). There were two restorations depicting early reptiles and early mammals (Elgin, Scotland and South African Karroo deposit, respectively) of 200 million years ago. Additionally, four restorations dealt principally with dinosaurian themes, or featured large Mesozoic reptilians. One of these was a *Triceratops* restoration published in the March 15, 1958 *ILN* issue, rather different from the more familiar Parker *Triceratops* pose we've become more familiar with in postcards and in 1960s BMNH publications. It was a wonder to behold these additional "Parkers," quite new to me after all these years! Like his other restorations I'd gained familiarity with, all of Parker's restored creatures retained a distinctively vibrant and engaging sense of true, purposeful 'liveliness,' which is perhaps the fundamental reason for why his art has proved so appealing. Although perhaps under-appreciated today, Neave Parker truly was one of our 'classic' paleoartists.

Note: (1.) A list of Neave Parker drawings purchased by the Ulster Museum and displayed at the 1993 "Prehistoric World of Neave Parker" exhibition (information supplied by Peter Crowther), with dates of *ILN* publication: *Ancient and Modern Animals of Africa* – June 19, 1954; *The 'British' Seaside in the Jurassic* - July 31, 1954; *Elgin 200 Million Years Ago* – January 26, 1957; *London 50 Million Years Ago* – January 14, 1956; *The South African Karroo: Origin of Mammals* – November 5, 1955; *Giant Crocodiles* – December 22, 1951; *Neave Parker and Dog Ghosts* – June 23, 1956; *Living Fossils: Extinct Ancestors* – July 27, 1957; *Latest Discoveries from Africa* – July 5, 1958; *Canadian Dinosaurs* – August 27, 1960; *Dinosaurs of the Gobi Desert* – January 9, 1960; *Chinese Dinosaurs* – February 11, 1956; *Trafalgar Square 100,000 Years Ago* – June 14, 1958; *Ancient Reptiles and the Sea* – February 20, 1960; *The Last Days of the Dinosaurs* – March 5, 1960; *Triceratops* – March 15, 1958. Neave Parker's paleoart paintings were also featured in Swinton's 1961 book, *Animals Before Adam* (London: Phoenix House, Ltd.), "including 8 drawings by Neave Parker."

FOUR

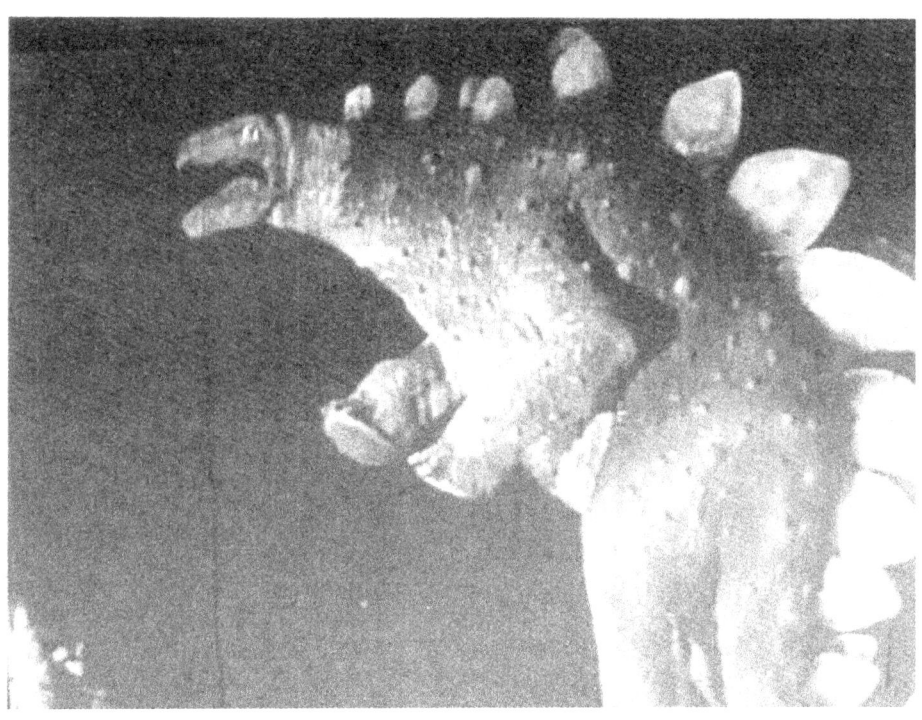

One of the earliest examples of dino-suitmation employed using 'trick' movie photography to make the bipedal Stegosaurus seem much larger—science-fictionally—than the puny human caveman shown at lower left. This was excerpted from a film short advertising Messmore and Damon's World A Million Years Ago display at Chicago 1933/34 World's Fair. See Chapters Fourteen (and Six) for more. (Rare test footage, no copyright stated)

Science Fictional Dino-Monsters

Who doesn't enjoy sci-fi dinosaurs of film and literature! I'm not a film expert by any means, but I do relish reconsidering features of movie-land dinosaurs with what is really known about them, scientifically. Thus, I provide you with this sampling of chapters which address not only the fakest-looking kinds of sci-fi dinos. Also, we'll delve into dinos that were conceived fictionally, with an intention of accurate portrayal, although the underlying scientific ideas have since been superseded. By the late 1970s, I'd written for two Earth science club newsletters, yet I really yearned to break into print in a 'real' science fiction fanzine. But what could I write about, given that my amateur interests orbited around dinosaurs and paleontology, subjects that really didn't seem to fit in anywhere at the time? Making matters worse, I hadn't really read very much on the topic either. Hope came though when I read an ad placed in *Asimov's Science Fiction* magazine about a new sci-fi fanzine named *Cosmic Landscapes* that Dan Petitpas was inaugurating to attract 'new writers.' After subscribing, I conceived a paleo-themed piece to whet Petitpas's appetite, which he reluctantly accepted. I hammered this out in our dining room using a clunky 1920s Royal "portable" typewriter. The earliest version of the article was eventually printed in a 1983 issue of that long defunct and forgotten *Cosmic Landscapes* under the title "The Lost Worlds of Science Fiction."

Only a few years later, I revised and expanded the article, forwarding a hard copy to D. Peter Ogden—then editor of *ERBANIA*, a fanzine devoted to all things concerning Tarzan and Edgar Rice Burroughs. I dismally lost hope of ever receiving a reply from Mr. Ogden, presuming he had no need of the piece. Then, completely out of left field in the *late 2000s*, I received issue no. 86 of *ERBANIA* (Spring 2003), containing my long-neglected article which had finally appeared unbeknownst to me a few years before, now under a new title,

"Geology in Science Fiction." Ogden also included a short, handwritten note expressing gratitude that he had re-discovered my old article to use. (Evidently, Ogden's home office desk was also a 'lost world' type place!) While I was thrilled to see the printed result, my elation ebbed when I saw several typos had been introduced. His or mine? Nonetheless, 25 years after my original submission, I was finally an *ERBANIA* author! (I followed up with a second piece for him — not included in this collection—shortly thereafter which was printed, this time in relatively short order.) I have retained my original title to the expanded version of the article here as Chapter Thirteen, while hopefully not transcribing too many fresh typos.

Chapter Fourteen, "Dinosaur Masquerade," brought me into contact with dino-phile Mark F. Berry, who later penned a Foreword to one of my books. I had begun reading his marvelous *Dinosaur Filmography* during a stay at my parent's home with my father who was undergoing chemo for lung cancer while my mother was away. And during that interval—thanks to Berry's book helping to keep my mind off woeful matters—the theme of this (now) particular chapter occurred to me, which I wrote and forwarded a few weeks later to J. D. Lees, editor of *G-Fan*. I included a 'plug' for Berry's book therein, and subsequently he was kind enough to introduce himself and thank me. Nice guy. Of course, now we know that Godzilla movies increasingly partake of those ubiquitous CGI effects, as opposed to suitmation. But the heyday of the live lizard acting as pseudo dino-monster is probably over.

The emphasis in Chapter Fifteen is on winged, flying paleo-monsters. Spotlighted is the discovery of a fossil 'bat-winged' *dinosaur* (with a mere two-foot wing span) that was recently described. (See *Prehistoric Times* issues nos.114, p.50, Summer 114, and 116, Winter 2016, p.50 by Steve Brusatte.) It is named *Yi Qi*.

In the case of Chapter Sixteen, my irreverent treatment of the monster "Agon" was of course intended as tongue-in-cheek, although it is true my copy of the DVD lacked English subtitles which limited my ability to follow the plot. Even though there are newer versions of the film with English subtitles, I chose not to alter this chapter for the present volume. This is perhaps especially because I was delighted to read that my fellow *daikaiju* scribe, Mike Bogue, had declared in a letter printed in the next *Scary Monsters* issue (no.78) that he thought my article was "lively and informative." Good enough praise for me!

Besides gazing at visual cues (e.g. restorations and reconstructions of dinosaurs), enlightened humans will *read* about prehistoric life in books, magazines and, now, digitized documents. Since the dawn of the dinosaur renaissance, there has been a plethora of print on paleontological topics. Far too much to analyze fully here. Arguably, the two most significant *popular* books of the 20th century dinosaur renaissance are Michael Crichton's *Jurassic Park* (1990), and Adrian J. Desmond's *The Hot-Blooded Dinosaurs* (1975). In between this pairing, many 1980s dino-aficionados might have declared that Robert T. Bakker's *The Dinosaur Heresies* (1986) warranted inclusion.

It was my father, Allen G. Debus, who inculcated me with a fascination for books (especially those crusty, older volumes) concerning paleontology. And in the wake of my *Dinosaurs in Fantastic Fiction* book, I continued to write a series of articles about prehistoric life as conveyed through short stories and novels—a 'novel' topic which, since the late 1970s especially, I've increasingly developed a fondness and appreciation for. I've selected several fanzine articles continuing my assault on prehistoric life in fiction literature, represented as Chapter Seventeen.

My first juvenile reading of a prehistoric animal fiction novel (e.g. "chapter book") was Robert Faraday's *The Anytime Rings* (1963). Little did I realize then that Faraday was a pseudonym for Bruce Cassiday, author of the first 'true' apocalyptic paleo-*kaiju* novel (-ization)—*Gorgo* (1960), fully addressed in my *Dinosaurs Ever Evolving*. After avidly absorbing *The Anytime Rings*, I moved on to Burroughs' *The Land That Time Forgot* ... then *At the Earth's Core*, Verne's *Journey to the Center of the Earth*, and Lovecraft's *At the Mountains of Madness*—all founded on paleontological themes. Eventually I was stricken with the creative notion that no one had really written critically about these and other titles before— and so I took on the task. One notes that as of circa 2010, the former frenzied interest in dino-sci-fi writing of the 1990s especially seemed to have been supplanted by *kaiju*-fiction novels and tales—undoubtedly influenced by the recent number of genre films, such as *Pacific Rim*.

Thanks to Justin Mullis for thoughtfully providing a PDF file of Frank Belknap Long's story (outlined here on pp. 180-181).

> These chapters originally appeared as articles in the following publications:
> Chapter Thirteen—*Cosmic Landscapes* no.6, October 1983, pp.20-21, 24.
> Chapter Fourteen—*G-Fan* no. 65, Nov/Dec 2003, pp.28-34.
> Chapter Fifteen—*G-Fan* no. 114, Winter 2017, pp.56-62.
> Chapter Sixteen—*Scary Monsters* no.77, Jan. 2011, pp.61-67.
> Chapter Seventeen—*Prehistoric Times* nos. 81-83, 86, 94, 113 (Parts 1 – 6); respectively, Spring 2007 (pp.22-23, 56-57); Summer 2007 (pp.22-23); Fall 2007 (pp.22-23, 27); Summer 2008 (pp.22-23, 27); Summer 2010 (pp.29-31); Spring 2015 (pp.36-37, 42).
> Chapter Eighteen—*G-Fan* no.115, Spring 2017, pp.68-75.

Chapter Thirteen—*Lost Worlds of Science Fiction*

During the 19th and 20th centuries science fiction and fact evolved together. Stimulated by revolutionary developments in paleontology, physical geology and evolutionary thought, science fiction writers attempted to fill many gaps within the body of scientific knowledge with imaginative tales of adventure and speculation. While clearly not their goal to provide scientific solutions, their presentations of prehistoric worlds and Earth's geological framework helped to popularize the emerging geological sciences. Interestingly, some of their imaginative stories reflect several actual differences between modern and Victorian geological understanding.

Sir Arthur Conan Doyle (1859-1930), highly revered as the creator of Sherlock Holmes, vividly captured the romantic spirit of Victorian paleontology and the stimulating atmosphere created by its leading scientific figures in his great adventure tale *The Lost World* (1912). In this story, two prestigious scientific adversaries, Professors Challenger and Summerlee, dispute each other's claims in much the same way two famous American paleontologists—Othniel C. Marsh (1831-1899) and Edward Drinker Cope (1840-1897)—actually disputed theirs over a century ago.

Cope and Marsh discovered many of the fossilized remains of dinosaurs we are familiar with today in Colorado and Montana during the 1870s. While Marsh, who was associated with Yale University, and Cope, affiliated with the Philadelphia Academy of Sciences, feuded bitterly over reconstruction, classification and priority over discovery of *fossilized* Mesozoic saurians, Challenger and Summerlee lock horns over whether *living* dinosaurs, and other prehistoric still exist on a remote plateau in South America. Summerlee is proven incorrect by Challenger, for soon after the explorers locate and scale the plateau, living *Iguanodons*, nesting pterodactyls and other prehistoric denizens are encountered. Some of these are identified by the scientists as creatures formerly inhabiting England's highly fossiliferous Wealden district.

In one frightening scene, an explorer is pursued by a giant flesh-eating dinosaur which leaps and hops after him. Doyle's dinosaur illustrates the antiquated notion that many bipedal dinosaurs incongruously hopped or bounded in 'kangaroo' fashion. Now more paleontologists believe that some dinosaurs may have been intelligent, warm-blooded creatures that were capable of swift, agile running.

The party triumphantly returns to London, following several perilous encounters with a merciless band of "missing links." Challenger's controversial premise—that extinct animals still exist in the land of "Curupuri"—is vindicated after he displays as evidence a small pterodactyl that had been captured during the adventure to a disbelieving throng gathered at London's Zoological Institute. The frenzy that seizes the audience recalls that which was occasionally inspired by Thomas Huxley (1825-1895) when he publically and often dramatically defended Darwinian thought during the latter portion of the 19th century by delivering stirring dialogue concerning ties between apes and man, or by presenting the undeniable 'missing link' between birds and dinosaurs known as *Archaeopteryx*.

Doyle's apemen are of particular interest now because he has recently been cited as a major perpetrator of the puzzling Piltdown Man hoax by researchers who re-examined certain key passages in his story. Doyle, it seems, may have prepared the fake fossil and later satirized efforts of scientists who tried to place it into the human evolutionary 'ladder' in *The Lost World.* (However today, evolutionary speciation through time should be viewed as a branching "bush," not a "ladder" of progress, sometimes with 'limbs' interrupted by extinction events.) In Doyle's story, the red-haired missing links may represent the Piltdown Man which was 'discovered' in 1913. The forgery was eventually uncovered in 1952 by Kenneth Oakley, who proved that the fossil was a fake using a fluorine dating technique. Eventually, South America – where Doyle's Professor Challenger encountered his living dinosaurs—became known as a crypt for many odd and gigantic fossil specimens, becoming known in some paleontological circles as the 'lost world.'

In Edgar Rice Burroughs' action-packed 1914 story, *At the Earth's Core* (1914), two explorers penetrate through five hundred miles of Earth's Crust in a mechanical drilling mole. They emerge on the inner-surface world of Pellucidar, making an astounding discovery that the Earth is hollow as its spherical shell rotates about a tiny radiant star. Primitive and bizarre life forms terrorize the explorers on this horizon-less world. Burroughs' explanation for why the evolutionary state of the inner crust is 'primitive' compared to the outer crust is because the innermost crust cooled more slowly after Earth's formation—thus, evolution was slowed.

Thereafter its surface became habitable for organisms at a much later geological 'period,' relative to Earth's outer surface. Many prehistoric beasts, including apemen (sagoths), and rhamphorhynchus-like mahars inhabit this strange world. (*Rhamphorhynchus* was a genus of winged pterodactyl that lived during the Mesozoic era.)

While drilling through the crust, the explorers notice that Earth's interior temperature never exceeded 153 degrees F, and that there is even ice present deep within the interior. Earth's radius was known to be about four-thousand miles in antiquity. (Eratosthenes 276-196 B.C.E.) However, it wasn't until 1913 when the American geologist, Beno Gutenberg (1889-1960), using seismic data, announced his startling determination that Earth has a 'core' with a radius of 2,100 miles.

Although Burroughs never related the existence of a world such as Pellucidar to seismological data, *At the Earth's Core* and its several Pellucidar sequels are fascinating considering that they were written at a time when much less was known about the tremendous temperatures and pressures in Earth's mantle. Interestingly, *At the Earth's Core* was published 57 years before Phase 1 of the highly publicized Mohole Project was completed, in which scientists actually probed Earth's interior by drilling miles into the crust.

The hollow Earth concept was popularized in 1818 by the U.S. infantryman, John C. Symmes (1780-1829), who held that Earth had open polar passages extending into the planet's interior. Scientific data for the Earth's average density and mass was determined by Henry Cavendish (1731-1810) in 1798. Symmes' conceptions became popular even though Earth's density could not be as great as determined by Cavendish if it were hollow.

Symmes was not the first to propose a hollow Earth concept. As early as 1692, renown astronomer Edmund Halley (1656-1742), of comet fame, suggested that Auroran light emerges from a polar passage. Nevertheless, Symmes' ideas proved so influential that the United States supported an American expedition to Antarctica in 1838 in order to investigate the polar passage theory. Although scientifically unfounded, it is apparent that Symmes may have influenced science fiction writers such as Burroughs and Jules Verne (1828-1905), who were able to populate subterranean worlds with prehistoric creatures … or Edgar Allan Poe (1809-1849) as well—who dispensed with prehistorical themes.

Even the English science fiction writer, H.G. Wells (1866-1946), in his story *First Men in the Moon* (1901), pictured the Moon's interior as a vast cavernous 'anthill' filled with insect-like Selenites. As in Verne's

Journey to the Center of the Earth (1864) a central sea resided at the Moon's core, and the diminishing effects of gravitational pull, which should be noticed as one hypothetically descends through the framework of a planetary body disappointingly were neglected. Furthermore, at the Earth's (or Moon's) 'hollow' core or central sea, gravitational force would not be experienced. The mechanics of this gravitational problem had been theoretically solved by Sir Isaac Newton (1642-1727).

French writer Verne presented two German scientists with a monumental challenge of tracing the footsteps of a great but fictitious 16th century scientist, Arne Saknussemm, through a volcano in Iceland to the 'center' of the Earth in his *Journey to the Center of the Earth*. This epic adventure tale is a masterful showcase of Victorian scientific thought.

Inspired by explorations of an associate, Verne raised the question of the origin of volcanoes in his story. Although the explorers descended through rock of recent volcanic origin, temperatures beneath the volcano never increase above moderation. In accounting for this unusual circumstance, Verne relied on researches of English chemist Sir Humphry Davy (1778-1829) to challenge the once highly regarded Central Fire theory, as proposed in Athanasius Kircher's (1606-1680) *Mundus Subterraneus* in 1678.

Volcanoes are now known to be geological features commonly associated with converging or diverging plate tectonic boundaries, and are no longer believed to be caused by the distillatory effect of an 'internal fire' or unaccountable exothermic, chemical reactions. Iceland is now known to be an insular hot-bed of volcanic activity owing to its situation on the mid-Atlantic Ridge system. Although German Alfred Wegener (1880-1930) was the first to hypothesize Continental Drift in 1932, connections between volcanism and sea floor-spreading were not realized until half a century later.

Eugene Dubois' (1859-1940) discovery of *Pithecanthropus* ("Java Man") in 1890 triggered vastly improved descriptions of 'apemen' in Burroughs' stories. However, Burroughs' fossil humans contrasted greatly with an enigmatical apeman fossil described in Verne's 1867 version of *Journey to the Center of the Earth*. When Verne's story was written, nine years after Charles Darwin (1809-1882) and Alfred R. Wallace (1823-1913) first published ideas about natural selection and the origin of species, the only fossil 'proto-human' known was Neandertal Man (1856). Although scientific controversy erupted over the matter of prehistoric humans, even after publication of Darwin's *The Descent of Man* (1871), it wasn't until

1913 when a French anthropologist set a convincing (though currently unpopular) standard for Neandertal restorations.

Upon discovery of the unusual hominid in Verne's story, an explorer remarks that Man might be well over 100,000 years old. Current estimates place modern Man's appearance at 40,000 years yet early human-type species on our 'bush' at well over three million years ago. Nonetheless this is an unusually bold position for Verne to have taken considering how unpopular Darwinism was then.

In another fascinating passage, an explorer lost in reverie imagines how the world developed since its primordial state. The dream traces in reverse sequence Earth's entire organic and physical evolution. "…I carried myself back to the far ages, long before Man existed, when the Earth was in too imperfect a state for him to live upon it." Verne's 'scriptural' period is not at all similar to that described by mid-19th century theologians, and in this passage appears influenced by doctrines of uniformitarian geologists.

Thus, we must certainly admire Verne's integrity in introducing non-conforming ideas about the origin of extinct animals, including fossil humans, and Earth's progressional development. In doing so, Verne (tacitly) acknowledged the principle of geological uniformitarianism, originally proposed by Scottish geologist, James Hutton (1726-1797) and popularized by Scottish geologist Charles Lyell (1797-1875). Uniformitarianism, which is at the foundation of modern geological thought, affirms that the gradual and uniform, rather than 'catastrophic' forces of nature dominate in the long run to produce most observable geological structures, and that the "present is the key to the past." Such ideology was of influential significance in the formulation of evolutionary thought because it excluded the role of supernatural forces from Geology.

Conversely, influences of Baron Georges Cuvier (1769-1832) are also very apparent in Verne's story. Verne's characters encounter many 'antediluvian' animals, often regarding them in deference toward the brilliant comparative anatomist. Cuvier adopted concepts proposed by Charles Bonnet (1720-1793) who believed that species could not evolve and that fossils were remnants of organisms existing prior to a geological catastrophe. Biblical floods were generally considered examples of catastrophes (although Cuvier did not invoke scriptural elements in his theory of the Earth). In *Journey to the Center of the Earth*, a central sea is regarded as the remnant of a former catastrophe.

Cuvier studied progression of fossil organisms from 'primitive' to more 'advanced' life-forms ranging through geological time. However, unlike many eminent naturalists of his day, Cuvier tactfully avoided controversial issues of species origination and 'special creations' when relating how the planet became repopulated after each catastrophic event. Cuvier was the most noted French scientist of his time, although his ideology concerning effects of geological catastrophes on progression of 'antediluvian' creatures, and immutability of species rapidly lost ground during his later years, turning instead to uniformitarianism and (increasingly) evolutionary thought. Ironically though, geo-catastrophism resurged in the latter 20th century after it was realized how the dinosaurs went extinct suddenly as a result of a large asteroid impact.

Treatment of evolutionary thought in science fiction—replete with prehistoric elements—perhaps culminating in Edgar Rice Burroughs' stories concerning the inadvertent exploration of a prehistoric southern Pacific realm known as Caspak in his 1918 novel, *The Land That Time Forgot*. Here, influences of Ernst Haeckel's (1834-1919) Biogenetic Law are evident. The Biogenetic Law held that developing embryos pass through stages resembling their evolutionary ancestors. In Caspak a mysterious, alternate evolutionary process gradually transforms prehistoric saurian and Cenozoic beasts into more geologically recent creatures, including human-evolutionary ancestors, throughout their *individual* lifetimes.

However, in 1901 A. Pavloc demonstrated that Haeckel's ideology couldn't be accurately used to describe growth stages of certain Jurassic fossils. Nevertheless, Burroughs' entry is an outrageously original idea for a science fiction story, considering that theories concerning development of organisms and origin of species are still hotly debated today. (Concepts of punctuated equilibria, in which speciation is thought to occur via a series of evolutionary 'jumps,' has superseded reliance on phyletic gradualism, in which speciation is thought to proceed more gradually while producing more intermediate fossil forms—'missing links.')

In the third installment of Burroughs' "land that time forgot" trilogy, *Out of Time's Abyss* (1918), when an explorer is savagely attacked and killed by a *Tyrannosaurus*, a comrade declares that the beast lived only six million years ago. Before refinement of radioactive age dating methodology, the age of the Earth, remoteness and duration of its several eras and geological periods, as well as the age of its extinct inhabitants were not precisely known. For many decades prior, the most reliable estimates for Earth's age had been determined from alternative data. For instance, that regarding the planet's rate of cooling—ranging up to 100 million years, using Lord Kelvin's (1824-1907) methods, only a fraction of the currently accepted measurement of 4.6 billion years. Such early estimates were inaccurate because contributions from Earth's continual, internal geothermal heat-releasing, radioactive processes were neglected. (The term "radioactivity" was defined in 1898.) Other estimates based on oceanic salinity, astronomical observations and sedimentation rates typically fell into this range as well.

It is now commonly believed that dinosaurs perished 66 million years ago, and that no human being has or ever will meet a genuine living dinosaur. Cognizant of such circumstances, during the mid-20th century, it has been the practice of contemporary science fiction writers to have our intrepid heroes either bridge this enormous expanse of time by traveling through it as in Ray Bradbury's classic short story, "A Sound of Thunder" (1952), or by traveling through outer space as in Anne McCaffrey's *Dinosaur Planet* (1978) to confront our beloved, but alien dinosaurs.

Scientific intrigue over puzzling matters such as the extinction of Neandertal Man and the extent of his contact with other types of prehistoric humans has prompted writers Jean K. Auel and noted paleomammalogist Bjorn Kurten (1924-1988) to explore prehistory during the Pleistocene. Their novels, *The Clan of the Cave Bear* (1981), and *Dance of the Tiger* (1978), respectively, demonstrated that general knowledge about human prehistory has considerably increased since the late 19th century through modern scientific investigation.

Author-scientist Stephen J. Gould (1941-2002), formerly of Harvard University, appropriately wrote in the introduction to Kurten's novel, "… these (science fiction) stories have a role in science. They probe the range of alternatives; they channel thought into the construction of testable hypotheses; they serve as tentative frameworks for the ordering of observations." While there is great truth in Gould's statement, the wonderful impact that these stories have on human imaginations cannot be taken lightly.

Although authenticity in Doyle's, Burroughs', and Verne's stories, for example, has considerably diminished through the years, many ideas in contemporary science fiction literature and film can be traced to them. Quite understandably, they're regarded as classics of this popular venue.

Through scientific investigations, the world has revealed many of its primeval mysteries. Scientists and laymen alike now have a much fuller comprehension than did our counterparts many decades ago of how our prehistoric world may have appeared and what it contained. An emphasis in science fiction of the past several decades may have shifted toward the vast reaches of the cosmos, but the prehistoric element in science fiction still attracts many readers who find solace and independence in the youthful mystical world of ancient times.

Chapter Fourteen—*Get Real: Dinosaur Masquerade*

'No Monsters (alive, dead, supernatural or undead) were harmed during production of scenes enacted in this film!'

Horrors! Here comes Godzilla, although this time cast as a real *live* Chuck-Walla Lizard adorned with pseudo-stegosaurian plates along its spine and tail, smashing through another miniature cardboard city set. Cameras roll, preserving its awkward sprawling gait and hissing tongue on film. Will this special effect visual cause us to cower in abject fear, or instead to die laughing? How many academy awards for special visual effects would Eiji Tsubuyara have earned using live lizards masquerading as 'Godzilla,' instead of his famed suitmation bipedal monster? Perhaps more than you'd think. Please consider.............

Dinosaurs embody a revered class of classic movie monsters. Claudine Cohen, author of *The Fate of the Mammoth,* noted how the earliest English horror novels as well as a 'taste for macabre ruins and graveyards' sprung from discoveries made by 18th and 19th century natural historians and paleontologists, such as Baron Georges Cuvier, who interpreted Earth's prehistory as a world wrought by the most awful and unimaginable cataclysms and deluges. Like those inspiring 'macabre' archaeological ruins, the earth could also be viewed as a 'ruined' relic. Thus, 'antediluvian' monsters, bones of which were disinterred from sediments representing these catastrophic episodes, seemed terrible and haunting.

Perhaps the primary reason why we so adore dinosaurs in fantasy films is because, perversely, we enjoy being scared. As Alfred Hitchcock noted, "Fear.......is an emotion people like to feel when they know they are safe." Intellectually, we know our safety is secured because the last of the real dinosaurs expired 65 million years ago. We watch, willingly, to be consumed in the story, realizing that we won't really be devoured. However, on-screen, right-in-your-face dinosaurs can still seem as threatening as any other kind of movie monster, or even more so because the mightiest of them can inflict more damage on a city or psyche than less powerful phantoms, werewolves and mummies. The most titillating elements in the dinosaur-as-movie-monster equation are that, unlike imaginary werewolves, vampires and reincarnated mummies, dinosaurs were/are real creatures, which thrived during a 180-million-year geological Era, and also that many were of monstrous proportions or of horrific temperament.

Because human actors do not anatomically resemble dinosaurs, swimming tylosaurs or flying pterodactyls, special effects are an essentiality of most dinosaur films. The dinosaur movie industry has been marked by four basic innovations in special effects. Namely, these are (1.) use of stop motion animation and cartoons (used most widely between the 1910s and the late 1960s), (2.) reliance on 'suitmation' (most often created for 1950s and 1960s 'B' films, but persisting into the present day), (3.) use of live reptiles (beginning in 1940) to which sometimes wire, rubber and cardboard prosthetics have been attached to render a dinosaurian 'look,' (4.) reliance on life-sized dinosaur sculptures, mechanical robots or props to interact 'live' on camera with human actors (a gimmick relied on throughout the 20th century), and (5.) most dramatically (beginning in the 1990s), computer-animated 'cgi' dinosaurs.

Usually, dinosaur movies are rated by how good the effects are. However, simply because the special effects seem mind-blowing, doesn't guarantee a 'four claws up' rating. Any dinosaur movie lacking that 'special' ingredient of effects wizardry would be a dismal affair indeed! In fact, even if the script is good and the actors turn in creditable performances, it's hard to make a decent dinosaur movie without rousing special effects. Then sometimes, because the effects are as awful as is the human acting, a 'cult-classic' is born. (Think of the 1962 Danish film *Reptilicus*, for example, which has some strange 'hold' over me every time it's televised.)

Aficionados of dinosaur films most highly appreciate stop-motion animation, if done convincingly, because it represents the greatest painstaking labor of hands-on, saurian love. Wouldn't it be great to play around on one of those miniature 'sets,' helping to make one of those films? Chief examples here include *King Kong* (1933), *The Valley of Gwangi* (1969) and *One Million Years B. C.* (1966), the one starring Raquel Welch. Much has been written about these highly regarded films and others especially those featuring the magical effects of Willis O'Brien, Marcel Delgado and Ray Harryhausen. After stop-motion, viewers laud the *Jurassic Park* era of computerized movie dinosaurs, due to their convincing nature, seamless perfection and on-screen fluidity. (Of course, full-sized dinosaurs and props were used to accent both these categories of films too.) "Suitmation" dinosaurs traditionally receive less critical acclaim, yet they are often memorable. But let's get 'real' for a moment.

In retrospect, most would perhaps agree that the least convincing movie 'dinosaurs' of all are created when live lizards and alligators wear dinosaur 'suits.' This is the often neglected, 'dinosaur impersonator' category. In his book, *Starring T. rex! Dinosaur Mythology and Popular Culture* (2002), paleontologist Jose Luis Sanz names this motley crew of would-be dinosaurs, "sauriodinosauroids." Sanz notes that:

> "These consist of innocent lizards (iguanas and monitor lizards, among others) and baby crocodiles to which rubber horns, spines, and dorsal fins are attached. This procedure, usually utilized in low budget films, was used repeatedly in productions of the 1950s and 1960s....it is based on incorrect basic suppositions. The idea is that present-day lizards are closely related to dinosaurs—or, to put it more bluntly, that the lizards of our countryside are no less than shrunken, pale shadows of former dinosaurs, relict forms of ancient lost splendor, and therefore have every right to portray their enormous disappeared relatives on screen." (p.115)

Live reptiles and lizards would appear to be sorry excuses for dinosaurs because, frankly, they don't anatomically resemble dinosaurs. Yet that hasn't prevented movie producers and special effects 'masters' from trying to pull a fast one on us, has it?

Perhaps few of you know the *first* and *last* films ever to use live reptile actors sporting 'dinosaur costumes' or otherwise posing as dinosaurs. The first screen appearance of a reptile outfitted in pseudo dinosaur guise, an alligator to which a nasal horn and wings were added, was in D. W. Griffith's *Brute Force* (1913). Don Glut described this apparition in *The Dinosaur Scrapbook* (1980). "The alligator... was ... heavily adorned with attachments. A horn surmounted the reptile's snout, and two sets of wings flapped atop its back, creating an effect more in keeping with a medieval dragon than a prehistoric monster." (pp. 118-119) For *Brute Force,* Griffiths also employed a snake in prehistoric animal disguise and the world's first complete dinosaur prop—a stiff, life-sized horned *Ceratosaurus.*

And the last film? That would be Don Glut's movie *Dinosaur Valley Girls* (1996). Okay—given that dinosaur expert Glut also employed several anatomically correct stop-motion dinosaurs in DVG, one might argue that his pet tegu lizard, 'Neecha,' really wasn't intended to be a 'dinosaur' in this film. In fact, in his *Dinosaur Filmography* (2002), Mark F. Berry claims that the last (made for TV) film employing the 'live-lizard strategy' was the forgettable *Island of the Lost* (1967). Regardless, I met the now deceased

Neecha shortly after his film debut, which is why I prefer to believe I had the privilege of petting the very last of the 'reel' scaly breed of movie actors to pose as a 'real' dinosaur. Neecha was on a rampage running through Don's Burbank home when we met there in 1995. (Two other reptiles appeared in *Dinosaur Valley Girls* - 'Big-Un', an Ornate Nile Monitor Lizard and 'B.C.'—Dons pet Boa Constrictor.)

Don described Neecha's performance in Chapter 3 of his *Dinosaur Valley Girls - The Novel* (1997, p. 88). Don cynically recalled, "We had to nudge the animal a bit before he moved the way I wanted him to. But Neecha did a fine job in his debut screen performance, causing us fewer problems than did some of the human talent." Admittedly, his grand "slow-motion speeding" entrance, filmed at 64 frames per second instead of the usual 24 frames per second to add 'gravity,' looks impressive.

Gentle 'nudging' aside, 'Dinosaur Don' wouldn't have harmed his beloved pet Neecha. But in the past, have overzealous producers been less scrupulous in recreating Cretaceous carnage and Mesozoic savagery using live reptilian actors? Well, there are rules for all to live by.

The American Society for the Prevention of Cruelty to Animals was founded in 1866 by Henry Bergh. Contemporaneously, in England there was a growing humanitarian movement at large to spare animals from suffering under the hands of non-medical biologists who surgically tested live animals to gain knowledge of anatomy. But British paleontologist Thomas Henry Huxley, for example, contested the humane society's view, espousing that as long as operations were conducted with the objective of gaining medical knowledge to prevent *human* suffering, then such experimentation should be condoned. Over half a century later in America, the use of live animals in fantastic films was viewed with skepticism. Squeamish film-goers didn't want to see animals brutalized on camera, right before their very eyes. Such humanitarian attitudes made an impact on the motion film industry through an organization, the Picture Society for the Prevention of Cruelty to Animals. By the 1960s, movie producers took pains to prevent the suffering of live animals used as 'actors' in films, including those wrestling, writhing reptiles masquerading as dinosaurs. As a final gag in the end credits to his *Dinosaur Valley Girls*, nodding to the PSPCA, Glut facetiously declared "No animals, extinct or extant, were harmed during the making of this motion picture."

Due to the successes of *Jurassic Park's* digital dinosaurs, the live-lizard strategy is nearly an extinct art form. And although many baby boomer traditionalists prefer stop-motion animated dinosaurs, purely for fun let's focus on how the live-lizard strategy was formerly employed to greatest effect, back when the rampaging flesh and blood 'dinosaurs' really were alive and very scary! For there were a handful of films effectively employing the 'live-lizard strategy'; foremost among this class of films were *One Million B. C.* (1940), *Journey to the Center of the Earth* (1959) and *The Lost World* (1960).

A major influence on Don's creation of *Dinosaur Valley Girls* was Hal Roach's *One Million B.C.* (1940). In fact, Don even named his pet tegu lizard after vocabulary used in the 1940 film to denote a reptilian dino-monster, or in cave-speak, a "neecha." *One Million B.C.'s* premise is that even if dinosaurs and, by definition, other huge reptilian 'prehistoric animals' didn't survive to our present, how cool it would have been had they survived into the relatively recent past, during caveman times!

One Million B.C. is a downright classic of epical proportions, even though it wasn't based on any famed fantasy writer's preexisting novel. Lon Chaney, Jr. starred as 'Akhoba,' (a year before Universal's *The Wolfman* hit the screens), and cave-babe actress—Carole Landis '(Loana') starred as Tumak's love interest. Tumak is Akhoba's son, played by Victor Mature. While the story as related by the human actors is similar to the 1966 remake, the approaches to prehistoric animal special effects contrasted considerably between the two pictures. While Ray Harryhausen created the utterly fantastic effects for *One Million Years B.C.* (1966) using stop-motion animation, Roach employed living animals including reptiles scaled-up to appear monstrously huge and shot in slow-motion, as well as one of the worst suit-mation dinosaurs ever filmed—a *T. rex.*

After being knocked out of a tree limb by a mammoth, really an elephant to which a shaggy coat of fur was applied by makeup artists, Tumak is found floating on a river, still alive. Now with the friendly Shell Tribe folk, Tumak, who was raised in a violent 'every caveman for himself' type setting—the Rock Tribe, learns table manners, politeness and customs, how to laugh and sing, how cave-writing is done, and most of all, how weapons of mass destruction—blade- tipped spears—can be made. Using a spear, Tumak kills the suitmation *T. rex*. Eventually, after a volcano erupts, the two tribes are united when Tumak and Akhoba rescue the Shell Tribe from the greatest neecha of all. This really huge 'dinosaur,' played by a rhinoceros iguana, is buried under an avalanche of rocks.

Several saurians appeared in the opening credits, including one which didn't make it into the latter portion of the movie, a Chuck-Walla Lizard impressively outfitted with two double-paired rows of stegosaurian plates, *and* a massive brow horn. While a publicity photo of this particular neecha beast can be seen in on page 179 of Don Glut's *Jurassic Classics: A Collection of Saurian Essays and Mesozoic Musings* (2001), the 'chuck-walla-saur' can barely be identified moving behind the film's opening credits. I wonder if its plates kept falling off during rehearsal. Not all the live animals who played dinosaurs in *One Million B. C.* (besides the human actor who played the *T. rex*) were reptiles, and not all the live animals appearing in this film played dinosaurs (such as the elephants in mammoth disguise). The *Triceratops* killed by Tumak was actually a pig, whose head was adorned with a great Halloween mask, that of a three-horned ceratopsian. And serving as another example, an armadillo with added brow horns played a way too huge-looking, armored 'glyptodont.'

In one dramatic scene, a gigantic *Dimetrodon*, really a dwarf alligator with an attached flexible sail, battles an enormous tegu lizard while Tumak and Loana cower under rocks in a crevasse. *Dimetrodon* wins, as we see blood oozing from the dying tegu lizard's jugular. Today such a grisly scene wouldn't have passed the American Society for the Prevention of Cruelty to Animal's (ASPCA's) scrutiny. In his *Dinosaur Scrapbook,* Don Glut stated: "Roy Seawright explained to me that the reptiles were starved for a few days, then brought onto their miniature sets. With a quick and mild shock from an electric prod, the animals would leap out at the nearest shocked creature - such as another equally hungry and shocked reptile." (p. 119) Hal Roach claimed he possessed certificates from the ASPCA testifying that no studio animals were mistreated. Nevertheless, the bloodier scenes were cut from the British release of the movie.

Any real *Dimetrodon* would have broken the bony vertebral spines in its sail the way Seawright's stage *Dimetrodon* is shown rolling about, overturning from back to belly, before eventually dispatching its opponent. Later, during the volcanic eruption sequence, a *Dimetrodon* and other dino-lizards avoid nasty burns from 'lava' on the miniature set by falling into fissures. 'Monstrous' sound effects were effectively created by mixing and dubbing-in lion, wild boar, elephant, dog, monkey and pig noises.

While the release of *One Million B.C.* predated my childhood, two later films, *Journey to the Center of the Earth* (1959) and *The Lost World* (1960) coincide with my dawning interests in paleontology. These two latter films are also associated in time and inspiration. Plotting of both movies departs liberally from the original novels written by Jules Verne in 1864 and Arthur Conan Doyle in 1912, respectively. However, for our purposes and for sake of brevity, let's just acknowledge that the prehistoric animals envisioned by Verne were not those featured in the 1959 production. Nor did Doyle's visions of Mesozoic wildlife resemble any of the creatures inhabiting the 'lost' plateau in the 1960 film. Movies based on both Verne's and Doyle's novels had been filmed previously, most notably a 1925 silent production of *The Lost World*. Based on the number of recent made-for-television remakes, these classics are certain to inspire further filmography.

Journey's 'sauriodinosauroids' have been regarded as the most convincing ever to use the live-lizard strategy. Except this is a fallacy because the animals special effects master L. B. Abbott intended for this film were a variety of Permian sail-backed creatures—genus *Dimetrodon* (like the dwarf alligator used in *One Million B. C.)*, which was not a dinosaur. Abbott's effects were introduced into the film, flawlessly. Dorsal sails made of foam rubber and wire were attached along the spines of 24, four-foot long rhinoceros iguanas. Then, to camouflage the attachment 'join' along their spines, the sails were bonded to painted cloth strips which draped over their sides. A most ingenious disguise! And totally convincing, even though Abbott's *Dimetrodons* appeared about 10 times larger than their fossil forebears. During the 'big bug' sci-fi 'mutation era' of moviedom, few people cared whether *Journey's Dimetrodon's* were way too big.

In Mark Berry's *The Dinosaur Filmography* (2002), we also learn that the cold-blooded iguanas remained lethargic on camera, unless a movie light warmed them up. But then after taking a few strides, they would dart away much faster than any 'gargantuan' 60-foot long creature could be expected to move. Actor Pat Boone who played 'Alec' in this film recalled that in filming these scenes, later composited with the reptile footage, he, James Mason who played Professor Oliver Lindenbrook in the movie, and the rest of the subterranean explorers had to react to:

> "... a guy on a ladder with a long pole with a rag on the end of it... and we had to cringe and pull back and brace ourselves against this guy on a ladder with a rag on the end of a pole. We felt like idiots." However, as Berry noted, "I defy the most-eagle-eyed nitpicker to spot one glitch in the composite of the men in the surf looking at the creature on the beach, or the shot of the explorers dragging their raft past the feeding pack. The old-fashioned optical mattes in the Dimetrodon sequence are as flawless as any state-of-the-art digital composite." (pp. 148-150)

At the film's climax in the sunken city of Atlantis, a giant lizard, this time a two-foot-long tegu lizard not identified by Lindenbrook as any particular dinosaur genus, threatens the explorers, right before the volcano jettisons them from Earth's 'center.'

In April 1960, both Abbott and James B. Gordon received Academy Awards for *Journey's* special visual effects. This was the second time such awards were granted for use of the live-lizard strategy, the first being in 1940 for Roy Seawright's use of photography in - you guessed it - *One Million B.C.*.

Okay—given *One Million B.C.'s* and *Journey's* striking special effects successes, I was surprised to learn that the next major motion picture to use the live-lizard strategy became so reviled by critics. After all, *The Lost World's* 'Irwinosauruses' scared the heck out of me at an impressionable age. Seriously! And it didn't matter that one of *Lost World's* characters Professor Challenger, played by Claude Rains who had evidently gone bonkers misidentifying quadrupedal lizards as 'dinosaurs.' Nearly every 1960s kid in the audience, including myself, knew what a *T. rex* looked like and it surely didn't resemble a lizard with spines! But still, certain scenes absolutely petrified me. So, you ask, "what's an 'Irwinosaurus'?" That's the crux of this movie's problem, as usually related.

Journey's box office successes provided incentives for next moving forward with a remake of the Doyle classic. At one time, stop-motion master Willis O'Brien had been invited to work on the project, but 'O'Bie' backed out after he realized that director Irwin Allen had decided to use "gagged-up lizards" (using 'Obie's' condemning term) instead of stop-motion puppet dinosaurs. Allen recalled how time-consuming were the 15 minutes of dinosaur sequences Ray Harryhausen labored to create for Allen's 1956 production, *The Animal World*. Allen recollected, "The process was just agonizing, getting back the footage in little pieces over weeks and weeks." Harryhausen groused, "...they did it the 'easy' way by gluing fins on alligators. They didn't look like dinosaurs, they just looked like something out of *Flash Gordon*."

Irwin Allen felt the live-lizard strategy had worked marvelously in *Journey*, so why wouldn't audiences accept another "dinosaur masquerade"? So instead, he turned to *Journey's* special effects master, L. B. Abbott. And not that it mattered, but at least two of *Lost World's* sprawling lizard-actors had previous 'dinosaur acting' experience, having starred in *The Land Unknown* (1957). The pieces all seemed to be falling into place for yet another cinematic marvel! Sigh! Although Allen's approach may have sounded reasonable at the time, today 20th Century-Fox's *The Lost World* is bemoaned as an 'only if' story, perhaps *the* film forever dousing the movie industry's interest in the live-lizard-as- movie-dinosaur strategy.

Instead of getting the most refined state-of-the-art, 1960s 'space age' era, stop-motion effects, in a dramatic scene where a pseudo-brontosaur battles a carnivorous 'dinosaur,' Irwin Allen merely delivered a monitor lizard outfitted with a ceratopsian-like frill fighting a caiman adorned with spines and a *Dimetrodon*-like sail. Commenting on the animals in this scene. Berry said:

> "Not content with mere back fins, the lizard dressers glued horns, plates, barbels, and even a floppy rubber neck frill on the critters; but the more prosthetics they added, the goofier the animals looked. The frill for example, makes its wearer look rather like some sort of ancient idol one might find inside a pyramid.... The specter of possible animal abuse always hangs over scenes like the big dino battle, which is not only distasteful but also a blatant copy of the famous fight in One Million B.C." (p. 249)

Lost World's "Fire Monster," (another finback reptile, incongruously identified by Professor Challenger as a *'T. rex'*) scene has always been a dino-monster favorite of mine. However, for all its faults, scenes in which a docile iguana with horns added, although scaled-up to look enormous while chomping on vegetation, and a nighttime scene in which an unseen 'dinosaur' stomps through the forest wreaking havoc, seriously, had me quaking with fear.

In defending his *Lost World* dino-creations, L. B. Abbott claimed the reptile actors "...naturally move the way monsters should." However, as Mark Berry countered "...the problem is not that lizards can't make serviceable monsters, it's that they can't make good dinosaurs." In fact, pre-production artwork for the film falsely advertised that 'real' looking dinosaurs would be featured in Allen's remake. As Don Glut remarked in *Dinosaur Scrapbook*:

> "Perhaps if the reptiles had gone unidentified as in One Million B. C., the picture might not have been so ludicrous. But no one, not even the youngest dinosaur buff in the audience, could rightfully accept it when Professor Challenger.... declares that a monitor lizard with a rubbery ceratopsian frill attached to its neck was a Brontosaurus." (p. 127)

The human characters in *One Million B.C.*, the first film featuring prehistoric animals since release of *King Kong* (1933), had no need to 'name' prehistoric fauna, other than by using the vague term 'neecha.' Furthermore, as Mark Berry noted in his entry on this film:

> "Far from apologizing for the lack of stop-motion, the film's promotional material actually bragged about it. The studio pressbook promised 'no animation whatsoever,' and an unknown copywriter came up with what must surely be the amazing ad line of all time: 'Actual living animals of a bygone age recreated and filmed by a new secret process.'"

Journey's *Dimetrodons* looked reasonably like *Dimetrodons*, their only flaw being that they were far too big. But *Lost World*'s prehistoric animals didn't look like any Mesozoic dinosaurs known to science. To make matters worse, Professor Challenger, embarrassingly asserted they were *actual* dinosaur genera. So maybe, penultimately, blame lies with *Lost World*'s script writers, not Irwin Allen's choice of special effects.

The live-lizard strategy has been frequently used since D. W. Griffith's introduction of the technique in 1913. Some of these attempts have been more successful than others. For instance, besides *One Million B.C., Journey,* and *Lost World* (1960 version), live lizards posing as 'pseudo-dinosaurs' made it into *Secret of the Loch* (1934), and a 1936 *Flash Gordon* serial. In *Flash Gordon*, 'sauriodinosauroids' referred to as 'ghastly monsters' battled for the first time on screen, four years prior to *One Million B.C.'s* dramatic reptilian combat scene. And dinosaur movie buffs might not recall that Ray Harryhausen even included a live, "gagged-up" iguana among his prehistoric *One Million Years B.C.* (1966) menagerie, a blunder which he now refers to as "....one of my, shall we say, rare miscalculations."

And then there are films like (sigh!) *King Dinosaur* (1955), which Mark Berry claims "...is a strong contender for the title of Worst Dinosaur Movie of All Time.... the special effects.... are astonishing in their inadequacy." I remember seeing this movie as a young boy, feeling cheated by these its titular 'king' dino-monster. And lest we forget the forgettable *Untamed Women* (1952), *Robot Monster* (1953), *and Women of the Prehistoric Planet* ((1966)... utterly awful films which also stagger the imagination in *their* inadequacy! Such offerings came about because by the 1950s, cheap productions such as these, burdened with ridiculous plotting, had begun pirating stock footage from *One Million B.C.*. This is at least one reason for why Hal Roach's 1940 film isn't—as revered as it should be today—as its award winning 'dinosaurs' have become sadly associated with abysmal movie fare.

Contrary to claims that a live iguana was actually burned in *Women of the Prehistoric Planet*'s explosion scene, Berry counters "....frame-by-frame examination of the sequence supports...that...the camera was stopped and the reptile removed from the shot just prior to the explosion, and the shots of it 'writhing' while ablaze look suspiciously like a rubber model being jiggled by the prop man." (p. 427) Irwin Allen also recycled 'Irwinosaurus' scenes into his televised productions, such as *Voyage to the Bottom of the Sea*.

While suitmation ruled the day in Japan from the 1950s through the 1970s and beyond, following their experiences with stop-motion dinosaurs in expensive productions such as *The Lost World* (1925), *King Kong* (1933), *The Beast From 20,000 Fathoms* (1953) and a nature documentary *Animal Planet* (1956), between 1940 to 1960, money-conscious American producers preferred the more economical and time-saving live-lizard 'detour.' Anyway, one can't trust a Chuck-Walla Lizard to destroy a miniature set of Tokyo exactly according to the script like a human actor wearing a dinosaur costume could. But in 'naturalistic' settings, steamy jungles or deserts, live-lizards could more or less convincingly play the role of dinosaurs, provided disbelief has been effectively suspended by a special effects team. It all depends on what you think you can get away with.

Although usually limited by expense and time, stop-motion was the earliest dino-monster technique to become widely used in major films. Whereas both live-lizard and suitmation (i.e. dinos played by human actors in dinosaur suits) strategies were American Hollywood inventions, fantasy dinosaur suitmation became refined in Japan. In Japan, beginning with *Godzilla* (1954), suitmation sizzled whereas in America it fizzled. This, however, isn't to imply that after Tumak killed *One Million B. C.'s T. rex*, live-lizards and stop-motion techniques were the only alternatives utilized by producers in English-speaking nations, because American and British producers continued to experiment with suitmation too.

Fairly decent films such as *Unknown Island* (1948), *The Land Unknown* (1957)—which also employed a live reptile, enlarged via optical effects, and *Gorgo* (1960) come to mind. But Asian dino-monster movie producers (e.g. Toho, Nikkatsu, and Daiei) have been wedded to suitmation ever since the heyday of the live-lizard techniques. For example, besides the Godzilla and Gamera pantheon of *kaiju* eiga, Tsuburaya Productions even used suitmation to portray *real* dinosaurs in *The Last Dinosaur* (1977). Meanwhile, in South Korea, a late 1960s suitmation-saur starred in *Yongary: Monster From the Deep*. By then, however, Japan's suitmation 'drago-dinosauroids' had reinvaded America, and, as some would argue, even outstripped the competition—stop- motion dinos as well as live lizard sauriodinosauroids.

'No,' I'm not professing which choice of special effects is best. What's most important is how and whether the dinosaur effects are *effectively* woven into the story. And, 'no,' I'm not one of those stop-motion purists. When done right and framed within a convincing story, suitmation really 'works' for me, and I can still watch *One Million B.C., Journey to the Center of the Earth* and *The Lost World* (1960) while enjoying live lizard 'talent.'

But c'mon, get 'real'! That 'gag' scene in *The Lost World: Jurassic Park* (1997) when, Spielberg's computer generated *T. rex* storms through downtown Los Angeles, prompting a fleeing Japanese businessman to say (in Japanese) "I left Tokyo to get away from all this!" doesn't 'work' for me because everybody knows that during their heyday, Japan's dino-monsters were (nearly) always 'suitmation,' (and rarely, to my knowledge, cgi, stop-motion or live-lizard) creations.

Chapter Fifteen—*Invention of Toho's First Two Flying Paleo-Monster species*

I first saw those fantastic giant Japanese dino-monster movies on the living room floor of our house back during the very early 1960s. That's when I saw Godzilla, Gigantis, Angilas, and two curious flying species metaphorically spawned in the Japanese postwar nuclear cauldron. These were Rodan, and another that never actually flew across our old black and white television screen – Varan. Yet Toho envisioned the latter two as flying creatures in their original filmic terrors. Guilty pleasures one and all! To me, next to Godzilla and Angilas, Rodan and Varan are by far the coolest-looking paleo-monsters emerging from Toho's inventive studios.

Over half a century ago, there were three Chicagoland shows where such sci-fi fare was regularly aired, *Thrillerama*, *The Big Show* and *The Early Show*. (*Shock Theatre* was long gone by the time we moved to the Chicago suburbs in August 1961, when 2nd grade was just beginning for me.) It would be relatively difficult discerning when exactly I was so avidly introduced to each of these films way back when, but a quick perusal of Ted Okuda's and Mark Yurkiv's *Chicago TV Horror Movie Shows: From Shock Theatre to Svengoolie* (2007), indicates that I probably saw *Varan, the Unbelievable* on the afternoon *Big Show* on November 18, 1965. I had assuredly seen *Rodan the Flying Monster* on one of these programs long before then because I was already very familiar with this monster when we saw *Ghidrah the Three-headed Monster* in the theater (featuring another fantastic flying titular creature).

I was a fully addicted fossil and dinosaur buff by then, and so it was quite clear from its flying abilities and general appearance that Rodan was intended to be some kind of gigantic, city-destroying pterodactyl. But whereas Rodan was designed along the lines of a pterosaur, Varan's closest living animal analog, based on its flying ability (which I wouldn't witness until many decades later on social media), may instead be the flying squirrel (or bat). But *Rocky & Bullwinkle* tittering notwithstanding, in their own right, pterosaurs may seem much cooler than flying squirrels. So, I'll suspend further discussion of Varan's flying scenes until after delving into those amazing pterosauria. Anyway, as we shall see, the Varan suit may have had an alternate dinosaurian derivation, one which I would not have known of during the early 1960s, inspired from an old restoration rarely reproduced in books today.

During my childhood, the largest pterosaur known to science was *Pteranodon*—the genus with that magnificent long crest protruding rearward from its skull. While its wingspan extended to an impressive 25-feet, paleontologists debated over its weight in life—approximately 50 pounds—during the 1970s. Just how efficient a flyer was this famous old fossil? Too heavy and it could never have flown; too little weight and it would have lacked sufficient musculature to take off and flap its wings. So a body having Rodan's estimated vast physical dimensions of 164-feet tall, a 500-foot wingspan and a 100-ton body mass strained every imaginable limit!

It defied any sense of reason that a creature the size and weight of a (20 million-year-old, post-Mesozoic[1]) Rodan, which hatched from such enormous eggs (100,000 cu. ft. volume), could fly. But when you are a mere tyke entranced by monster scenes on a television broadcast showing the tornadic wind storms generated by those powerful wings ... do you really care? Over a decade later, in 1971, remains of a much larger pterosaur, later assigned the name *Quetzalcoatlus,* were discovered. Adults of this genus had a 40-foot wingspan and may have weighed up to several hundred pounds. So it seemed as if pterosaurs evolved in size through geological time, as such genera had evolved by the Mesozoic Era's final "Period," the Cretaceous. By the mid-1970s, corresponding to a cultural phase in science known as the "dinosaur renaissance," it was also becoming generally accepted that pterosaurs were energetic warm-blooded flyers and soarers. The last of the lot perished 66 million years ago, leaving birds (avian descendants of certain dinosaurian lineages) to conquer the skies.

Paleontologist and paleoartist Mark P. Witton provides an excellent discussion of many genera of real pterosaurs in his 2013 book, *Pterosaurs: Natural History, Evolution, Anatomy (*Princeton University Press). Pterosaurs enjoyed a 150-million-year history of diversification, reflected among an intriguing range of wing and head shapes, mouth and dental apparatus and body sizes, evolutionarily adapted for a disparate variety and range of paleoecological lifestyles and habitats. While today, the most popular if not utterly fascinating genus might be the largest group, the "azhdarchids" (e.g. including *Quetzalcoatlus* which in adulthood stood as tall as a modern giraffe), the filmic giant monster Rodan was clearly based on and assigned to smaller genus *Pteranodon* (with a 20-foot wingspan), species r*odan* (as stated in the film). Another clue is that Rodan's name in the Japanese film title, "Radon" (i.e. *Sora no Daikaiju Radon*), as noted by J. D. Lees, is derived from letters in the genus name *pteRAnoDON*.

Most likely, the shape of Toho's flying *Pteranodon* models was influenced by the image of this genus as painted by famed paleoartist Rudolph F. Zallinger in his "Age of Reptiles" mural for the Yale Peabody Museum (completed in 1947). As stated in a past issue of *G-Fan* and in my most recent 2016 book, at the time, by 1954, Toho technicians were familiar with Zallinger's Age of Reptiles painting. Zallinger had also painted a marine Mesozoic setting which could also have been influential, titled "The Reptiles Return to the Sea," in which several *Pteranodons* soar and hover over large fish and gigantic aquatic saurians swimming below. This painting (also) appeared in a contemporary ("pre-Rodan") issue of *Life* magazine and so it remains possible that Toho made this restoration available to their monster designers.

In one scene in the 1956/57 film, scientists assessing evidence for a paradoxically gigantic *flying* creature even refer to a watercolor restoration of the *Pteranodon* shown on camera. However, as noted by J. D. Lees in his excellent 2007 *G-Fan* (no.79) article:

> "... early concept drawings from the planning stages of *Rodan* show that the monster was originally conceived as more bird than reptile. It seems the ancient link between reptiles and birds, *Archaeopteryx*, was more of an inspiration for Rodan than was *Pteranodon*. Drawings show Rodan with feathers on his wings and along his sinuous tail. Had spfx director Eiji Tsuburaya proceeded along those lines, Rodan might well have ended up looking a lot more like the creature from *The Giant Claw* …Early concept art showed a feathery, more birdlike *kaiju*." (p.53)

A recognizable publicity still highlights a 'transitional' lightly feathered Rodan maquette with thinner wings, indicating that the avian analog had gone fairly far in planning stages. And yes, there was a diminutive *Archaeopteryx* visible in Zallinger's Age of Reptiles mural too. For the record, as of 2011, little crow-sized *Archaeopteryx* has been reclassified as a "deinonychosaurian theropod" dinosaur (*not* a bird).

Once erroneously viewed as leathery "bat-winged" analogues or even as "dragons of the air," pterosaurs, which were lightly furred warm-blooded creatures in life rather than bare-skinned cold-blooded reptilians, are now more correctly regarded as efficient flyers with sturdy wing membranes strengthened with tear-resistant micro-fibers. Fueled by the dinosaur renaissance mission of rediscovery and scientific scrutiny, over 100 pterosaur genera have been described. Imagine how many *other* monstrously immense pterosaurs could be designed today for a Toho film on the wealth of such scientific data!

Unsurprisingly, Rodan appears recognizably different within various scenes in its first movie appearance (and thereafter as noted by Lees), as well as in associated promotional advertising. Why is this? In its debut, we first spy an enormous hatchling; later in the film two winged adults cavort through the skies. After landing, their wings drape toward the ground from a staggering height. Rodan's "hero suit" (or should I have said a "Haruo suit") designed for wear by Haruo Nakajima, has not a single crest, but twin, paired horns projecting rearward from the back of its skull. When worn, the suit stood upright, bipedally (contrary to how paleontologists now believe many pterosaur genera moved) and flaps its huge wings, sweeping city structures and vehicles aside into demolished rubble. Three recurved clawed fingers are embedded in the suit extending midway from the wings, which appear leathery, bat-like or rather resembling a Dracula cape. It is difficult delineating any convincing elongated "wing finger" extending outwardly from those recurved claws though. (Toho's "Fire Rodan" introduced in 1993 appears more pterosaur-like, albeit with bat-like finger bones instead. Ray Harryhausen also committed this error with a pterosaur in *One Million Years B.C.*, but did anyone really mind?)

The original full-sized Rodan suit has numerous spiky protrusions visible semi-symmetrically in rows over its otherwise scaly looking, ridged abdomen (in an odd way, anticipating the light body "fur" noted on several pterosaur genera decades later), and a tail resembling either a uropatagium (a wing membrane extending between the legs and tail), or (more likely) a rounded tail vane, relatively short in proportion to its body size. As is the case with certain genuine pterosaurs, yet unlike *Pteranodon*, Rodan has short teeth within a beak often used to 'hen-peck' giant monster foes. While no egg known to science could have the huge dimensions (e.g. estimated to be 100,000 cubic feet) of Rodan's "reptilian" egg discovered in the cave under the mine by protagonist Shigeru (Kenji Sahara), yes, pterosaur (small) eggs and even well-developed embryos within eggs have been recently uncovered and studied by paleontologists. So Toho got that part conceptually correct too. For his amazing 2015 cover illustration published on the cover of *Mad Scientist* no.30, artist Mark Maddox's rendering of Rodan in flight earned him a Rondo.

When flying, however, and viewed from above or below, Rodan's wing contour appears even more 'pterosaur-like' than in scenes where the paleo-monster is shown standing while flapping its wings to cause destruction. (In fact, in the Showa era films, Rodan seems to flap its wings more vigorously while standing upright than when soaring at supersonic speeds through the air.) Dramatic flying scenes were created using suspended props, as opposed to suitmation involving a human actor inside. J. D. Lees noted that while design of smaller Rodan models and props used for flying scenes was more closely aligned to the shape of genuine pterosaurs, of course, the 100-pound suit worn and operated with such difficulty by Nakajima absolutely had to conform to the human body. "There was a suit for actor Haruo Nakajima to perform in, and in the same scale, a flying version fitted with a forward-pointing head and wider spread wings. The second Rodan was designed to be held aloft on wires for scenes such as when Rodan flies over Fukuoka ... A half-size flying model was created, and also a half-size upper body that was operated like a puppet. Three flying models were constructed for panoramic shots or when the monster is flying at a distance." (Lees, *op. cit.*, pp.53-54)

That juvenile Rodan in the cave has diminished wings that do not yet appear equipped for adult flying, although in the American 1957 version of the film, it is suggested that the Rodans were fully grown at birth, thus precocial, and could fly immediately after hatching.

Overall, using this assortment of models and suits an impressive, disbelief-suspending illusion of real pterosaurs menacing modernity was crafted for Toho's first monster movie installment involving a flying creature. Then, after an 8-year long hiatus, American monster movie fans applauded Rodan's triumphant return in 1965's *Ghidrah the Three-headed Monster*, although the 'pterrific pterosaur's' first filmic appearance was its finest. But let's instead move on to Toho's second, more incongruously flying beast – Varan, perhaps Japan's first "triphibian" monster. (Like "Gappa" it can walk on land, live under water and even fly.)

Varan, the Unbelievable (1962) is rarely televised today. And the original 1958 Japanese version, considered by critics to be far superior to the American version, is unavailable on DVD. So if you've only seen the former film then you probably are unaware that the monster Varan flew in some scenes edited out from the original for American audiences. Interestingly, there could be dual zoological influences underlying conceptual design of the monster suit, one dinosaurian and the other mammalian. Intrigued?

My first published contribution to *G-Fan* was a Letter appearing in issue no. 61, where I suggested that Varan's suit may have had a dinosaurian derivation. I'll insert a segment from that piece: "I am suggesting here, without proof, that Varan could have also been inspired by artist's images of one of the first Mesozoic reptiles to have been categorized as 'dinosaurian,' … *Hyaleosaurus*, as depicted by mid-19th century paleoartists does have an appearance much like that of Varan!" (p.28)

A decade and a half later, an intriguing Facebook entry popped up on my screen one morning. Here, an individual had noted how a rare 1886 image from a French popular paleontology book showing a dinosaur standing up to brace itself against a modern 6-story building (see front cover) resembled Varan because it had many needle-thin spikes spread over its back and tail. Without going into too many details here, I replied to this FB remark in noting that this was instead a fanciful image of a *Stegosaurus* as once known, "borrowed" from a prior illustration appearing in *Scientific American*. (In fact, my FB posting is the 'nugget' which triggered the idea for this article.) The 1886 restoration more closely resembles a bland Toho "anguillasaurus" than Varan, let alone *Stegosaurus*.

But *Hylaeosaurus* as originally restored appeared much closer in anatomical design to a mid-20th century Varan than did very early perceptions of *Stegosaurus*. Like the real ornithischian analog dinosaur *Hylaeosaurus*, Varan usually moved on camera (crawling) as a quadruped, but (analogous to another dinosaur genus – *Iguanodon*) may be regarded as a "facultative biped," as in early scenes where it emerges from the lake, walking upright. A bit of further clarification concerning suit construction is needed before moving on toward Varan's flying capabilities, excised from the American version of the film.

Peter H. Brothers, an admirer of Varan's suit design, commented in *G-Fan* no. 59 that "… Baran (sic) isn't so much a prehistoric reptile as an ancient and malevolent god come to life… The costume was …in its own modest way, revolutionary, with its impressive carapace and bulging muscular structure … foreshadowing the Teenage Mutant Ninja Turtles and even Gamera." (p.23) Martin Arlt in *Mad Scientist* no.22 adds (based on recollections of Toho technician Keizo Murase): "It is unclear who actually designed Varan. … Varan was decorated with a series of ornamental spines that ran down his midline, from head to tail. The spines were meant to be transparent, which would give them a kind of luminescence on film." These were made from clear vinyl spray hoses, "cut diagonally to create a sharpened end." (Arlt, pp.4-5) The spines, which are elongated and thin, and not shaped as broader triangles (like the Rhedosaurus'), were embedded into a tough dorsal carapace whose bumpy texture was achieved by affixing peanut shells into the clay prototype. (Arlt, pp.4-5) In fact, throughout the entire history of scientific paleo-restoration ("paleoart") this spiky Victorian (iguana-esque) *Hylaeosaurus* is the *only* dinosaur bearing such arresting morphology. Likewise, Varan also is a unique, one-of-a-kind, fictional paleo-monster.

Arlt also mentions that 3 versions of Varan were constructed for the film. Haruo Nakajima performed in a full body costume; a smaller hand puppet was used in facial close-up scenes, and a half-size scale prop was made for those flying scenes rarely seen by western audiences. The membranous 'wings' were not attached to the full body costume – thus making Toho's paleo-monster seem, to me, even a bit more "hylaeosaurian." So perhaps, analogous to protocol in 1956's *Rodan*, scientists in the 1958 picture should have renamed Toho's creature, *Hylaeosaurus*, species – *varan*.

Benjamin Waterhouse Hawkins ranks as the mid-19th century's greatest paleoartist, best known for sculpting his large statues of dinosaurs as perceived by paleontologist Sir Richard Owen. These went on public exhibition on the Crystal Palace grounds at Sydenham in 1854, where they still stand along with several other prehistoric animal figures: one of the restored dinosaurs was a spiky looking fellow named … *Hylaeosaurus*, at the time only known and described on the basis of fossil scrap. In examining the image of *Hylaeosaurus* as formerly restored by Hawkins, *particularly* at its dorsal series of 'crystalline' spikes arranged from head to tail, doesn't this rather fanciful image of a real dinosaur rather vaguely resemble Varan (*sans* wings)?

Varan's warty, pebbly-looking armored "carapace" (i.e. sculpted with aid of peanut shells), extending dorsally over the upper ribcage area, actually provides a more 'ankylosaurian' feel to the monster. Interestingly, in modern, late 20th century, circa 1980 dinosaur renaissance era restorations, *Hylaeosaurus* appears less spiky, but much more armored 'ankylosaurian' than when compared to Hawkins's original conception. Varan's body proportions differ from *Hylaeosaurus'*, but then, in spite of the scientific derivation, so does Rodan's relative to its precursor, *Pteranodon*.

It's certainly a stretch (again without *any* proof), but possible that Toho monster suit designers relied on Hawkins's concept of *Hylaeosaurus* (which itself also appears rather like a modern Galapagos Iguana) when they conceived Varan. Had one of Toho's technicians visited Sydenham before or after the war? A subliminal association perhaps? Was there a photo of Hawkins's models in an available book or magazine somewhere? We may never know. So is the jury still out? Is my origin theory for Varan's monster design …ahem, "un-believable"? Considering that Varan was originally conceived as a *flying* creature too, my idea may not quite … get off the ground … after all, because *Hylaeosaurus* was 'merely' terrestrial.

At 50 meters long, it would seem inconceivable that Varan could fly, let alone on thin yet voluminous, wing membranes that would have made Rocky the Flying Squirrel (debuting in 1959) pride of his pack. (Boris and Natasha would have never caught him!) Varan's wings, regarded as "almost pointless" by Mike Bogue in an online "American *Kaiju*" movie review, extending from below the monster's wrists toward the tops of its ankles, were constructed from a plastic sheet. Given a possible hylaeosaurian paleo-monster 'bauplan,' then what might have been a conceivable zoological derivation of those wing add-ons? This time think 'mammals,' not dinosaurs or pterodactyls. As incongruous as it may seem, Varan's side wings most resemble and are highly derivative of those parachute-like organs spread by gliding mammals, such as flying "lemurs" (which are not true lemurs), and of course flying squirrels of which there are 44 species. Hardly "monstrous" in size or bearing, such mammals are generally small – usually a foot long or less – weighing only a handful of pounds. Their wings (known as patagium) stretch from wrist to ankle; some have patagia that envelop the hind legs along the tail. (Bats are different – they have finger bone-reinforced wings.) With such structures, gliding mammals can sail over 200 feet from tree to tree.

Whereas Varan's "pointless" flying was readily removed without remorse from the 1962 Amercanized film, Rodan's flying abilities were the absolute point. By then, on *Rodan's* taloned heels, American audiences had just been treated to another flying menacing manifestation in 1957's *The Giant Claw,* having alien avian affinities (vaguely resembling the 'archaeopteryx-ish' Rodan prototype). Several years later, analogous to Varan, Reptilicus' atrocious flying scenes were also excised from an American version of the titular 1962 film, due to their "awful execution," as opined by Jeff Rovin. (Yet they remained in the tie-in *Reptilicus* novelization.) Shortly thereafter Toho scored heavily with one of monsterdom's most popular alien foes—the winged Ghidrah, or aka King Ghidorah.

The first paleo-flying monster menacing (to minor degree) modern civilization may have been Professor Challenger's heron-sized pterodactyl transported back from that "curupuri"-infested South American plateau in Arthur Conan Doyle's novel, *The Lost World* (1912). Clearly though, since the mid-1950s, giant flying *daikaiju* and other filmic monstrosities, founded upon real creatures known to science, often flock to center stage within the science fictional/horror arena.

Note 1: Not to be hypercritical, but the scientist team in *Rodan* errs in several respects. They suggest incorrectly that the pterosauria lineage harkens back possibly even to the Carboniferous period. And they date a Rodan egg shell fragment at 20 million years using carbon-14 age dating analysis, which would have been useless given such a remote age. Of course, that cited 20 million years date corresponds to the Miocene epoch, well over 45 million years *after* the Late Cretaceous, when *Pteranodon* thrived. This would make the genus a 'crypto-survivor' during the age of Man, unknown in an intervening ~70 million-year fossil record.

Key References:

Martin Arlt, "Varan the Unbelievable! The Last of Toho's 1950s Monster Films," *Mad Scientist* no.22, Fall 2010, pp.3-9; Martin Arlt, "Rodan the Flying Monster," *Mad Scientist* no.30, Winter 2015, pp. 18-29; Peter H. Brothers, "The Monster Varan: Toho's Black and White Sheep," *G-Fan* no.59, Nov/Dec 2002, pp.14-25; J. D. Lees, "Rodan: Evolution on the Wing – (Second in a series)," *G-Fan* no.79, Spring 2007, pp.52-57.

Chapter Sixteen—*Sleuthing the Giant Mysterious Monster*

Many are the variety of giant Asian monsters, starring in films of such varied quality. Just when we think we've seen or at least heard of them all, especially those from that hallowed early 1960s era, when wee 'monster kids' were in the making, another emerges from the radioactive gloom. Ever hear of the "Atomic Dragon"? Isn't that Godzilla, you might reply. Well, true, but here I refer to *"AGON, the Atomic Dragon."* What's this one about … well uh, it's about a giant atomic monster. Here's one that went mostly under the radar, unmentioned either in Donald F. Glut's popular *The Dinosaur Scrapbook* (1980), or his *Classic Movie Monsters* (1978). Jeff Rovin didn't list Agon in his *Encyclopedia of Monsters* (1989). Fortunately, sources available on the Internet today can help demystify Agon, although not completely. For instance, *daikaiju* expert, Mr. August R-AGON-E provided a very short synopsis at the Classic Horror Film Board, and of course there's the International Movie Data Base (or IMDB.com). There are a couple short clips posted at YouTube. But for starters, let's 'screen' the film, which is rather 'Agon-izing' to do given that (as of 2011) available copies (DVD/VHS) lack dubbing or English subtitling. Well, maybe lack of detail herein will prompt 'Agon-istic' behavior among those of you thirsting for more.

Agon is most certainly an obscure monster, yet much better than the Korean "Yongary." In body outlines, it sort of resembles a cross between Godzilla and Gorgo, with subdued, low-ridged triplicate row of dorsal osteoderms (e.g. "Plates"). We see this early in the film as the approximately 150-foot tall monster is shown destroying a building and blowing fiery breath at the puny humans below. Here I use the word, "film," when actually AGON was intended as a four - part television serial. It was produced in 1964, possibly by Nippon Denwa Eiga. The original Japanese title was *Maboroshi - No daikaiju Agon* (or, *Agon the Mysterious Giant Monster*). The story was written by Shinichi Sekizawa and Kozo Uchida. Mr. Ragone states interestingly that a reason cited for why it wasn't aired until 1968 is because Toho claimed copyright infringement on the story that seemed similar to *Gojira* (1954). However, Ragone goes on to say that this isn't true. The real reason is that Agon's sponsor "dropped out." Agon was directed by Norio Mine and Fuminori Ohashi—who designed the monster costume. The actor who wore the Agon monster suit was Etsuji Higashi. All four parts of the mini-series are available on a DVD offered by "Atlas Films." I watched my copy with avidity, even though I couldn't follow what was said by the actors (as I do not speak Japanese). But we can still follow along as the film follows a traditional giant monster 'formula.'

First we see a nuclear explosion. Then, following an incident along the coast, after a landslide causes a car to crash, reporters, police and a nuclear scientist are alerted to the scene. They find a mysterious cave and detect radioactivity nearby. There's also hint of a strange growling sound prevalent. Turns out there's nothing of particular interest inside the cave, but then after the Geiger counter suddenly pegs, something boils up out of the waves offshore. Horrors—it is a giant dino- monster, Agon! We can even see its neck pulsating under the jaw as it breathes (a nice effect unseen in the early Godzilla films).

The main reporter ("reporter # 1") and the handsome heroic looking guy who is probably a detective drive together in a sports car to the National Atomic Energy Center where they talk to the lead scientist and meet an attractive female scientist. At one point—that pivotal time in all monster movies where the scientist (Prof. Ukyo) has some explaining to do—we hear the reporter say "Atomic Dragon" in English. Next the "Danger" button flashes and the scientists run to the seashore where they detect considerable radiation with their geiger counter, announcing Agon's reappearance. This time Agon comes ashore and, naturally, people flee.

Next comes a tension-filled scene where the pretty lady scientist's ankle is pinned under a fallen tree while Agon relentlessly approaches. She falls into a crevasse as the ground splits under Agon's powerful stomping feet. Meanwhile, the air raid siren blares, as Part 1 of the serial comes to a dramatic edge-of-your-seat moment! In Part 2, the girl is rescued by reporter #1, as Agon advances upon the National Atomic Energy Center installation, toppling buildings and other structures. Agon exuberantly stomps on (toy) cars, destroys what appears what appears to be a nuclear reactor and then gets smothered in radioactive smoke. Later, reporter #1 flirts with the pretty lady scientist at the hospital.

Now comes that traditional scene where the scientist provides further elaboration on what's going on, and it's really too bad that I can't understand what they're saying. It seems as if the lead scientist is theorizing that the dino-monster Agon was sleeping blissfully on the sea floor, only to be awakened by detonation of a nuclear bomb. I'm sure there's much more to it than simply that, however. (Well, maybe not too much.) Next, predictably, the military general character throws in his two cents, and soon we see air force jets soaring overhead. A military offensive is evidently in the making.

Agon rises up out of the waters while tanks roll into position along the coast. Jets attack, firing missiles; Agon breathes fire at a passing jet, which crashes. Cannon fire and tanks do not stop mighty Agon. Next, in homage to Ray Bradbury and America's filmic 'Rhedosaurus,' Agon symbolically destroys a lighthouse! Curiously, the detective rolls an atomic facility vehicle possibly containing some radioactive material inside ... downslope into the ocean - effectively 'feeding' the monster. The giant monster fishes the van out of the surf. Then Agon disappears under the waves. Part 2 has now ended.

Part 3 takes up the tale of a couple of crooks who eventually get their dues at the end of Part 4. It's unclear exactly what they're trying to finagle, but at one time they steal a tube marked "U" (possibly for 'Uranium'?). I won't recount their goings-on here in detail. Then the dino-monster resurfaces and bites a row boat with a young boy (who has appeared before in the first two parts) trapped inside. In its mouth, Agon carries the boat with the boy inside along the shore. It isn't clear if he's holding the lad hostage, or if Agon intended to rescue him from the bad guys, however inadvertently. In any case, with the boy in harm's way, the military cannot launch a full-scale attack even though Agon remains on the rampage, smashing a bridge and inflaming other structures. Incongruously, the boat remains in Agon's deep set mouth. A rope ladder is lowered from a hovering helicopter so the boy can escape from Agon's jaws of certain doom, but agitated Agon flames the helicopter. The air raid siren blares.

Now Prof. Ukyo explains important stuff to reporters, one of whom—Eureka!—suddenly has an idea, helping to save the day. Meanwhile a second helicopter arrives, this time carrying a box dangled underneath on a rope, causing Agon to drop the boat rather carefully to the ground so that the boy isn't unduly harmed. Agon is lured by the helicopter's tempting package. The two bad guys (who as we see may be drug smugglers) reappear, who find themselves being dragged along on Agon's immense tail. They steal the helicopter after it lands, attempting to taunt the monster with the package. But Agon uses his fiery breath to destroy them inside the helicopter. (My mere guess is that the box contains radioactive material.)

We notice that Agon's suit is visibly on fire, and so the dino-monster heads out to 'sea,' presumably to douse the flames so the actor inside isn't hurt. Next the narrator cuts in probably with stern moralistic tones. Alarming music blares as Agon's head disappears below the waves. Agon is destined never to return.

Reactions to this film posted at various places on the internet overall aren't too condemning. Some people say the movie is 'fun' and entertaining, and I concur. The online store I bought my DVD, from "Gotta See DVDs", claims that Agon has "...great experimental electronic music and sounds ebb and flow with the natural ambience of organic location noise over the continuing visual blanket of surreal and interesting sepia tone colored images ... one of the those great, electronically pulsed, audio/visual cinematic fever domain dreams that are fun to experience on dark evenings in dark rooms." Well, yeah - but it's more fun watching Agon smashing those cardboard buildings.

Just a few more 'bullets' before I go, once more courtesy of Mr. Ragone's online blogs. *Agon, the Atomic Dragon* is considered 'parallel' to another show, *Ultra Q*, aired from 1964-1965, which is regarded as Japan's answer to Rod Serling's *The Twilight Zone*. Agon is an early example of a *daikaiju* devised for television, although not the first of its kind. That honor may instead go to a curiously titled children's show, *Marine Kong*, which survived 26 episodes broadcast in 1960. Marine Kong was about a giant robotic dragon controlled by villains. And it seems the Agon monster suit may have been re-customized for another show even before the 'Atomic Dragon' four-part serial aired. For Agon's costume was embellished further by Mr. Ohashi who, for example, added saber-tooth fangs. The resultant, a shape-shifting extraterrestrial monster from Planet Pal named "ARON," then starred in episodes 13 to16 of *The Space Giants* (1966 to 1967).

Nope—Agon isn't as good (or cool) as Godzilla. Ratings wise, if *Gojira* (1954) is a "10" and *Yongary* (1967) is a "2", then "Agon" is a definite "4" even if I can't understand what the actors said. I just wish someone would get around to adding English subtitles eventually.

Then I would watch this film 'agon' & 'agon.'

A Hamilton & Co., Ltd., London, 1951, no copyright stated (A. Debus collection).

Chapter Seventeen—*Spoiler Alert! Dinosaurs in Fantastic Fiction*—"Extras"

With any article there must be a 'catalyst.' In this case, after many decades, the catalyst was something I became personally reconnected to two years ago that really caused memories to flood. Following the decease of my younger brother, Rick, a geologist - OK - stop! I'm, not trying to bring you all down here ... well, but what this catalyst turned out to be was an old LP children's record that my niece returned to me as we were cleaning out my brother's place. This was Golden Records' "A Day In the Life of a Dinosaur" (1964). It still had the cardboard album cover more or less intact. Most intriguingly (and appropriately for our purposes here), "A Day in the Life of A Dinosaur" is a *time travel* story involving dinosaurs. I hadn't seen this album in well over 40 years.

The two main characters were "Bronty" whose voice was actor Phil Foster's, and the space traveling "Terra-naut" played by Peggy Powers. It was a simple idea, 'fostered' by NASA's then escalating space program successes. Through internet searches, I've since discovered that CD and MP3 versions of this recording are now available, so I won't elaborate. But to my 10-year old dinosaur-infatuated mind—then so impressionable as well in the 'famous monster' arena and sci-fi venue, "A Day in the Life of a Dinosaur's" songs diabolically created an indelible mind worm. Rick and I listened to this record so many times that the lyrics became grooved (not unlike Time's 'spiral') into my brain. Surprisingly, even after so many years, our copy of the record wasn't too badly scratched at all. Did I really want to experience all this again, after so many decades of mind worm 'healing'? I tucked the record away, reconsidering every now and then, until mid-November 2009 when my wife finally fixed our old turn table. And so I listened, reminiscing. To what extent did this record fortify my budding interests in the ever-popular time travel element of science fictional dinosaurs? I can only imagine, but probably a lot.

In "A Day in the Life of a Dinosaur," a Terra-naut launches in a big Rocket. Guided through communications with Time Machine Control Center, the Rocket is whisked backward at a rate of over 100,000 years per minute to the Jurassic "era." Via "special radio" equipment, Terra-naut remains in direct contact with Control Center, allowing us to hear all the adventures plus the eight songs she sings with the amiable "Bronty." Along the way, we learn that cousin *Stegosaurus* is a scholarly sort, who has written the "Dinosaur's Who's Who," (perhaps THE original dinosaur encyclopedia). And we also hear early 1960s reflections about dinosaur extinctions (e.g. glandular diseases and mammals eating dinosaur eggs). It's a neat little sing-along type of story, even for adults to listen to!

Indelibly, skilled writers portray enduring prehistoric scenes with words, one thousand of which—according to adage—add up to a single paleoartist's picture. That's more or less a premise of my (then latest) book *Dinosaurs in Fantastic Fiction* (2006). Due to contractural space limitations, therein I omitted some references to genre novels, especially if they seemed somewhat derivative in nature, thematically repetitive. So, I'll use this welcome opportunity to introduce several sci-fi novels and lost world-ish stories which either slipped through the cracks of my book, for which discussion was necessarily trimmed, which missed deadline for inclusion, or which I discovered later on. Here we'll stride through science fiction's hallowed extra lost world literary devices or time travel inventions, permitting humans to stand alongside 'living' examples of prehistoria—dinosaurs and other prehistoric creatures.

Nearly two centuries ago, there was no prehistoric fiction to speak of, because the very idea of a "prehistoric" past was newly fast emerging and in flux. Earliest forms of anything bearing semblance to a deliberately contrived "fictional" prehistory came either in the form of contemporary cartoons incorporating life restorations of prehistoric animals, or via speculative imaginary passages (published in books), often composed by amateur paleontologists or men of medical or scientific training who were infatuated with fossils they had collected. Sometimes visual artistry melded with elaborately written compositions, embryonic versions of the sort of paleoimagery we're most familiar and comfortable with today. Usually, however, such excursions from mainstream Geology were merely vignettes, scene-setters, or "pen-pictures" (as Charles H. Sternberg referred to them) rather than completely fleshed-out fiction stories. Eventually, skilled, visionary writers (e.g. Jules Verne) couldn't help themselves from writing fiction novels based on developments in the then newest, startling science – Paleontology. So first let's explore a bit more about dinosaurs of space-time and parallel worlds. A favored form of sci-fi (or 'sf') literature melds dinosaurs and other prehistoric animals with time travel.

In the Epilogue to *The Time Machine*, H. G. Wells's narrator suggests that his time traveler protagonist may, on another exploration of time, have been...

> "...swept back into the past, and fell among the blood-drinking, hairy savages of the Age of Unpolished Stone; into the abysses of the Cretaceous Sea; or among the grotesque saurians, the huge reptilian brutes of the Jurassic times. He may even now—if I may use the phrase—be wandering on some plesiosaurus-haunted Oolitic coral reef or beside the lonely saline lakes of the Triassic Age."

However, despite his expert training in paleobiology under 'Darwin's Bulldog' Thomas Henry Huxley, Wells never wrote the intriguing story suggested in his Epilogue. That task was left to other writers, although surprisingly left unaccomplished for decades.

One of the earliest to tackle the dinosaur-time travel theme was "Richard Marsten" often a pseudonym for mystery novelist Evan Hunter who, however, claimed *not* to have written the story which I'll introduce here - *Danger Dinosaurs!* (1953), a young adult's novel.

Marsten's minor masterpiece was written on the heels of Ray Bradbury's classic short story—"A Sound of Thunder" (1952). Borrowing key elements from Bradbury, *Danger Dinosaurs!* is distinctive. Alex Schomburg's front cover design—showing some sort of gangly, humpy-backed theropod (spinosaurid?) menacing a pair of hunters—gets the heart pounding. Delving in, we soon find that the story is about a hunting expedition to the Jurassic Period, one hundred million years ago. Incidentally, *hunting* of cryptozoological or otherwise prehistoric wild game has been a traditional sf theme since Arthur Conan Doyle introduced sportsman Lord John Roxton in *The Lost World* (1912).

Protagonist Chuck Spencer is a hired guide for Dirk Masterson's unsavory hunting party which enters the Jurassic via a 'time-slip.' However, Chuck's (and his older brother Owen's—named for 'Sir Richard Owen, perhaps'?) instructions are that dinosaurs and other prehistoric wild game are *only* to be photographed, not killed with guns. In explaining time-slip theory, Owen uses a simple (yet now antiquated) analogy—the grooves on a phonograph record. "Assume that the outermost groove of the record is the past. And the groove nearest the center is the present. When you play the record, the needle travels from past to present..." But if powerful machines—like those wielded by Chuck's and Owen's employer—nudge the needle to 'skip' outward, then the needle has slipped into the past. To avoid interfering with the future, time-slipping into any period inhabited by mankind is legally prohibited. To limit impact of the time travelers' anachronistic presence and therefore minimizing the chances of altering the future by effecting the past, time-slippers are protected from their surroundings with a transparent stasis field dome holding wild game at bay. In seven days, the time-slip device will transfer those reentering the rendezvous relay area back to their present time.

Rather than merely photographing dinosaurs, Masterson has an ulterior motive. After 'accidentally' ramming a jeep through the force field, Masterson begins shooting stegosaurs and pterosaurs with guns he's stowed away for the trip. Consequently, Owen who is crushed in a 'brontosaur' stampede caused by Masterson's recklessness is quickly erased from memory, because if he died (100 million years) before he was ever born, then he *never existed*—an interesting plot 'thought.' Intriguingly, 'wounded' space-time heals itself by eliminating Owen's 'former' existence. Soon, even Chuck's personal memories of his brother have faded, as an alternate, bent 'history' is charted for survivors.

After encountering a pair of geologists who were lost in a prior mission to the wild Jurassic terrain after discovering uranium deposits, unscrupulous Masterson reveals his true colors—what he's really after. At all costs, he covets those valuable deposits for himself! But Masterson righteously gets what he deserves. He's eaten by a lunging Allosaur... although later after the survivors return to an altered present his (metaphoric) 'record' has disappeared from reality.

Although often the case—as in Bradbury's "A Sound of Thunder," not all time dino-treks end up in the Cretaceous. For in one curious neglected entry, Roy Sheldon's (pesud. for British novelist E. C. Tubb, b. 1919), *Moment Out of Time* (1952), time travelers are projected into the swampy Jurassic instead. Despite leading cover art implications, in Sheldon's novella there are but few dinosaurs actually encountered. Nor are there intriguing time paradox matters to resolve. No, Sheldon's is more of a psychological study, as an unlikely combo of mobsters and scientists find themselves exiled from the year 2050 on a one-way projection into the past. This happens, or so the engineer explains, because "…the Giron-flow phenomenon is still very imperfectly understood." Well, uh okay—I always had troubles with that gadget too.

While the time travelers do have two bloody encounters with different kinds of dinosaurs, a twenty-foot long, speedy 'kangaroo' variety, and a twelve-foot long carnivorous type with alligator-like heads, these are readily subdued with their energy blasters. In fact, the travelers seem mainly menaced by the unscrupulous mobsters within their party, and more grossed out by the swarm of enormous cockroaches and dragonflies which greedily consume corpses the humans leave in their wake through the jungle, than by the few encountered dinosaurs. Interestingly, Sheldon's is one of the first travel-through-time stories involving dinosaurs (although not nearly as memorable as Ray Bradbury's contemporary classic, "A Sound of Thunder").

Einstein himself probably would've approved of Lisa M. Graziano's and Michael S.A. Graziano's 2006 novel, *Hell Creek*. Perhaps borrowing thematic elements from George Gaylord Simpson's posthumously published novella, *The Dechronization of Sam Magruder* (and possibly even ideas spun in Douglas Lebeck's *Memories of a Dinosaur Hunter*), the Grazianos' minor classic relies on a chronal accident which projects three (and a half - *eew*!!) people and a faithful dog backward 65 million years in time. Particle physicists working on a matter transferral project are stunned when their experimental device actually moves objects through space-time, meaning the farther the object moves in space (and the bigger the object is), the further backward in time it goes. Then, after a paleontologist is summoned to identify strange beetles which have been materializing in, then later vanishing from bottles subjected to experiment, all 'hell' really breaks loose. Now the abject castaways begin a death-defying quest to find a reversion point one thousand miles away where they might be re-projected into their normal present.

Along the way they're confronted by sickle-clawed raptors, *Triceratops* herds, and an elderly, half-tamed, battle-scarred *T. rex* named 'Corla.' And they also find a 'caveman' living there, indicative of space-time travel's probabilistic nature. It seems the chronal device has cast copies of themselves from the present back to up to four *other* Late Cretaceous time slices too, each perhaps separated by hundreds or a few thousands of years. Rather paradoxically, the Rex-taming caveman is the last 'descendant' of two of the time travelers 'copies' who were projected a century or so earlier into the Cretaceous than are the novel's protagonists.

As billed by *The Best Reviews*, John Varley's 2005 novel *Mammoth*, is "Terrific ...H. G. Wells meets *Jurassic Park*." While I wouldn't go quite that far, its premise is immensely intriguing--once more involving what turns out to be a most peculiar time machine, this 'time' discovered alongside a 12,000-year-old frozen mammoth specimen interred with a frozen woman and a man who is *wearing a wristwatch*! For us paleophiles, this is definitely 'can't put it down,' 'you had me at hello,' material. Nobody ever really figures out how to operate or control the time machine, which is an unlikely looking device consisting of a strange box containing scores of marble-sized ... well, 'marbles.' But when it works unexpectedly, fortuitously, the device casts two human characters backward into the North American Pleistocene. Then they suddenly return to Wilshire Boulevard along with a herd of mammoths, most of which are gunned down by policemen; two survivors are placed in a greedy billionaire's extravaganza - "Cenozoic Park," where the great extinct mammals of yesteryear are either animatronically restored, or (in the case of the two mammoths) living. Ultimately, it seems as if the box of 'marbles' warps time into a locus or loop, linking the mammoths' normal space-time with modernity, looping the lives of the two protagonists who are fated to exist in both. Good story!

Alan Dean Foster's short story, "Pleistosport" (*Prehistoric Times* no. 64). Foster relates the tale of sportsman Thackeray who travels back into the recent Ice Age to bag a woolly rhinoceros. In a twist from how such stories usually proceed at plot level, the transporting 'Chronovert' device can only project people back up to five million years, but not longer, thereby obviating the possibility of traveling to the Mesozoic era. Thackeray, overly confident and armed with the best rifle, equipped with a heat sensor, that money can buy is outsmarted by a stealthy *Smilodon*. "For the first time he began to think of the saber-tooth as a possible danger instead of an unmounted display." Ironically, the big cat, even greater hunter than Thackeray himself, proves to be his awful undoing.

By the early 1990s science fiction Grandmaster L. Sprague de Camp found himself extending the adventures of his famous time-traveling raconteur Reginald Rivers. Rivers first entered the annals of sci-fi literature with Sprague de Camp's famous 1956 short story, "A Gun For Dinosaur." By 1993, the author had crafted sufficient tales concerning time travel into prehistory, ranging from the Triassic to the Pleistocene Periods, constituting an engaging short story collection—*Rivers Of Time*, with a smart looking cover painted by paleoartist Bob Walters. Included was a yarn titled "The Synthetic Barbarian," first appearing the September 1992 issue of *Asimov's Science Fiction Magazine*.

"Synthetic Barbarian" was about a slightly deranged man named Standish who with partner Hofmann, yearns to hunt dinosaurs in the past, just like the cavemen did. It takes considerable effort for Rivers to explain that cavemen never could have hunted dinosaurs. So eventually they travel with their various weapons in Rivers' safari company's Time Chamber back into the North American Oligocene, in hopes of taking a brontothere trophy (now regarded as Late Eocene). Upon their arrival, Standish dresses as he thinks a 'caveman' should—hence the word in the story title "synthetic." Anachronistically and unlike any genuine caveman, Standish also carries with a high-powered crossbow. Soon he is covered in bug bites and has a serious sunburn, since, crazily acting in caveman persona, he's doffed most of his clothes. After harrowing incidents with a charging hippo-like rhinoceros *Metamynodon*, and later an entelodont "terminator pig," Rivers realizes that Hofmann and Standish may be settling an old feud here in the past regarding a woman. He needs to get them outta there before someone is killed or seriously injured by mistake, or deliberately. So by the time they spy a magnificent foraging brontothere, because they've each gained sizeable trophies already, the ostensible purpose of their hunting trip becomes anticlimactic.

One of the most intriguing published dino-time travel tales of recent years is James F. David's *Footprints of Thunder* (1995). Here, a cultist mix of fringe scientists, their groupies and smart college kids band together in predicting when the end of the world as-we-know-it will occur, based upon their interpretations of old manuscripts (i.e. "Apocrypha" of the Babylonian Zorastrus) and bizarre stranger-than-fiction type natural phenomena that would have appealed to *X-Files'* FBI characters Scully and Mulder, or Agent Dunham in *Fringe*. Such a smattering of evidence suggests an acceleration of unexplained activity, which according to the fringe scientists' calculations, will lead to world upheaval—although at the outset they don't realize the nature of the impending cataclysm, or what it will entail.

Basically, what happens next is that space-time becomes "quilted" over itself, although not quite so seamlessly. Vast geographic provinces of the Earth are 'replaced' by areas originating in the Cretaceous Period, that is, according to the (former) historical continuum. Quilting causes a breakdown in the fabric of space-time, threatening causality. Entire U.S. cities vanish overnight, apparently into thin air—replaced with steaming jungles and a most unusual fauna. A major tsunami in the Carribbean washes out the American gulf coast, forcing a horrified family to ride a mother sauropod toward shore, while it is savagely attacked, steadily ripped apart and eaten alive by marine reptilian predators. And whatever happened to those modern 'replaced,' populated territories of the globe that vanished into thin air?

While the President's team of experts strive to solve the problem, plotting to reverse the effect, most of this novel is about the characters' adventures in these newly grafted, uncharted prehistoric 'territories' that have arrived in a new, altered '20th century.' What caused Time to unravel?

The catalyst seems to be U.S. and Soviet hydrogen bomb detonations of the 1950s, which inadvertently created micro-black holes at the centers of nuclear fusion. Unbeknownst at the time, these explosions triggered time wave disruptions with resulting space-time displacement. Based on new calculations performed by the President's physicist, the only way to reverse the problem (i.e. neutralize the effects of space-time manifestations caused by earlier explosions) would be to detonate a huge nuclear bomb over Oregon's Cretaceous 'park.' But not everyone supports this risky and implausible solution.

Problems with the space-time continuum only escalate in David's 2006 sequel, *Thunder of Time*. David fuses fears over dangers posed by atomic weaponry with prehistoric specter, thus linking the ever popular 'dinosaurian time travel' and traditional 'war-with-dinosaur' themes. This novel explains the reasons and consequences for the disastrous time-quilting warp (a concept involving "orgonic energy," which I won't delve into here to avoid completely spoiling it for you). Visits to the Zorastrus Temple in the steaming Yucatan jungle (which is a time conduit wormhole linking future and past ages across parallel universes), adventures at the Lunar Flamsteed Crater, and immense nuclear explosions at a secret government installation in Alaska combine to make this a thriller! And there are also several memorable dinosaur-related sequences. As the climax builds (rather complexly), with multi-verse *reality* at stake, vicious *Velociraptors* and a Tyrannosaur threaten the mission at every labyrinthine bend! Like I said, a real page-turner.

Besides that aforementioned 'A Day in the Life....' recording, time machine gizmos coupled to prehistoric life also have entered the children's literature arena. Another giant of science fiction, Lester del Rey, delighted younger readers a half century ago with his *Tunnel Through Time* (1966). This is a time machine tale (reminiscent of Irwin Allen's old *Time Tunnel* television series), where something goes wrong with a physics experiment intending to "offset gravity." The Past remains continually alive (in the past), yet attainable using a machine that "contracts time" within a circular ring wherein time moves incredibly fast although in reverse. But there's always a problem with "slippage," that is, when the actual target in past time is missed due to systematic measurement error. And so two high school-aged boys and a paleontologist project themselves back into the Mesozoic, working their way 'upward' in time when the time machine's circular ring sequentially reappears, allowing passage, leaping through geological ages toward the Present. After observing a variety of Cretaceous reptiles and dinosaurs (while braving a close encounter with battling *T. rex* and *Triceratops*), they wend their way into the Cenozoic Ice Age; from there into the warmer Late Pleistocene. Returning to the Present, they conclude the world isn't yet ready for time travel.

My parents ordered this next short novel for me as part of a scholastic books school order. I couldn't simply resist the front cover come-ons, "Adventure in the Time Machine," or the descriptive promotion, "Two boys learn the secret formula that takes them back into a world where time began." Yes, this was my first-time travel novel, and it involved prehistoric life. Well, except there wasn't really any sort of time *machine*, per se. Rather, this next story relied on a curious disbelief-suspending gimmick for traveling through time, which to any fourth grader would have seemed marvelously plausible. "Robert Faraday's" (pseud. for Bruce Cassiday) *The Anytime Rings* (1963) is essentially an adventure tale of two boys and their, er, well.....'dog.' There was never a *Lassie* episode quite like this though.

In "Anytime Rings," two young adventurers, Jon Martinson and Dick Petterson, are enjoying themselves at the circus, when they spy a broadside advertising 'Colonel Sarno's Mystical Menagerie,' billing "Amazing Monsters Unseen Anywhere Else on Earth Today! Come visit the Zoo Inside. We dare you!" Greeted by the shadowy Col. Sarno, the boys soon see animals in a display pit lit by blazing torches. There they behold a 'Jurassic Park' in miniature—living specimens of *Triceratops, Ankylosaurus* and a scaly-skinned *Cynognathus*—which they immediately recognize from their dinosaur book readings. Despite the boys' accusations that the creatures must be fakes, because they died out millions of years ago, Sarno insists upon his integrity. When the armored dinosaurs fight below a pole supporting a torch is toppled, and a fire swiftly envelops the tent canvas.

Only the boys, Sarno and, as we soon discover - Cynognathus, survive the blaze. Then Sarno reveals his secret, allowing him to 'go then.' His "anytime rings," are engraved with Albert Einstein's famous equation, $E=mc^2$. Science so advanced that it's magical. Anyway, the boys grab a pair of their very own 'anytime rings,' and Sarno instructs them how they should be properly used, without really explaining *how* they work. Importantly, Sarno cautions, "Do not try to find the secret. Just use the rings for the fabulous powers they possess - and believe in them... Be certain also that the ones you take with you can be trusted in other places and in other times. You have a great responsibility." Furthermore, "... upon entering *then*, you will have instant comprehension and command of the language."

Next, they find poor dog-like *Cynognathus* wounded yet alive, which they intend to escort back to its proper geological period using their anytime rings. So, as Sarno instructed, they utter the catalytic phrase, "All the king's horses and all the king's men couldn't put the world back together again," setting off 'instantaneously' for ... one million years B.C., which as Jon instantly realizes, unfortunately, isn't the right temporal destination whatsoever, as the Triassic 'dog' lived 180 million years ago instead. But it's too late and soon the boys find themselves standing in a Pleistocene jungle.

Their harrowing adventures begin. *Cynognathus* flushes out a huge Mammoth from the foliage, while a stone arrow shot by a bow that barely misses injuring Jon or Dick. Next the trio are found by a golden-skinned cave boy, who has "...an almost Neanderthal look to him..." The cave boy, named Fee, takes them to his rustic village of the People of the Leopard situated at the base of an active, smoldering volcano. Fee's father is the Chief of this tribe, who is plagued by an underling, power-hungry High Priest Zumi, who mistrusts the boys. This tribe is in conflict with the neighboring People of the Zebra, who we learn Zumi has had dastardly dealings with.

Pitted against Zumi's sacred wolf-dog, loyal *Cynognathus* emerges a true champion who rescues the boys from peril. Ostensibly requiring proper sacrifice to appease the gods, vengeful Zumi nearly roasts the boys in a lava heated lake. This deed is thwarted by the Chief just as the volcano erupts. Their only way to escape the pouring lava is through the rival People of the Zebra's village, but their safe passing will only be allowed if *Cynognathus* (now named 'Nogg') can defeat a vicious saber-tooth tiger in a bloody battle-to-the-death. Following a great monster duel, not only does faithful Nogg come through once more, killing *Smilodon*, but the tribes make peace after all. When it comes time to leave, the two Chieftains decide to retain Nogg as their sacred animal. Arriving back in their present, sans Nogg, the boys realize their travels were authentic, not resulting from post-hypnotic suggestion. And Sarno is there too, ready to enlist the boys in his continued pursuit of historical artifacts. One wonders what their parents would say to this proposal!

Curiously several years before release of *Back to the Future* (1985), readers who were so lucky to find the promotional advertising, purchased copies of Lewis S. Brown's quaint, *Yes, Helen, there were DINOSAURS* (1982). The 143-page soft cover was written and profusely illustrated in black and white by Lewis S. Brown, who chose of all things a 1965 Volkswagen Beetle for a time machine, equipped with a "Time Motor."

This motor evidently runs on sheer "...refined and liquefied imagination"; it cruises steadily at 100 million years per hour. (A far cry from a DeLorean, indeed!) Brown stated on a website that his "...story is designed to be interesting to young people, but would be of interest to all ages. I chose a Volkswagen Beetle for a time machine because readers would know the size of one, and illustrations of animals and plants are to scale with the car, giving an excellent idea of the size of things." While Beetles were still common then, that isn't the case anymore.

Brown wrote his book following the onset of the 1970s Dinosaur Renaissance, yet prior to the vast unfolding of 'everything dinosaur' that swiftly materialized during the 1980s and beyond. This is an entertaining yarn, reflecting the state of contemporary dinosaurology. Uncle Homer's destination is the Late Jurassic Period (i.e. 138 mya, which would now be considered Early Cretaceous). Along the drive, he teaches his niece, the titular Helen - with her little dog "Pepper," about paleontology. As they pass backward through time they spy other more recent fossil animals and plants dating from successively older epochs of the Cenozoic Era. They pass into the Cretaceous Period "80 million years ago" without a hitch from sudden tumultuous chronous waves of catastrophe (i.e. - such as caused by asteroid collision or devastating volcanic eruptions - contentiously debated in 1982). Brown (via Uncle Homer) reflects and instructs principally upon the warm versus cold-blooded physiological dinosaur metabolism debate. Brown evidently was an early convert to then controversial views of paleontologist Robert T. Bakker.

Arriving in the Jurassic, unsettling things unfold. They're chased by a hungry ceratosaur, and Uncle Homer is bitten by a furry-looking pterodactyl. Eventually other dinosaurs battle before their eyes, and of course, mischievous little Pepper gets into trouble with indigenous locals. An Allosaurus shot with magic bullets falls unconscious atop their Beetle, miring it in the mud. They have their worst struggles on "Great Rock III," otherwise known as "New Speculations Island" where fast running dinosaurs are all endothermic. Uncle Homer samples the blood of a feeding sauropod, finding it to be a "true endotherm." In the end, however, were their adventures merely a dream? Well, maybe not as Helen finds a small blue and yellow *Archaeopteryx* feather in her pocket. Proof? You decide.

Dino-time travel tales can work on several levels. Sometimes a veritable "Wellsian" machine isn't required to go backward. Often one's own mind or imagination, or an enhanced dream-like state provides the needed passport into prehistory.

For the 21st century researcher, the older the book is, the rarer it probably is also, especially if the book wasn't a 'best-seller.' One such volume is John Mill's *The Fossil Spirit: A Boy's Dream of Geology* (1855). Like others written since, Mill's story does not rely on a Wellsian time machine device, but instead a spiritual, supernatural mode of 'viewing' prehistory - in this case, "transmigration of souls." No, I haven't yet managed to get my mitts on a copy of this rarity. All I know about it comes from Ralph O'Connor's *The Earth On Show* (2007). It would seem that mid-19th century stories based on a theme of time travel 'journey' into the past, revealing past geological ages, weren't quite as uncommon then as one might surmise. (For instance, Verne's *Journey to the Center of the Earth* qualifies as a quasi-sort of time travel tale.) The "dream" in *Fossil Spirit* is related by a Hindu fakir who has experienced the prehistoric past and its denizens firsthand, as a series of formerly living fossil animals leading to the present. For instance, the fakir's spirit formerly occupied an *Ignanodon's* body, lived life as a plesiosaur, and later inhabited a giant ground sloth's body. Only by virtue of the fakir's "supernaturally lengthened memory" are we able to comprehend "...the different animal forms he had assumed as his soul had transmigrated through geological history." This is perhaps not unlike imaginative passages in Gideon Mantell's *Wonders of Geology* (1838), in which "...Mantell (imitating the *Arabian Nights*) introduced a long-lived extraterrestrial traveler who had visited the earth at different epochs, and who had reported on the changes he had seen." (Quotes from O'Connor, p.366.) Certainly, during the mid-1850s and 1860s—revolutionary decades in the history of evolutionary-paleontological understanding—many writers were attracted to the literary realm of possibilities in 'unveiling' Earth's past ages, sequentially. *Time travel* was a popularly employed motif and science fictional gimmick even then.

We note the oldest means of faux time travel are still resorted to. For two of the most 'time-honored' means are the transmigration of souls approach used by John Mill in his aforementioned *Fossil Spirit*, and the "waking dream," as in Jules Verne's famous 1864 passage in *Journey to the Center of the Earth*. (See chapter one of *Dinosaurs in Fantastic Fiction,* or my article "Reframing the Science in Jules Verne's *Journey to the Center of the Earth*" in *Science Fiction Studies,* no. 100, vol. 33, Nov. 2006, pp.405-420, for considerably more on Verne.) Such early conventions offer means of "time travel," sans machine, utilizing the mind instead. We just can't let go of them. These literary devices remain employed today as in passages incorporated into Mark Jacobson's 1991 novel, *Gojiro,* and Sam Enthoven's 2008 novel *TIM: Defender of the Earth.*

Rare are the occasions when professional paleontologists write fiction, but when they do it's usually about imaginative time excursions into paleo-history. For the conclusion of his 1999 popular book *Time Machines*, University of Washington paleontologist Peter Ward crafted an elaborate "pen-picture" titled "Cretaceous Park," (not to be confused with Robert Bakker's 1994 fictive piece). Ward reconstructed the Late Cretaceous Pacific Ocean off the coast of North America. Departing in a time machine from June 15, 2222, Ward's time traveler recedes to 76 million years ago. Donning scuba gear he dives into the ancient sea to observe marine life, predaceous mosasaurs and, Ward's specialty—ammonites. As evinced from his detailed reconstruction, ultimately Ward concludes, "The informed imagination... may be the best time machine of all." A marvelously crafted vignette!

Dinosaur guru & expert, Don Glut's 2006 novel, *Dinosaur Valley Girls* invokes a time travel talisman, known as a 'kaza,' found in an archaeological dig. Except rather than projecting the time traveler back to the Mesozoic Period, protagonist Tony Markham is instead whisked off to a land of one million years ago co-inhabited both by dinosaurs as well as modern-aspect cave people (especially female cave-babes). Obviously this prehistoric realm is more inspired by the 1940 Hal Roach film *One Million B.C.*, rather than real science. Granted three wishes from the magical *kaza*, first Markham travels back into the alternate paleo-world setting known as Dinosaur Valley, where he meets his love, beautiful blond Hea-Thor.

Time travel elements are intriguing as Tony seems drawn into a cosmic loop, mystically bound by the *kaza's* powers. His destiny is the Dinosaur Valley which he's aware of through dreams and nightmares experienced in *our* present. He first suspects their reality after seeing artifacts (including the kaza) validating those dreams in a museum's paleontological archive. Then, although having experienced a week's worth of adventures in the stone-age past, he returns to his present only *5 minutes* after having disappeared in a puff of green smoke. Tony's simple explanation for his dramatic reappearance, as he plans his fated return to Dinosaur Valley is, "Only five minutes? Oh, right. Time. Space. It's all relative." What would Einstein say?

Aliens interacting with dinosaurs either on Earth or in outer space have been tied to time travel as well, as we note in a classic 1937 story. In

P. Schuyler Miller's "The Sands of Time" (1937), a time traveler named "Terry Donovan" relates his adventures into the Cretaceous Period to a bone-digging field paleontologist named Professor Belden. Donovan has constructed an Egg-shaped time machine situated in a laboratory, operated by lots of cables, switches and dials. Deducing that "time is coiled like a spiral," Donovan has demonstrated that by cutting 'across' to an adjacent coil of the spiral it is possible to travel backward in time, jumping approximately sixty million years per coil. He declares, "Some other age in earth's history lies next to ours, separated only by an intangible boundary, a focus of forces that keeps us from seeing into it and falling into it. Past time - present time - future time, side by side. Only it's not two years, or three, or a hundred. It's sixty million years from now to then, the long way around!"

Travel in a machine across *spiraled* Time into Prehistory has been relied on by several writers of paleo-fiction (such as in Marsten's aforementioned story). While Miller may not have invented the idea, his is the oldest use I have encountered of this device. So while the field paleontologist is content to dig up the bones of a fossil "Trachodon," as 'proof,' Donovan shows him photographs of an assortment of *living* dinosaurs he has witnessed after sliding backwards to the coil of spiral time immediately (and conveniently) adjacent to our own—the Cretaceous. But before forever disappearing into the past, Donovan tells Belden about his greatest adventure there.

Most curiously, there are warring factions of people from another world living alongside the dinosaurs. Donovan intended to bury a box containing pure radium into the Cretaceous beach sand, that would be dug up in rock outcrops existing in his present (e.g. 1937), following his return. This would permit an exact calibration of the length of time, existing across spiral time. But he is chased by two carnosaurs, one of which he kills with his gun. Wounded by the first, Donovan is about to be devoured by a second, which is obliterated by a ray gun-toting woman (whose human, yet extraterrestrial nature is left unresolved). Several years following Donovan's disappearance via the "Egg," Belden discovers both his and the girl's sandaled footprints in Cretaceous bedrock. These are criss-crossed by a three-toed tyrannosaur trackway; the three sets of footprints "vanish at the water's edge." Without elaborating here, P. Schuyler Miller's story bears striking analogies to Arthur C. Clarke's later and better known1950 story, "Time's Arrow."

Another in this vein was rather obscure author 'Philip Barshofsky's,' "One Prehistoric Night," published in the Nov. 1934 *Wonder*

Stories. The names Barshofsky and Philip Jacques Bartel were evidently pseudonyms for M. M. Kaplan, a "mystery man" of whom quite little is known regarding his writing career. Michael Ashley had reprinted Barshofsky's dino-tale in a marvelous anthology he'd edited in 1974, *The History of the Science Fiction Magazine, Vol. 1, 1926-1935*. Barshofsky, like several writers before him and since, contemplated what might have happened if aliens (in this case Martians), visited Earth during the Mesozoic. You see, Mars is a dying planet, and only its neighbor Earth can supply the vital and bountiful resources their (insectoid) race needs to avoid extinction.

But Jurassic age Earth is a savage world, and its cruel, unintelligent (i.e. dinosaurian) "blood mad" inhabitants far too numerous and powerful for the beleaguered colonists to withstand. Ultimately, "stupid earthly creatures had preserved the world for earthly intelligence to come aeons later." "One Prehistoric Night" must have been inspired partially by Alexander M. Phillips' 1929 analogous tale, "Death of the Moon," especially in consideration of a curious line added by Barshofsky. For on this '(one) prehistoric night,' "Even the impassioned moon gasped in awed surprise." But Barshofsky was most certainly influenced by RKO's most famed 1933 film release, as we read in another curious passage reflecting the carnage, when, momentarily, an allosaur attacking a spacecraft hull "… resembled King Kong atop the Empire State Building."

So in the case of aliens invading Mesozoic Earth, sometimes the noble dinosaurs emerge victorious (as in Barshofsky's 1934 entry) while on other occasions they lose – totally big time, as in Russian vertebrate paleontologist Ivan Antonovich Efremov's 1948 story, translated as *Star Ships*. Efremov not only was one of the primary investigators of the Mongolian fossil beds during the late 1940s, but also was a prolific writer of geologically oriented science fiction; at least two of his tales involved dinosaurs. In *Star Ships*, however, mass extinction of the Late Cretaceous dinosaurs is the theme. When humanoid aliens land on Earth intent on extracting radioactive elements from the mountain crusts to power their great spaceships, they're menaced by dinosaurs, resulting in a wholesale onslaught of our native fauna. Yes—the K-T boundary (or as it's now more properly called, the "Cretaceous-Paleogene" mass extinction). Efremov used his factual knowledge of dinosaur "graveyards" as proof of the former slaughter, yet adding an additional fictionalized discovery made by the paleontologists – the skull of a strange (*alien*) humanoid with its accompanying advanced tools found within the fossilized dinosaur 'cemetery.'

With or without a time travel device, the linking of outer space and dinosaurs remains a popular sci-fi theme. In 1953, the year prior to when Toho Productions first introduced Godzilla to movie-goers, a British writer, John Russell Fearn, issued a pair of novels intended for adolescent readers, which involved dinosaurs erupting from a Jurassic "pocket" within the Earth, followed by an alien invasion. These were titled, *A Thing of the Past* and *The Genial Dinosaur*. Fearn, a prolific writer (and author of the authorized Universal Studios' 1954 novelization, *Creature From the Black Lagoon* – although therein writing as "Vargo Statten"), instead used the pseudonym "Volsted Gridban" for his two dinosaur books.

In Fearn's *A Thing of the Past*, dinosaurs spew forth from within a basalt fissure where they've been preserved by volcanic "neogene gas" since the Jurassic Period, over the English countryside, attacking London (less than a decade before "Gorgo" and "Paleosaurus," both of filmic fame, later did so as well), In *The Genial Dinosaur*, the only remaining dinosaur, affectionately named Herbert (a truly gigantic, carnivorous form of *Diplodocus*), inadvertently thwarts an alien invasion. Turns out that on the home planet from which the invaders traveled, dinosaurs are considered highly sacred. Thus, needing a male dino-specimen to 'seed' their waning female dino-population, the alien leader graciously accepts Herbert conditionally, in exchange for halting the Earth invasion. Certainly a curious pairing of fantastic tales for younger readers!

S. M. Stirling's sterling alternate history novel, *The Sky People* (2006) is an homage to Edgar Rice Burroughs, as references to the latter's classic paleo-sf are peppered throughout. In Stirling's scenario, Venus actually is the swampy, watery world envisioned by astronomers during the 1930s. Strangely, living dinosaurs and Neanderthal men have been discovered on Venus! Since the 1980s, both EastBloc and competing American settlements have sprung up on our 'younger' sister planet where only the fittest, brilliant and highly trained selected people have come to study Venusian fauna. Due to Venus' slightly lower surface gravity and higher air density, some differences are apparent though. For instance, insects and pterosaur species are absolutely huge; the ecosystem seems "far too energetic." Venus appears to be some sort of parallel-Earth. The DNA of Venusian animals is uncannily similar to terrestrial creatures. Is the theory of evolution wrong then?

Geologists note that before 150 million years ago, during an anaerobic era and before deposition of a distinctive organic boundary layer, there's little evidence for life in the Venusian rock record. Then, rather suddenly, fossils analogous to those found on Earth pop up in deposits characteristic of an aerated planet. It's as if Venus had been *terra-formed*, and that possibly the planet may have been intended as a wildlife 'zoo' for a master celestial race of beings, the Lords of Creation who may have 'seeded' life on Mars and Venus (and Earth?). However, in *The Sky People*, the essence of the story surrounds heroic Marc Vitrac and his party of earthlings who set out on a quest to rescue an EastBloc Russian cosmonaut whose rocket crash-landed in the wild country. Braving parasite-infested titanosaurs, vicious sabertoothed cats, feathered raptors and carnivorous giganotosauruses, ultimately Vitrac—sort of a literary cross between Burroughs' 'Tarzan' and Carson of Venus—falls in love with 'Teesa,' a beautiful Venusian native woman. Vitrac commands the respect of Teesa's modern-aspect tribe, then victoriously rides a huge horned dinosaur they've tamed into battle against the savage oppressive Neanderthals, before foiling an EastBloc terrorist plot.

So are we done yet? Well, while we're on the subject of Venus we find many are the tales of Venusian dinosaurs that have sprung from the pens and typewriters of ye' olden authors! Two additional tales worth briefly mentioning here are Edgar Rice Burroughs' 1937 novel—*Carson Of Venus,* in which Earthling Carson Napier (a Tarzan-like hero) encounters Venusian seas teeming with gigantic reptilian monsters. One of these, the immense "Rotik," even has a periscope-like appendage attached to its noggin. Burroughs also populated terrestrial landscapes of Venus with other strange dino-creatures—the 6-meter long, crocodile-snouted "Vere" and "Kazars" (parrot-beaked and feathered denizens) have cameos, in his engaging *Lost On Venus* (1933). There are also "beast-men," viewed as "evolution's first steps from beast to men," underscoring the supposed parallel evolution taking place on our sister planet.

Another Venusian example is Kurt Mahr's (i.e. Klaus Otto Mahn's, 1934-1993) segment of a 1972 story, "The Venus Trap," continued within volume 17 of the *Perry Rhodan* series, translated from German into English by Wendayne Ackerman, and edited by Frederick Pohl and Forry Ackerman. In one segment, the Perry Rhodan party, fleeing bad guys, are menaced by a pterodactyl with a 100-foot wingspan. In another passage another team tromping through a Venusian jungle are attacked by a ponderously slow-moving tyrannosaur, referred to as a "Tyrex." Note that the dinosaurs in such tales are merely posed as "paleo-props," and are not at all central to the plot.

And speaking of Venusian tyrannosaurs …

Scotty, beam us back to 1953 again, at a time when Venus was popularly (yet not so scientifically, anymore) regarded as a warm, steamy planet, and when the future of manned space-rocketry seemed linked to atomic power! This time though "Carey Rockwell's" (pseud. for Joseph Greene, Richard Jessup and possibly other writers) novel *The Revolt* on *Venus* summons the courageous crew of the *Polaris* to the rescue, after vanquishing a Venusian *Tyrannosaurus*. This title was no.5 of eight in the *Tom Corbett, Space Cadet* novel series intended mostly for adolescent boys—spinoffs from the early 1950s televised program. The series was inspired by the writings of Robert Heinlein. Although from today's perspective a lush, watery Venus viewed as a planet where parallel fauna evolved lagging remains millions of years behind Earth's stately 'ladder' of evolutionary progress indeed appears to be an 'alternate' universe.

Corbett's Venus is replete with huge saurians, including living *T. rexes*! With blasters and paralo-ray pistols at the ready, Tom Corbett and his pals discover that hunting a *Tyrannosaurus* for sport turns out to be the key toward thwarting a revolt of Venusian Nationalists, sworn enemies of Solar Guard. In a dramatic scene, while lost in the great jungle, spaceman Astro observes a Rex battling a 100-foot long, 5-foot thick snake. "With three coils wrapped around the tyrannosaurus's body, the snake was trying to wrap a fourth around its neck and strangle it, but the monster was too wily. Rearing back, it suddenly fell to the ground, its weight crushing the three coils around its middle. The snake jerked spasmodically, stunned, as the tyrannosaurus scrambled up again. The ground trembled and branches were ripped from nearby-by trees. All around the jungle had been leveled. Everything fell before the thrashing monsters." With smashing scenes like that, is there any wonder why I was a fan of the novel series during the early 1960s? Eventually, the *Tyrannosaurus* is killed, the revolutionaries are defeated in an epic space battle and order is restored throughout the solar system. If you ever plan to return to Venus, given its dinosaurian-like horde, then "Spaceman's Luck" to you!

Frank Belknap Long, Jr.'s short story, "Exiles of the Stratosphere," printed in the July 1935 issue of *Astounding*, concerned "beings with lighter-than-air metals who live in our upper atmosphere and occasionally plague us with their outcasts, e.g. those titular "exiles." After many generations of living in stratospheric domes, protagonist "Lutation" descends to the crust of the now wild, savage Earth, risking contamination from planetary toxins from which stratospheric denizens no longer have immunity, to destroy several treacherous exiles. Belknap Long describes the planetary surface thusly: "Since man's ascent to the stratosphere reptilian evolution had swung again upon an upward curving arc. Creatures resembling the gigantic saurian of the Jurassic age moved once more through the lush jungles and by the rims of the plague-infested marshes. This was the dark and awful crust…" In midst of his mission, Lutation rescues a female from a pursuing "Trallator"—a 100-foot long dinosaurian with "foreshortened upper limbs," dispatching the dino-monster using a vial of toxic 'animalcules.' The tale is reminiscent of British geologist Charles Lyell's mid-19th century theory of a cyclically evolving Earth, in which a futuristic 'Mesozoic Era' might someday return.

Certainly, by the late1930s, the lost world - island concept had become the most popular fictional framework in which crypto-dinosaurs and other prehistoria could be merged with explorers from modernity. By then, this ever-popular theme had been rehashed numerous times in fiction novels, short stories and in films. While dino-philes revere this most popular and enduring form of genre (especially in movies), some of the contemporary literature examples may have become 'lost' themselves along the wayside. Here are several examples.

When Willis Knapp Jones' short story "The Beast of the Yungas" was published (*Weird Tales*, Sept. 1927) paleontologists such as Henry F. Osborn viewed the great sauropods as predominantly *aquatic*, inhabiting marshes and swamps to relieve stresses on their supposed fragile limbs. (It's since been determined that they were terrestrial animals.) When stalwart Winslow, an intrepid world explorer, is asked by a charming lady at a dinner affair, "Haven't you ever been afraid?" he relates a Bolivian adventure where, not he, but a fellow named Manion became literally frightened to death. "It's not a pretty story… I believe I'd better not tell it." But guests press for details, and so Winslow concedes. Winslow and Manion believed they were on the trail of a living *Diplodocus*, which "…lived ages ago during the Pleistocene period before the tyrannosaurus came along and killed them off." The expedition party tramps through muck and steaming rain showers into the deepest jungle where they spy a hidden valley. Climbing down a vine ladder, Manion decides to spend the night on the valley floor. Then, as Winslow relates confidentially to a gentleman—sparing the awful details from the ladies who would be doubtless upset to hear—brave Manion died, evidently paralyzed by fright. But this animal, *Diplodocus* or whatever it was, didn't eat Manion's flesh. Because the monster must have been herbivorous and, judging from its tracks, huge, "So I think he may have seen such a prehistoric animal …"

While today, Jones' choice of the 87-foot long *Diplodocus* as dino-monster may seem rather odd, bear in mind that during the 1920s people were simply astounded by the largest dinosaurs' enormity. *T. rex* hadn't yet become a national icon. Furthermore, by then *Diplodocus* had become a cosmopolitan dinosaur thanks to Andrew Carnegie who had several casts of his exceptional museum specimen made and delivered to museums in Europe, Mexico and South America. Also, Jones' story mirrored a chain of historical events, as unsubstantiated rumors of living dinosaurs still living in an African Congo lake setting (Rhodesia) had triggered Carl Hagenbeck's cryptozoological expeditions of the early 1900s. And certainly Arthur Conan Doyle's literary successes with *The Lost World*, serialized in 1912, may have inspired later works such as Jones' "The Beast of the Yungas," which may well be the prototypical 'Reginald Rivers' type yarn as stylistically perfected by L. Sprague de Camp decades later (e.g. - through his masterful time-travel writings as "A Gun For Dinosaur"1956).

James Blish, one of sci-fi's hallowed names, wrote an account of an African expedition as it perhaps really might have happened long ago. (Wink!) *The Night Shapes* (1962), be it fiction or skewed documentary, concerns an early 1900s expedition into the deepest, darkest region of the African continent, resulting in the discovery of living dinosaurs! Kit Kennedy, white denizen of the Congo region most familiar with local natives, has an affinity for the wild backcountry and a reputation of being mindful of jungle lore and customs. So when a team of European scientists embarking on a mysterious safari (i.e. which turns out to be a search for the source of illegally mined radioactive pitchblende ore) invades this unforgiving, infested turf, he's propositioned to become their guide. Along their fateful journey, they encounter the famed *Mokele-mbemba*, or the legendary "night-shapes"!

Expedition leaders are captured by hostile natives who, enslaved near the swampy river inhabited by night-shapes—which turn out to be living dinosaurians, mine the deadly ore. A developing dilemma, however, is that if the river valley of the night-shapes is divulged to the outside world, then it would be plundered and exploited by white invaders and hunters. So (not unlike Tarzan), Kit Kennedy comes to the rescue, thwarting the dinosaur-hunting expedition when it inevitably arrives, with fear of the unknown, of course aided by a conveniently timed, body-trampling dinosaurian stampede! Kit perceives the Valley of the night-shapes not as a "zoological garden," but "… a place where old and deadly animals still live, animals that (the modern) world couldn't even tolerate." But the 'shapes' are also metaphorical for mankind. For, "They can never die. The night-shapes aren't animals or men, or demons ... They're the ideas of evil for which those real things only stand. The real things are temporary. They can be hunted. ... But the shapes are inside us. They've always lived there. They always will."

These days, alternate histories are all the rage in science fiction. So is steam-punk. But when you stop and think about it, aren't most fictional tales 'alternative' or parallel to reality anyway? Of course, the essence of an alternate history story is always 'what if' certain key events in history had been changed either slightly or even a great deal such that the protagonist must confront significant changes in the altered universe setting? Invoking a modicum of quantum physics—as we may have gleaned from those *Back to the Future* films, alternate realities are exactly what one might expect to find in parallel world settings, which as *BTF* characters Doc and Marty showed us, typically involves time travel. Which is why space (and time) alternate universe dinosaur tales merit 'space' (and therefore time) here.

Johnathan Green's *Unnatural History* (2007) is an adventure taking place in the era of Pax Brittania. Here, Great Britain's dominion over the world has persisted into the late 1990s under Queen Victoria's rule, well, that is, a robotic/bionic version of the 19th century queen that's been kept alive for nearly a century, artificially. Britain has conquered space (yet strangely relies on steam powered machinery for domestic use). In one gory scene, because dinosaurs haven't entirely gone extinct in this alternate universe, terrorists known as the Darwinian Dawn bomb the Challenger Enclosure in a London Zoo causing the fleeing dinosaurs to stampede. A heroic, if not peculiar gentleman/detective/adventurer named Ulysses Quicksilver dispatches a great 'Rex-like' "Megasaurus" on the loose before it can do further harm gobbling up citizens left and right. To Ulysses the terrorists' view of how history should be transformed is unnatural, yet dinosaurs are an apt metaphor for how the Darwinian Dawn views the Queen's rule. For a chief terrorist declares... "Magna Brittania is an outmoded dinosaur, a creature of a bygone age that should have become extinct long ago. Just think of me as the agent of evolution." Ulysses eventually rescues Queen and Country in her darkest hour from a horde of prehistoria de-evolving through "acute genetic regression." At its core, *Unnatural History* revolves around terrorists who are fed up with the enduring political status quo. They're intending to take down the Empire using a chemical formula concocted in a mad scientist's laboratory causing humans to genetically mutate into prehistoric forms. A band of escaping mad-slasher convicts are exposed to the chemical, transforming into a horde of savage apemen who take over the Crystal Palace (which was never destroyed in war), just as a zeppelin armed with more de-evolving chemical bombs crashes through the glass roof. Earlier, the scientist who invented the chemical suffers accidental exposure. He transforms all the way back down evolution's 'ladder' in a series of scenes and encounters staged with Ulysses Quicksilver ... from an apeman, into a horrifying reptilian 'lizard man,' then a frog-like amphiboid, and finally a gelatinous mass of cells. This is the fate for all those coming into contact with the de-evolving chemical. But ultimately Quicksilver saves the day.

While the fictional device of using scientific, biochemical lab methods to reanimate extinct, prehistoric creatures predates this next entry, Edgar Rice Burroughs' "The Resurrection of Jimber-Jaw" (*Argosy*, Feb. 20, 1937), in turn, predates *Jurassic Park's* biotech dino-monsters by over a half century. For many decades "Jimber-Jaw" was sort of a 'lost' tale, as Burroughs' scholars focused on his novels, made more available through paperback mass reprinting since 1963. And "Jimber-Jaw" is a gem, absolutely one of his best stories. A scientist named Slade who crash-lands in Siberian tundra spots a 50,000-year-old Pleistocene cave-man frozen in a wall of ice—much like the frozen woolly rhinos and mammoths which do turn up occasionally in the frozen north. Using the latest scientific techniques (coupling blood and posterior pituitary fluid transfusions), he resurrects the cave-man, whom they name Jimber-Jaw.

Subsequently, he's introduced to modern civilization. But this turns out to be like taking King Kong back to New York, as it all ends tragically, with Jimber-Jaw unaware of how much time has elapsed since his first 'death,' or, most significantly, the whereabouts of his long-deceased mate, Lilami. You may find "Jimber-Jaw" included in a trilogy of Burroughs' lost tales, *Tales of Three Planets* (1974).

In such bioengineered dino-stories, there's often a basic foundation of applied factual knowledge (i.e. the fact that prehistoric animals of various known genera, as reconstructed by scientists, did formerly live on Earth in fantastic geological settings). Then this foundation usually becomes melded with a satisfying kernel of current scientific research and discovery, leading to wildly ensnaring, what-if, disbelief suspending scenarios.

In the case of Kenneth Robeson's *Doc Savage: The Time Terror* (*Doc Savage Magazine*, Jan. 1943), the kernel of then recent contemporary truth and wisdom stoking our senses, thus permitting us to 'accept' the scientific gimmick when unveiled, happens to be laboratory research into vitamin biochemistry (as was being conducted extensively during the 1920s through the 1940s). Therefore, although not the first example in its hallowed lineage, in a way "Time Terror" vaguely presages those recombinant DNA/molecular biology dino-monster sci-fi stories of the 1980s and 1990s, like Crichton's *Jurassic Park*.

"Robeson" (an alias for Lester Dent), had been paired with prehistoric animals a decade earlier in a better-known novel titled *Land of Terror*, or your basic, run-of-the-mill prehistoric lost island tale; an island destined to become destroyed in a climactic volcanic eruption. But as framed in Time Terror, Robeson takes us further, this time to a jungle plateau situated in (of all places!), counterintuitively, the Arctic. But the premise of this fast-paced story involves far more than man's reaction to prehistoric inhabitants found in yet another "lost" world.

Land of Terror occurs during World War II time, and the 'bad guys' are military Japanese who covet a new chemical discovered by an American chemist, Calvin Western, who has isolated it (akin to the vitamin research happening in the scientific world then) purported to accelerate evolution! They've captured the chemist and so heroic Doc Savage and his team strive to rescue him before the Japanese learn the secret chemical formula. It seems the chemist had first discovered the chemical which halted evolution on the lost jungle plateau, as evinced by the presence of a few marauding prehistoric mammals and a few dinosaurs. But Western goes one further and after identifying the chemical dissolved in the plateau's circulating fresh water that halted evolution, was able to also synthesize another "opposing" chemical, a "vitamin-like substance," that will *accelerate* evolution by enhancing overall "adaptability" to the environment. If the Japanese gain hold of this latter chemical, then they would become the world's master race! Fortunately, Doc Savage and his team of specialists, with Calvin Western, are far more egalitarian-minded, thus thwarting the dastardly plot.

Robert Wells' 1969 novel, *The Parasaurians*, is another take on the dinosaur hunting safari theme (first popularized by Ray Bradbury's 1952 short story, "A Sound of Thunder.") "Parasaurians" also touches on the time-honored 'war with dinosaurs' theme. If there's a dinosaur (or another variant of 'prehistoria') running rampant somewhere, then why is it that man simply must shoot it? And if they're already extinct, then how we yearn to restore them to life so we can kill them with our guns, threatening them with (another) extinction.

Robert Wells rather anticipated some of Michael Crichton's ideas in *Jurassic Park*. In "Parasaurians," dinosaurs are resurrected using genetic code preserved in frozen ovoid remains discovered in fossil deposits exposed by retreating icecaps, extracted with a dash of test tube biochemistry, then properly analyzed using "Mattterling's equations." In Wells' case, however, the beyond-robotic yet "animated" dinosaurs, known as "parasaurians," typically are non-living mechanized reconstructions (reminding readers today of those 1980s Dinamation models). That is, except for most recent technological advancements and unlawful enhancements permitting two breeding "species" (i.e. *Stegosaurus* and *Tyrannosaurus*) to propagate in the wild (not unlike what happened with those raptors in *Jurassic Park*) – all thanks to a meddling scientist who surreptitiously delved into Frankensteinian territory: the forbidden mysteries of life.

So in the year 2173, recruited by Megahunt Chartered (Inc.), an eclectic business catering to those who desire hunting the most unusual and dangerous big game conceivable (and only those wealthy enough who can afford it), Prof. Ross Fletcher signs up for a hunt. The dinosaurs, which for all practical purposes appear genuine, are distributed throughout a tropical island named El Pais. He witnesses the "birth" of a parasaurian *Triceratops*, startlingly becoming animated after a human tech breathes into it. "The audience ... held its breath. The acolyte below exhaled his into the tube. Man and beast of the prehistory of his own planet were connected there in a shocking reversal of evolution." The safari tragically ends with an engaging tyrannosaur chase.

Intelligent dino-reptiloid humanoids have long been staple creature features of fantastic fiction. However (as I've discussed elsewhere) their scaly presences began to proliferate during the mid-1950s following invention of the hydrogen bomb, or more specifically, following the Soviet Union's catch-up development of its own fusion bomb. And they've enjoyed common place status in fantastic fiction since the late 1970s, although especially so after Dale Russell (and Ron Seguin) introduced the "Dinosauroid" in 1981. Accordingly, two other tales will be briefly mentioned. Stuart Vaughn Stockton's *Starfire: The Mending, Book 1* (2009), is about an intelligent dinosaur civilization existing in the stars, projecting a medieval 'sword of thrones' type ambiance. Here, the dinosaurs interact much as people would in a fantasy adventure. And a very familiar giant dino-monster has even taken on distinct anthropomorphic tendencies, as in Greg Cox's 2014 novelization of the movie script for *Godzilla*.

But there's too much to outline here and my time (and space) drawing short. And so harking back to this article's 'catalyst,' in the words of Rosalind Van Gilder and Victor Millrose, who scripted *A Day in the Life of A Dinosaur*, I close with these words - "... so thank you, my friends, for a wonderful day in the life of a dinosaur ...."

Chapter Eighteen—*Big, Fierce, Extinct ... Radioactive: Godzilla's Essential Formula*

During the "dinosaur renaissance," when postmodern understanding of dinosaurs and other prehistoric animals went through such profound scientific upheaval, in answer to a commonly posed question, 'why are dinosaurs so popular, especially with children', psychologists arrived at the "archetypal answer." Dinosaurs, they noted, are "Big, Fierce and Extinct." (I will refer to this as the "BFE" formula.) Yet the essence of this idea became quickly challenged by scholars Stephen J. Gould and William J. T. Mitchell.[1] Well, one can readily see that BFE as a concept wouldn't be restricted to only dinosaurs. Not all dinosaurs were "big" or "fierce, for example. Many realize they aren't even completely extinct because birds *are* dinosaurs.

And BFE could apply to many other kinds of prehistoric monsters, yes, including *fictional* varieties, which ordinarily should be 'extinct,' but have rematerialized in sci-fi stories and movies. BFE was already an important 'paleo-monster' theme of a major wave of American "dinomania" proliferating from circa 1900 into the mid-1930s. But World War II quelled this early sense of fascination, and so by the early 1950s, BFE became amended and augmented in popular culture by a new equation: "BFER" = *Big Fierce, Extinct ... Radioactive*! This revision clearly applies to Godzilla. Let's briefly outline evolution of the BFE concept as it transformed into BFER during the 20th century. And so, first, how did public fascination focus on the 'Big-ness' factor in regard to our favored paleo-monsters? Here, I straddle that mystical, time-honored domain between real and fanciful dinosauria.

"B" - *Enormous!*

The immensity of paleo-monsters proved utterly fascinating during the 19th century both to fossilists and laymen alike. During a pre-World War II American 'wave' of dino-popularity, we note a profound preoccupation with their great sizes, especially as manifested by those colossal, long-necked sauropod dinosaurs. While there had been stories of giants in literature about for decades (e.g. *Gulliver's Travels*) and some gods were often viewed as gargantuan by the ancients, in the case of extinct monsters, the great sizes of several genera was indisputable. True—there were intriguing paleo-mammals to ponder such as the enormous Mastodon and Mammoth, or giant ground sloths, gigantic "lizards" like *Iguanodon* and *Megalosaurus*, proved eminently popular in the later Victorian age.

By the late 1880s fantasy artists began rendering early images of over-sized dinosaurs (as then known to science), invading modern urban centers, as we see for example in "Stablo's" art, thus introducing a theme later to become rampant throughout the 20th century. His 1886 illustration published in a life through geological time book by Camille Flammarion showcased an over-sized, pensive *Iguanodon* (as then scientifically viewed) displaying customary behavior later 'adopted' by a number of famous, increasingly pugilistic 20th century prehistoric monsters, particularly from 1925 onward—both in film and literature. A quarter century later, Jules Lermina vivified this striking scene in passages of his novel, *Panic in Paris*, of which more shall be stated shortly. However, the atmospheric opening to Charles Dicken's *Bleak House* novel (1851) suggested (without visual illustration) the earliest merging of a dinosaur (*Megalosaurus*) in a modern city – London.

Universal appeal of potentially menacing dinosaurian BIG-ness truly expanded with discovery and announcement of the largest of the clan—the great long-necked sauropods, of which spectacular specimens were disinterred and 'resurrected' with great hoopla in North America and Africa between 1877 and 1915! Following their scientific description, as disseminated into the mainstream, the pop-cultural era of the HUGE dino-monster had begun.

Let's briefly note how striking imagery of the truly giant prehistoric monster entered public consciousness during the turn of the 20th century, leaving widespread lasting impression through arresting museum displays and early fantasy art. Three kinds (genera) of dinosaurs principally became associated with the public's concept of sheer, utter reptilian massiveness: *Brontosaurus*, *Diplodocus*, and the mysterious "Gigantosaurus" – later named *Brachiosaurus*. In a 1898 *New York Journal* article, an artist depicted how the (by then already described) *Brontosaurus* would appear gazing into an 11th story window of the New York Life Building. This illustrated a news story announcing a new discovery, purportedly the "most colossal animal ever on Earth just found out west." The newspaper image evidently electrified steel magnate and philanthropist Andrew Carnegie's visions, thus resulting in another paleontological expedition which led to discovery of the nearly complete skeleton of the Jurassic *Diplodocus*. Carnegie lost no time in having plaster replicas of this new fossil specimen prepared and sent to museums in Europe, Mexico and South America. Meanwhile, *Brontosaurus* went on to 'star' in several classic films and stories of the early 20th century (i.e. *The Lost World* and *King Kong*). Then, discovery of another record-setting dinosaur from the African bush of Tendaguru, also documented in vividly illustrated newspaper accounts, further fueled imaginations.

This was that historical period when newspaper headlines often reported discovery of the latest "most gigantic" of all land animals & dinosaurs. But the genus found in Tendaguru Africa truly tipped the scales, holding the size record for decades. A partial specimen of this monster had already been collected in Colorado in 1900, which was described by paleontologist Elmer Riggs as *Brachiosaurus altithorax* in 1903. (Two resin skeletal reconstructions were prepared for the Field Museum of Natural History during the 1990s, one of which may be seen at the United Airlines terminal at O'Hare International Airport.) A German expedition then found a far more spectacularly complete skeleton in Tanzania (former German East Africa), along with several incomplete specimens—permitting a complete composite reconstruction. Some of the material excavated between 1908 - 1914 was initially referred to as "Gigantosaurus," but later named *Brachiosaurus brancai*. (By the late 1980s, G. S. Paul christened the African variety *Giraffatitan brancai*.)

The African giant measured 80 feet from tip of tail to its snout, its head waving 40 feet above the ground at the end of its long neck, all strung together on very heavy bones. Given that its discovery coincided with the atrocities of World War I, while its preparation began during a period trending toward World War II, understandably it took several decades to mount its titanic skeleton in Berlin, by 1937.[2] But during the intervening period, before Brachiosaurus's image was established concretely, people reveled in the anatomical uncertainties, envisioning the bulk and enormity of this fantastic Jurassic giant. For example, W. D. Matthew (American Museum of Natural History vertebrate paleontologist), overestimated the great "Gigantosaurus's" length at 130 feet long (considerably longer than the sleeker, 90-foot long *Diplodocus*). And, he opined, the monster's weight supported on its relatively weak limbs would have prevented an existence living outside of swamps where it would have been buoyed by water. Matthew's speculations reflect the public's infatuation with the sizes of certain dinosaurs a century ago. Artist Vincent Lynch illustrated Matthew's 1914 *Scientific American* article ("The Largest Known Dinosaur"), showing an awesome "Gigantosaurus" walking along a city street, flanked by skyscrapers of the time while excited crowds yield to the towering behemoth's pounding footsteps.[3]

Ulrich Merkl, author of the splendid *Dinomania: The Lost Art of Winsor McCay, the Secret Origins of King Kong, and the Urge to Destroy New York* (2015), claims cartoonist Winsor McCay (1867-1934) played a significant role in our (then) embryonic fascination with gigantism (particularly of the sauropod variety!), that is, toward the imagined destruction of our cities. It seems that the American Museum's *Brontosaurus* (i.e. formerly, formally *Apatosaurus*—which itself remains a valid genus) skeleton, mounted for display in 1905 deeply influenced McCay's vivid 'extrapolations' of nature gone awry, creating an oft-friendly, gentle brontosaur character, featured in his famous cartoon, "Gertie the Dinosaur" (1912, 1914)—the very first dinosaur movie, and a sequel, "Gertie On Tour" (1921) which may have never been released or distributed commercially. As noted by Merkl, Winsor McCay generally seems to have been fascinated with relative sizes of *giant* creatures which he featured in cartoons, human forms as well, ambling about through New York City. Merkl also suggests that McCay's sauropod dinosaur at loose-in-the-city-themed comics may have influenced others to take up the calling later in prehistoric dino-monster 'attack' films. While he makes a plausible and interesting yet circumstantial case for McCay having possibly had some sort of subliminal bearing on Merian C. Cooper's *King Kong*,[4] such evidence is either lacking, or at best 'second derivative' for inspiring Toho's *Gojira*.

When it comes to Godzilla, Winsor McCay's influence may be glimpsed too, but vastly diluted 'genetically' like a great, great paleo-parent. One must be cautious to avoid the fallacious argument that simply because something came before does not necessarily mean it caused something to result later that might be similar or seemingly derivative in nature. (The Latin term for this is, "*post hoc ergo propter hoc.*")

During the 1900s through 1920s, for editorial reasons or general amusement, the popular press often printed cartoon images of non-dinosaurian giant animals, bugs, robots, and other beasts romping through metropolitan areas. So the BIG-ness motif was very much already out there before *King Kong*'s debut. Dinosaurian hugeness was also metaphorically linked to the idea of obsolescence (and thus extinction), as noted in contemporary editorial cartoons. And although sauropods were dauntingly gigantic or even in some cases fearsome (like the brontosaur in RKO's *King Kong*), they were usually not portrayed as fiercely as other dinosaurs: rarely were they imagined to be as titanic or towering as Godzilla.

The horrific idea of sauropods trampling over humans (and buildings) thereafter became a staple ingredient of sci-fi films and published fantasy stories through the 1930s.Thereafter, sauropods attained their popular zenith when the Sinclair Oil Company linked dinosaurs with fossil fuel products in a spectacular display at the 1933-34 Chicago World's Fair, adopting the image of a brontosaur for its gasoline station brand logo. Meanwhile, besides the animatronic 'robots' exhibited by Sinclair and Messmore and Damon at the 1933/34 Fair (which my father witnessed as a boy and recounted for me later), life-sized statuesque restorations of sauropods began appearing around the world. Such sculptures were constructed by paleo-image makers in Rapid City, South Dakota, Calgary, and Hamburg, Germany, massive labors of love proliferating through the mid-20th century, globally into the modern day. With these huge monstrosities to stand next to and safely marvel at, the relatively puny size of humankind is poignantly apparent.

Those old dino-giants of the Roaring Twenties have been vastly superseded in recent times. A recent record holder for size, *Argentinosaurus*, was eclipsed by yet another South American superlative sauropod, whose mounted skeleton is displayed in New York's American Museum of Natural History, thought to have grown up to 130 feet long (i.e. *twice* as long as *Brontosaurus*!), a whopping 77 tons! Regardless of size, in modern times, sauropods usually only rate as second shrift paleo-news relative to any new information concerning everybody's (current) favorite dino-monster – *T. rex.*

"F" - *Ferocious!*

Dinosaurian hugeness was certainly something to behold firsthand, yet human psyche was also increasingly captivated by the ferocity of certain Mesozoic monsters. By the early 1930s, in the pop-cultural realm, public fascination for dinosaurs had generally eclipsed that of other prehistoric creatures, such as the magnificent paleo-mammals. Furthermore, the BIG dino-monster attractor was supplanted by virtue of the presumed *ferocity* of other species. Although Sinclair's "Brontosaurus" proved a popular attraction at the Chicago World's Fair, far more imposing was their life-sized pairing of *Tyrannosaurus* and *Triceratops*, aggressively staged as in

Charles R. Knight's famous (then) recently competed Chicago Natural History Museum mural. While McCay's art may have been widely disseminated via the newspaper medium, Knight's prehistoric life paintings were more riveting, memorable and everlasting. The frightening to behold 'Knightian' *Tyrannosaurus rex* was also a gigantic saurian, even though its size in adulthood couldn't compete with that of the sauropod clan. *T. rex* steadily rose to the forefront of dino-monster lore and popularity during this interval, thanks to the American Museum's skeletal 'Rex' reconstruction completed by 1911, Knight's keen artistry of the monster and a filmed segment from RKO's *King Kong*,

According to scientists and artists, who portrayed scientific interpretations on canvas in sculpture, and in early sci-fi films, *Tyrannosaurus* and its evolutionary flesh-eating cousin *Allosaurus* were of much fiercer disposition and temperament than any of the sauropods. As W. D. Matthew proclaimed, early American Museum dinosaur skeletal displays were intended to showcase, realistically, scenes "millions of years ago when reptiles were the lords of creation, when 'Nature red in tooth and claw' had lost none of her primitive savagery."[5]

Trampling through and efficiently demolishing cities, a task which fictional sauropods couldn't perform as dramatically or as menacingly as a 'prehistoric' snarling, sharp-fanged beast on the prowl also divulges an innate fierce disposition. Besides McCay's foundational gentle "Gertie" flicks, later cartoons instead featured vicious theropodous dino-monsters as titanic frightening menaces, terrorizing primeval landscapes or citizens of a doomed city. Max Fleischer's 1942 "Superman" episode, *The Arctic Giant*, and the "Rite of Spring" segment from Disney's 1940 *Fantasia* effectively convey this monstrously transformed, savage "persona."

So a menacing "fierce" *fantasy* dino-monster must satisfy a two 'layer' definition. First, you have its potential to cause wanton destruction in a metropolitan area due to its great size, damage which could collaterally be inflicted by a roving, although usually not so viciously inclined, "gentle" giant, sauropod. Secondly, in contrast, there's the innate, savage rage of the blood-thirsty, carnivorous kind, which could be witnessed either in the primeval jungle or in our cities. In the arena of pop-culture (perhaps fueled by horrors of two world wars?), appeal of the wayward, bewildered sauropod lost or misplaced in modernity, eventually became preempted by the ferocious bestial pseudo-theropod; rapaciously ravaging formidable jungle prey, destroyer of cities, emblematic of pending global warfare.

Prehistoric creatures, especially "fierce" dino-monsters, fuel our "archetypal" senses because they conjure Darwinian 'Nature red in tooth and claw,' savage imagery. Although one item of advertising art for 1925's *Lost World* movie highlighted a theropod on rampage in a city, this eye-catching scene was not portrayed in the movie. (Instead it was a wayward stop-motion animated brontosaur that wanders through London, collapsing a bridge.) Then in 1933, RKO's *King Kong* featured that fan-favorite, famous scene where a paleo-monster, Kong, brutalizes a bloodthirsty tyrannosaur to the death; later the "prehistoric ape" battles his way across New York City. By then, Knight also had completed three of his most famous and frequently reproduced paintings exemplifying ferocity in the prehistoric world: a pair of tussling "Laelaps" (1897), and two later depictions each showing *Tyrannosaurus* squaring off versus *Triceratops*: the classic prehistoric showdown. Furthermore, in fiction, dino-monsters also countenance undying ferocity when humans are devoured. Arguably, however, heyday of the utterly soulless, *Fierce* dino-monster didn't materialize until the Cold War's onset.

It is interesting that throughout his aforementioned book, Merkl shows how Winsor McCay wielded great potential to disseminate the general idea of *prehistoric* monsters plundering cities through his art, as in later scenes from classic movies such as 1925's *The Lost World*, and *King Kong* (1933). Although, as noted, the idea of a live dinosaur simply situated in, yet not shown ravaging, a modern city is not a "visual" that he invented, having been superseded by at least three 19th century images, two of which appeared in editions of Camille Flammarion's grand 'evolutionary epic' *Le Monde Avant La Creation de L'Homme* (1886). Essentially McCay was thwarted by Fate, an "injustice of history."[6] For example, his Gertie-like "Dino" strips of 1934 (highlighted by Merkl in his new book), involving another huge sauropod awakened from a rock quarry that invades New York City were never released because of his untimely death.

Conceptual similarities do exist between several of McCay's cartoons and notable sci-fi scenes filmed thereafter. In referring to the "lowly status" of newspapers, Merkl laments how McCay's dino-themed newspaper cartoons "… were condemned to brief lives as disposable consumer goods." Yet, he contends, "On the other hand, they reached so many readers …"[7] Perhaps rightly, he credits McCay with producing the first animated cartoon of a giant monster attacking a modern city, a ~12-minute feature titled "The Pet" (1921), available on YouTube. Here, the giant monster is not a dino-monster, but a strange *daikaiju*-like 'puppy' that swiftly grows to gigantic proportions before succumbing to a squadron of biplanes and dirigibles. This tale, introducing the "giant monster attacks metropolitan city" theme in film, smacks of a later, Venusian dino-monster's exponential growth spurt in Henry Kuttner's 1940 story, "Beauty and the Beast," a sci-fi tale which may have in turn influenced Ray Harryhausen's 'Ymir' in *Twenty Million Miles to Earth* (1957). But, as already stated, what came before may not necessarily have caused something later to happen.

Besides sci-fi films, pseudo-dinosaurian ferocity was also fictionalized in literature, with many works appearing prior to 1945. Although several fictional examples were addressed in my *Dinosaurs in Fantastic Fiction* (2006), a French novel recently translated into English worth considering is Jules Lermina's aforementioned *L 'Effrayante aventure* ("*Panic in Paris,*"1913, English ed. 2009). Half a decade earlier, the April 1, 1905 ed. of the *Chicago Sunday Tribune* carried 'news' of a death struggle waged between a *Diplodocus* and *Tyrannosaurus* in downtown Chicago atop the old Montgomery Ward building tower on Michigan Avenue. (See *G-Fan* no. 112) Note the metaphorical 'conflict' between the BIG and FIERCE here. In the outcome of this savage battle, both monsters plummeted to their inevitable … *Extinction.*

So that familiar dino-monsters run amuck in a city, *fighting each other* theme was introduced over a century ago in literature. It would be another four decades, however, before the first pseudo-dinosaurian battle fought between a pairing of giant monsters (unleashed by the hydrogen bomb) in a major city (Osaka) would be dramatically staged on film, as witnessed in Toho's *Godzilla Raids Again* (1955). *Gigantis* triggered an onslaught of giant paleo-monster 'rallies' such as *King Kong vs. Godzilla* and *Ghidrah, the Three-headed Monster.* Today, in the modern cultish era of battling "*daikaiju,*" this apocalyptical theme has proven to be incredibly popular, both in film and literature. And it's really taken off considerably further since the day of those not so long ago issued *Dinosaurs Attack!* trading cards (Topps, 1988), invoking a parallel universe where humanity had not just one dinosaur to tangle with, but a ferocious saurian 'army' attacking out of time. Merkl suggests that a psychological urge to (symbolically) destroy New York City in comic strips (such as McCay's) or later in films (such as in *King Kong,* etc.*)* lay in underlying technophobic, luddite tendencies: rejection of urban life and its pollution. Hence, perhaps, movie shorts like *The Pet* and collectibles like those gruesome Topps trading cards.

While Robert Bakker's famous restoration of a sprinting *Deinonychus* (1969) heralded onset of the Dinosaur Renaissance, it was Michael Crichton's cunning raptors in his *Jurassic Park* (novel 1990; film 1993) that put those vicious *Velociraptors* on the pop-cultural map. In 2001,

Jurassic Park 3 featured a very cool sail-backed *Spinosaurus* proving more formidable than even *Tyrannosaurus*. As we know though, none of Crichton's bioengineered creatures should be dwelling in the present day. They already had their chance during the Cretaceous and 'failed.' Has nature gone awry?

"E" – *Extinct*!

People young and old alike, regardless of gender enjoy seeing "live" dinosaurs, fully restored, imbued with plausible movement; it's an enchantment. According to W. J. T. Mitchell, the dinosaur:

> "… is welcomed with open arms by the movies. From the first animated dinosaur feature (*Gertie the Dinosaur*) to *Jurassic Park*, the dinosaur stands at the cutting edge of animation. This is not … because the figures of dinosaurs are more complex or difficult to animate than … any other animal. It is because … no one has ever seen a dinosaur (or) … seen one move. … The cinematic image of the animated dinosaur shows us something that can be seen nowhere else…"[8]

After all, they're extinct, therefore harmless when restored. Right?

As alluded to above, in a well-known *Natural History* magazine column, Stephen J. Gould reiterated a psychologist's opinion that dinosaurs became popular because they're "big, fierce, and *extinct* – in other words, alluringly, but sufficiently safe."[9] (my italics) The idea of their extinction had already been allied to "monsters" of long ago, as exemplified in the title to the Rev. H. N. Hutchinson's popular 1893 book, *Extinct Monsters* (London: Chapman & Hall). Because dinosaurs are extinct, they may be considered safe. Really? Well not quite.

For perverse reasons that perhaps only psychologists could elucidate, humans enjoy witnessing, or *imagining* disaster on many levels. From large asteroids colliding with Earth, apocalyptical gamma ray bursts from space, to giant robots toppling city skyscrapers—by now it's all been graphically portrayed at the movies, and on the SYFY and History networks, for example. Of course what is emblazoned on our retinas from the television screen cannot physically hurt us ("It's only a movie; it's only a movie!"). We realize that there is only a small probability of such events actually happening someday. Yet in many cases the broadcast "disaster" is a metaphor or a 'hidden meaning' within the script representing something dreadful that is trending in reality which could harm us, or already is. One must simply pierce fiction's outer veil to reveal the real condition.

So does extinct really imply "safe' because 'they' can't get us in our comfy chairs? Well as in many of our prehistoric dino-monster films, the disturbing dilemma is of an anachronistic nature. The fact that living, in the flesh 'prehistoric' creatures have been resurrected and are existing with man signals a disquieting, perhaps apocalyptical imbalance in Nature. Such 'extinct' creatures should *not* be here, and their presence presages an unnatural pending sense of doom, angst. For example, think of the *Jurassic Park* series (i.e. Frankensteinian use of genetic power, overpopulation – also as in case of 1998's TriStar *Godzilla*) or *Gojira* and *The Beast From 20,000 Fathoms* (fear of the then often tested H-bomb), or *Godzilla vs. the Smog Monster* and *Gargantua* or the recent *Shin Godzilla* (uncontrolled ecological havoc).

Any intelligent discussion of life's history through geological time must eventually embrace the touchy topic of extinctions in the fossil record. Prior to our enchantment with the BIG-ness of popular paleo-monsters and their Ferocity, was mankind's absolute amazement with their former existences in past geological ages, so long before the modern era, as divulged by scientifically minded men such as Georges Cuvier (who dwelled on their catastrophic extinction), Gideon Mantell (who pondered the first truly popular dinosaur – *Iguanodon*) and others. During the earliest intellectual phase, stemming back two centuries ago, large extinct "reptilians" such as the marine ichthyosaurs and plesiosaurs and winged pterodactyls captured imaginations. As geological sciences matured, it became understood that there were discrete "periods" and "eras" in which certain extinct organisms lived and were 'restricted' to, beyond which traces of their once living forms were no longer recorded as fossils. Dinosaurs became recognized for their hallowed ranking in the immense scale of geological time.

Until the beginning of the Cold War, dinosaurs as known to the public were most often regarded as icons from the geological past. Thanks to startling displays in natural history museums and paleontology books written for laymen by paleoartists and scientists, on a pop-science level, apart from fantasy films and cartoons, dinosaurs were usually referred in reference to that Mesozoic Era spectrum of the geological time scale (which was constantly being refined, especially with radioactive age dating techniques). Dinosaurs also personified obsolescence: paleo-poster children of the Extinct. It was difficult to think of them without pausing to reflect upon when in the context of Time they once dwelled. To a large extent, their signifying 'purpose' (as *the* most popular set of extinct organisms felled by Nature), was to reflect on geological time and extinction – the sobering fact that it does indeed happen, and that it is pending for all life.

That time-honored 'procession of life' theme with iconic 'saurians' guiding through the mists of Time pervaded prominently throughout popular geology from the early 19th century on up through the mid-20th century. As I've shown previously (in *G-Fan* no. 98, Winter 2012), Godzilla's suit design was (perhaps principally) founded on a *T. rex* restoration featured in *the* 20th century's most famous (American) painted panorama showing life's progression through geological time (i.e. Rudolph F. Zallinger's "Age or Reptiles" mural at the Yale Peabody Museum). As outlined in Part One of my 2016 book, *Dinosaurs Ever Evolving,* thanks to popularized science, by the 1930s this geological evolution 'idea' was simply 'in the air' then, *zeitgeist.*

Dr. Yamane in *Gojira* theorized that Godzilla's origins lay in a discrete geological period, the Jurassic. So why was such a creature alive today? At least the traditionalized life through time theme more or less placed dinosaurians properly in geological time where they were supposed to stay. But what happens when they would step out of that 'place' … in horribly mutated forms?

"R" - *Radioactive*!

The aforementioned, *Panic in Paris* may represent the earliest 'cause & effect' merging in fiction of a radioactive substance ("vrilium") with menacing paleomonsters. A machine charged with this vrilium inadvertently bores through subterranean Paris, into Tertiary strata, where our heroes encounter frozen prehistoria from the Jurassic and Pleistocene. These readily thaw and attack the city, although there are few casualties. When Lermina's story was first serialized, 1910, civilization wasn't yet in a "cold war" period; radioactive bombs didn't exist. Also description of the attacking, revivified horde illustrates this was still a pre-*T.rex* period of paleo-fascination, when mammoths, megatheria, *Brontosaurus* and iguanodonts were pop-culturally regarded as 'equals.'

Four decades later, Cold War era hypothetical notions, later publicized by authorities such as Otto Schindewolf, Daniel Cohen, Isaac Asimov, Carl Sagan and Adrian J. Desmond, that radiation emanating from distant supernovas may have exterminated dinosaurs or other extinct organisms, bore no fruit scientifically. In the science fictional realm, however, radiation arguably did have significant ramifications when it came to resurrection and metaphorical destruction of dino-monsters. A roving radioactive dino-monster should be dead or totally 'zombified' due to the absorbed radiation. How Godzilla copes with such massive radiation doses accords well within the bounds of a (giant) *mysterious* beast, affirming its status as chief *daikaiju*.

On page 9 of the slightly revised, reprinted 1982 Preface to his superb *Keep Watching the Skies – 21st century edition* (2010), Bill Warren states that in 1950s sci-fi flicks, the radioactive gimmick wasn't significant. Warren claims:

> "... although the 1950s did tend to be worried about atomic warfare, radiation in SF films wasn't a means of expressing this fear, probably not even unconsciously. It was just a way of originating an unusual or interesting menace, or explaining one already conceived. Radiation was used to explain many wonderful things, from giant insects to walking trees to resurrecting the dead. This was not a form of nuclear paranoia, merely cheap and simple plotting."

My opinion? I think Warren is correct for most B-movie cheapies, he downplays radiation's significance on public paranoia in a number of other important films, but is wrong in the grim case of 1954's *Gojira,* where the H-bomb threat theme is pivotal, nearly palpable. Consider that memorable scene, for example, on Odo Island, when Dr. Yamane observes a crackling Geiger counter analysis within Godzilla's footprint; a radioactive trilobite—until then known only in the fossil record—is found embedded in one of these prints thus linking radiation with the inevitable. In Godzilla's visage, radiation is an unnatural manifestation of man's inhumanity.

I've addressed the Cold War period in two prior *G-Fan* articles (issues nos. 105 and 109). *Gojira* is a cinematic outcome of two nuclear strikes ending World War II. More directly though, Toho's masterpiece reflects what historically became known as the "Lucky Dragon incident" of 1954, resultant of US hydrogen bomb testing in the Pacific. The latter event fortified widespread fears of the Hydrogen bomb (much more powerful than any atomic bomb) and its potential for widespread radioactive fallout contamination.

Such fears of global annihilation were more subdued yet still reinforcing for American audiences in the 1956 version, *Godzilla, King of the Monsters*. Meanwhile, by then, Americans had become mindful of Russia's development of their own hydrogen bombs. Would a 'world war 3' suddenly erupt between the two rapidly emerging superpowers? Civilians were building bomb shelters, just in case. Not unlike Godzilla, itself a radioactive 'saurian' manifestation of the proliferating H-bomb threat, Dr. Serizawa's "oxygen destroyer" also becomes metaphorical for the hydrogen bomb. In total war, a super weapon is needed to counter the menacing product of other super weapons.

Paleontology is a highly visual science, and our enchantment with such imagery has fueled public fascination with dino-monsters, especially in the cinematic realm—movies. We may even discern a basic twofold outline in "paleoart," that revered, scientifically oriented practice of visually and realistically reconstructing extinct life forms, pertaining to (1.) a traditional life-through-geological-time edifying theme, then (2.) transforming into visual imagery conforming to a "dinosaur renaissance" phase (which I believe had its embryonic beginnings during the 1950s as geologists began reconsidering causes of mass extinctions – including volcanic eruptions and radiation-spewing supernovas). However, prehistoric animal (and dino-monster) images have always been introduced on many levels of media, purpose and sophistication, intended for a wide demographic. Yes, those plastic dinosaur toys and even your old 1964 Aurora Godzilla model kit, for example, would qualify as 'paleoimagery.'

So, when considering the more widely encompassing, pop-cultural arena of paleoimagery (of which paleoart is but the scientific subset), we must acknowledge another prominent branch, or 'trend,' spawned in the Cold War era. This influential category corresponds to dino-monster imagery reflecting extinction concerns, such as conveyed via early entries in the Godzilla movie series where exposure to H-bomb radiation proves foundational, and including other apocalyptical examples of 'Godzimagery' such films have inspired.

Conclusion:

Now let me take you back to our early 1960s living room in the Chicago burbs, where as a young lad I sat watching agog & amazed at my first viewing of *Godzilla, King of the Monsters* when it played on an afternoon movie program. Staring wild-eyed with utter fascination at this invincible, great-finned biped with the haunting roar that chilled my soul

…. a creature never to be reckoned with that was clearly …. Big, Fierce, Extinct, and yes certainly Radioactive ….

Then & now, coolest dino-monster ever!

Notes: (1.) Stephen J. Gould, "Dinomania," in *Dinosaur in a Haystack: Reflections in Natural History* (New York: Harmony Books), 1995; W. J. T. Mitchell, *The Last Dinosaur Book* (University of Chicago Press, 1998), pp. 9-14. (2.) Gerhard Meier, *African Dinosaurs Unearthed: The Tendaguru Expeditions* (2003, Indiana University Press). (3.) W. D. Matthew, "The Largest Known Dinosaur," *Scientific American*, Nov. 28, 1914, pp.443, 446-447. (4.) In the matter of Kong, Merkl states (p. 205), "Surely there are too many parallels here for this to be a coincidence." Also, in lamenting the "lowly status of the comic strip," (pp. 191, 251), he generally "… cannot prove it …." (p.129). See Ulrich Merkl's, *Dinomania*, 2015. (5.) Ronald Rainger, *An Agenda for Antiquity: Henry Fairfield Osborn* (The University of Alabama Press, 1991), p.158. (6.) Merkl, p.105. (7.) Merkl's quotes and perspectives are expressed on pp. 191, 205, 251. (8.) Mitchell, pp. 62-63. (9.) Gould, "Dinomania," p. 223.

FIVE

Josef Pallenberg's 25-foot tall statue of the Iguanodon in Hamburg, as shown on the cover of the April 8, 1911 Scientific American. See Chapter Twenty-one for more.

Statuesque Dinosaurs

Nearly four decades ago, on the first day of my new job, I noticed a dinosaur postcard pinned on the wall of a co-worker's cubicle. This was a card not then accounted for in the 'Debus collection,' nor one I'd seen before. It was a card showing a life-sized *T. rex* battling *Stegosaurus* on a rocky prominence, photographed in gory, reddish haze, as displayed at Disneyland's "Primeval World." It was a beacon-like card, beckoning fabulous vistas to my future, while linking to that co-worker's past—a fateful joining of world-lines. Yes, that card certainly proved enchanting for more reasons than one. Never underestimate the power of curious association. For that day I made the acquaintance of one, Ms. Diane Schlitz, and as the story goes, one thing led to another, and, well she—i.e. the person whose cubicle in which that postcard resided—later became my wife in late 1982!

Dinosaurs are perhaps most properly witnessed when they're restored as life-sized statues and sculptures. Surely, the largest of their clan is most strikingly perceived this way. I've written previously about such models usually arranged in theme parks, including several significant examples in 2002's *Paleoimagery*. Life-sized dinosaur statues and sculptures are being made by paleoartists and paleoartisans around the world at a pace never before anticipated, thanks both to the public's welcoming reception as well as a host of new chemicals and materials that facilitate such major undertakings. In the following chapters, favoring primarily what has come before, I chose to commemorate several of these works and their makers in three disconnected pieces, published in *Prehistoric Times* magazine.

In early 2000, my once co-editor—Gary Williams—with whom the fanzine *Dinosaur World* was published between 1997 and 2000, had contemplated writing a comprehensive book focusing on life-sized dinosaur statues and 'parks.' Unfortunately, this prospect fell through by the fall of that year, as I elected to move on from *DW*, and Gary and I parted ways. Regardless, a book solely on that topic, printed a decade and a half ago, would be hopelessly out of date today, as there are so many more examples of life-sized prehistoria to see in institutions and outdoor theme parks scattered around the world. Beyond those examples mentioned in *Paleoimagery*, for posterity and for those desiring additional info, here's a bit of an update.

> These chapters originally appeared as articles in the following publications:
> Chapter Nineteen—*Prehistoric Times* no.112, Winter 2015, pp.42-43.
> Chapter Twenty—*Prehistoric Times* no.80, Winter 2006, pp.52-53, 56.
> Chapter Twenty-one—*Prehistoric Times* no.106, Summer 2013, pp.41-44.

Chapter Nineteen—*Prehistoric Monster Memories*

How refreshing to learn of Chris Kastner's "Backyard Terrors" life-sized dino-sculptures, situated in Bluff City, Tennessee.[1] Here's someone involved with PT who not only builds remarkable dinosaur attractions, sharing his process, while reaching out to Facebook friends and readers of this magazine. What's more, he's talented, carrying on a fine artisan tradition—one that truly fascinated me half a century ago, and for reasons to be unveiled below still does.

So, let's begin during the late 1950s by when I had certainly known of those magnificent creatures known as dinosaurs. And so I probably also learned from somewhere that, allegedly, dinosaurs and other extinct beasts were 'only' to be encountered in museum displays – usually as skeletons, or as glossy Plates reproduced in books. Only? Then my Dad had related his own experiences at the 1933-34 Chicago World's Fair, with its famous "World A Million Years Ago" and Sinclair dinosaur exhibits. And from his "dinosaur scrapbooks" compiled during the mid-1930s, I was aware of the concrete dinosaur statues in Rapid City, South Dakota. A few years later I heard about yet another Sinclair display, although this time planned for a New York World's Fair scheduled for 1964. I wanted so badly to see this new Fair with its life-sized dinosaur restorations in person, but while my parents wouldn't travel to New York, my grandparents went (sniff!) without me.

At the Field Museum of Natural History I'd witnessed sculptor Frederick A. Blaschke's life-sized "Titanothere" and Dawn Horse displays, as well as Leon Pray's impressive life-sized sculptural rendering of *Andalgalornis* – well, nearly a "dinosaur." These were superb, only serving to whet my appetite for how truly stupendous a full-sized, colossal World's Fair 'brontosaur' sculpture would be—if I could ever see one. Meanwhile, I was also learning that dinosaurs could also be startling, mainstay roadside attractions. A true slice of Americana, that is, outside museum grounds— realm of 'true' paleoartists. If only I could manage to see some of these life-sized dino-statues … anywhere– supremely difficult when you're not quite an adolescent, relying on others to drive you there, anywhere!

Perhaps the first such, to me, large rudimentary representations of "dinosaur statues" I ever saw were displayed outside near a street corner at a company lot across from Old Orchard shopping mall in Skokie, Illinois. I vaguely recall they resided on what may have been some kind of a lumber company; a few wooden "dinosaurs" were constructed out of logs. One was bipedal and another was shaped or formed somewhat like a horned *Triceratops*. Yes—very crude looking constructs, but my grandparents who sometimes took me along when they shopped at Old Orchard would point out the "dinosaurs" every time we'd go there (during the late 1950s and early 1960s). I still recall noticing them on into the early 1970s after I'd received my driver's license, although by then, they were in dilapidated condition (then surely a refuge for carpenter ants).

Occasionally, my parents enjoyed traveling with my grandparents to the Wisconsin Dells area during the late summer and fall, and this is probably where I witnessed my first 'real', 'in-the-flesh' life-sized dinosaurs! Lo & behold! Along the Wisconsin River, across from the main drag, there was a dinosaur park known as "Prehistoric Land." The prize viewing was of a 70-foot long "Brontosaurus" sculpted in concrete by artist Dick Day (during the late 1950s). Alas, the dinosaurs disappeared in the fall of 1989, when the property was redeveloped into a mini-golf course.

From books my dad owned, by the early 1960s, I had seen photos of Josef Pallenberg's German, pre-World War 1 dinosaur statues. And of course by then, from Edwin H. Colbert's books I knew of Benjamin Waterhouse Hawkins' large, long-since antiquated 19th century prehistoric animal restorations for both the Crystal Palace Exhibition, still present on the grounds at Sydenham, as well as New York City's ill-fated "Palaeozoic Museum." But although our family had lived for many months at various times between 1959 and 1971 in Cambridge and London while my dad was on sabbatical, rather infuriatingly we never went to see Hawkins' Victorian dinosaurs.

So these and other visions seemed tantalizingly unobtainable, such as my reaction to a new arrival posted from the Calgary Zoo—a postcard showing "Dinny" a huge brontosaur statue erected in 1935, and photo slides of other cement dinosaurs residing there, documenting a segment of my grandparents' 1963 vacation trip. Then, ditto, another frustrating factoid entered my life from afar in the form of a brochure advertising a "Prehistoric Forest" situated in the Irish Hills of Michigan. Would I ever (be allowed to) see these magnificent replicas?[2] Gee—being a kid was getting tougher all the time.

But then at last I learned that *the* 64' Sinclair dinosaur models were traveling to certain destinations throughout the midwest during the summer of 1966. They made a stop in Chicago Heights where we drove to see them, the dinosaurs standing triumphantly atop flatbed truck trailers. Finally I was able to experience a small facet of a genuine World's Fair! A cast of the Sinclair "Trachodon" was sold to Brookfield Zoo, where it still resides.

During a trip to Grand Canyon Arizona and Utah in August 1972 with my grandfather, on our way to Bryce Canyon, day we were attracted to a roadside display of dinosaurs (not unlike those I'd seen in Chicago Heights). This was in Orderville, Utah, and the recreations displayed outside in the blazing summer heat were every bit as sensational and as "accurate-looking" as the 1964 Sinclair dinosaurs. As we hiked around admiring the fiberglass constructs I noticed an older proprietor on the premises, who as I later realized may have been the creatures' creator himself – Elbert Porter – who'd been laboring on this project since 1959. Years later, in 1991, I rediscovered these same dinosaurs outside the Utah Field House in Vernal, Utah, to where they'd been moved in 1977. On that latter occasion I was also able to purchase a wonderful, informative booklet written by Mr. Porter about the making of his dinosaurs titled *Dinosaurs Return* (1982) in the gift shop.

During college years, paleontology temporarily fled from my radar screen, in light of other scholarly (and some not so 'scholarly') pursuits. Interest in life-sized prehistoric animals remained dormant, the flickering flame re-lit briefly one evening in 1973 when I had a close encounter with remnants from the World A Million Years Ago exhibit in Chicago's Oldtown. A few years later in 1976, during a December trip to Los Angeles with a girlfriend, I reveled in the life-sized sculptural recreations of Pleistocene mammals in Hancock Park near the George C. Page Museum (La Brea Tar Pits) cast in high durability ornamental cement by sculptors Herman T. Beck and Joseph L. Roop during the 1930s, augmented by 1960s additions sculpted in fiberglass by Howard Ball. During this same trip I also witnessed more of Ball's magnificent 1960s handiwork—more dinosaurs formerly displayed at the New York World's Fair, then resituated to Disneyland's "Primeval World." Perhaps there was still hope: the 'thrill' evidently wasn't gone!

Well, after this interval, I obtained a copy of Don Glut's, to me— highly influential, *Dinosaur Scrapbook* (1980), with its emphasis on dinosaur parks and life-sized dinosaur recreations, several of which I later saw in person during the 1990s. But during my paleontological re-awakening of the late 1970s and early 1980s, I also visited the fiber-glass

"Perry Mastodon" sculptural restoration display for Wheaton College in Illinois. (Two decades later I found myself interviewing Richard Rush for *Dinosaur World* magazine.) Rush had managed not only the Perry Mastodon recreation, but had also directed a magnificent life-sized diorama of an anatomically correct *Tyrannosaurus rex* feasting upon a *Triceratops* corpse, placed on display in "The Third Planet – Earth," at the Milwaukee Public Museum in 1983. Then a few years later, along came those once ubiquitous, never-stop-motion Dinamation models!

The dinosaurs *had* indeed returned, and so had my profound interest in them. Since then I've been fortunate to marvel at many life-sized dino-recreation exhibits scattered around North America. Besides Glut's precocious *Dinosaur Scrapbook* focused on dinosaur popular culture, one may read more about these and other dinosaur parks in two 2002 McFarland publications, Appendix One to Glut's *Dinosaurs: The Encyclopedia – Supplement 2*, and in my own *Paleoimagery: The Evolution of Dinosaurs in Art*, especially Chapters 30 and 31. Evidently also afflicted with the dino-building bug, fellow minded Mr. Kastner's exertions completes the vicious, red-in-tooth and claw cycle.

Notes (1.): See Mr. Kastner's article in *Prehistoric Times* no.110, Summer 2014, pp.36-37. (2.) The 'spoiler' is that, yes, during the late 1990s we finally did see them, and eventually *other* amazing dino-constructs displayed in several western states—California, Nebraska, Utah, Colorado and Oregon.

Chapter Twenty—*Calgary's Prehistoric Zoo*

When it comes to dinosaurs, western Canada is a chief point of interest. Not only were their fossilized remains found there in spades during the early pioneering years of dinosaur collecting by stalwarts such as Charles H. Sternberg and Barnum Brown (who both wrote popular accounts about their Canadian dinosaur discoveries), but also in more recent years by paleontologists such as Phil Currie. One of western Canada's most popular paleontological attractions has for many years been Calgary Zoo's "Prehistoric Park." Surprisingly very little has been written about this popular park in the many dinosaur fan-zines and publications first emerging during the 1980s era of dino-love. And so for its history one would turn to the writings of Donald F. Glut, who enlightened us in his *Carbon Dates* (1999), *Dinosaurs: The Encyclopedia* (*Supplement 2*) (2002), and *The Dinosaur Scrapbook* (1980). Another book, David Spalding's *Into the Dinosaurs' Graveyard: Canadian Digs and Discoveries* (1999), offers a bit more on the Park's early displays as well as further details on the history of the Calgary Museum which featured fossil bones as far back as 1911.

No, I've never visited the Park, so I'm not an expert. But relatives have, and through their travels marvelous postcards and photographs arrived from afar. My grandparents' Prehistoric Park visit of 1963 aroused early interests after I received their postcard showing 'Dinny the Dinosaur,' which I still have. They wrote, "Dear Allen: We saw this monster this afternoon. He was over two times as high as your house..." By 1963 I had transformed into a major 9-year old dino-phile. But who wouldn't, knowing there were dinosaurs twice as high as my house roaming around out there in the northwest! Recently, I located my grandparents' old photographs and projector slides, stored in the basement. I'd never seen these images previously, showing Calgary's dinosaur sculptures including the *Stegosaurus* and a few others (shown here). More on those shortly.

Here, I plan to cautiously outline a history of the Park, admittedly and apologetically conducted at 'long-range.' But I'll also foreshadow an interesting development concerning the Park's primeval inhabitants. Lacking the Park's complete story and realizing that others were then possibly researching who might do the job more competently, I avoided the temptation of writing about Prehistoric Park in chapter 31 of my *Paleoimagery* volume. Now, at this belated juncture, I can fare at least reasonably in this digression—so here goes.

While the practice of completing life-sized dinosaur restorations for public display would appear to be a British-European tradition—considering Benjamin Waterhouse Hawkins' Crystal Palace dinosaurs and other exhibited Victorian antediluvians, and Josef Pallenberg's century-old dino-sculptures for Hagenbeck's Hamburg Zoo, for example—many of you wouldn't have known that the oldest dinosaur park of sculpted prehistoric animal restorations placed on *permanent* display in North America was Calgary's "Prehistoric Park," which even slightly antedates Rapid City's "Dinosaur Park"—the oldest such park built in the United States. However, there were other American challengers to the throne such as the life-sized, mechanized dinosaurs from the 1933/34 Chicago World's Fair, and Hawkins' never completed "Palaeozoic Museum," circa 1868 -1870. Today there are many such parks scattered around the world, many of which I've had the good fortune to see; an amply illustrated book really should be written about them all someday.

Collectively from Glut and Spalding, we learn that Prehistoric Park was inspired by Pallenberg's Hamburg dinosaur models, as, in 1932, Lars Willumsen, the Calgary Zoological Society's newly appointed director, convinced Society president Dr. O. H. Patrick that a similar park should be constructed in Calgary to memorialize Canada's rich fossil reserves. Then, sculptors and engineers were invited to work on the ambitious and to say the least 'mammoth' undertaking. Glut mentions sculptor Charlie Beil (1894-1976), engineer Arne Koskeleinen, and Finnish artist John Kanerva (b. 1883). Beil is most renown for his skilled bronzes of western scenes, sculptures of cowboys and plains indians. In 1973, Beil was honored with the highest medal which can be bestowed upon a civilian; "Member of Order of Canada" for his artistry. And in 1968, he received a Doctor of Laws from the University of Calgary. Of Koskeleinen who scaled-up Beil's miniature models to full-size, there is less readily available information; further information concerning Kanerva's contributions is coming to light. Paleontologists such as Charles M. Sternberg, Barnum Brown, Loris S. Russell, Charles Gilmore and W. E. Swinton offered technical expertise as to the dinosaur designs. In 1957 Swinton even proclaimed that Prehistoric Park's models were the most accurate figures of prehistoria yet sculpted.

Most likely owing to its gargantuan size, the final dinosaur statue of the *original* grouping was "Dinny," completed in 1935. Situated on St.

George's Island, Dinny was fashioned as a super-sized, 107-foot long and 35-foot-tall concrete *Apatosaurus*. It weighs a remarkable 120 tons! According to Glut, Dinny was unveiled to the public on August 9, 1935. Fifty-two years later, Dinny the Dinosaur would be officially designated as an "Alberta Historical Site.' Spalding differs, however, claiming that Dinny—declared the Park's official mascot in 1959—was instead introduced to the public in 1938. The first dinosaur statue to actually be constructed was a scaled-up *Chasmosaurus* with many to follow over the next two years. The prehistoric menagerie lineup would steadily increase over two decades, until circa 1957 (stated by Glut to be an approximate date of the Park's last 'original' addition), when the *first* dinosaurian wave then boasted a total of 50 statues.

An interesting short DVD made in 1958 is available from the National Film Board of Canada titled "Craftsmen Young and Old," (for $20.00) of which less than 4 minutes is devoted to Mr. Kanerva, "The Man Who Made Monsters." The black and white film shows grade school children touring the Park, while the aged-looking, cigar-smoking, short and pot-bellied Kanerva demonstrates how concrete is applied over wire mesh to flesh out dinosaur bodies, and then subsequently plaster as 'skin' texture is slathered over the concrete. The film was most likely made during the late 1950s, as the narrator states that "upwards of 30" life-sized statues had been completed, "after a quarter century." We also learn that Kanerva was inspired to build prehistoria during the 1920s after his interest and curiosity had been aroused from a "silent movie" involving prehistoric life. Before proposing to build the statues he researched by reading every book he could find on paleontology. Interestingly, the film shows one statue was at an in-progress stage; Kanerva lifts one happy child aboard an impressive *Uintatherium* model.

Besides glorious Dinny, Glut also published photographs of the original *Plateosaurus, Allosaurus* and *Styracosaurus*. To me, some of the lesser sculptures generally seem similar in quality to those of Wisconsin Dells' now defunct Enchanted Forest. But Dinny is spectacular, worthy of commemoration!

Calgary Zoo's Prehistoric Park opened to the public on August 25, 1937. Glut provides a listing of prehistoria originally placed on display in his *Dinosaur Scrapbook,* which I won't repeat here for sake of brevity. Originally, the display was not arranged in an obvious chronological sequence. But that matter was remedied in 1956 when, facilitated by the Alberta Society of Petroleum Geologists, edifying alterations were made, "...so that the visitors could trek through time. Forklifts and a Royal

Canadian Air Force crane rearranged the models, and they were repaired and repainted." (Quote from Spalding, p.106) Quite an undertaking; several must have cracked or fractured during movement. The ever-popular life-through-time portrayal now began in the Pennsylvanian coal age, extending through the Mesozoic and into the Pleistocene.

An old, undated card shows what sort of mischief younger zoo patrons would have gotten themselves into at Prehistoric Park. The card is actually a postcard-sized 'Hunting License,' authorized by the Ancient Order of the Most Grand and Phoney League of Dinosaur Hunters for 25 cents. A line entry was added to identify "'Next of kin,' in case 'Dinosaur catches you." On the front side three of the models are visible, including the aforementioned *Corythosaurus*.

A 1992 visit to Calgary by my geologist brother and his family piqued my curiosity concerning the park's history. This is because photos my brother took there didn't seem consistent with those I was more familiar with appearing in Glut's *Dinosaur Scrapbook*. Instead of the odd assortment of often quaint-looking, mid-20th century stylistic statues, I spied what appeared to be a number of Paul Jonas sculptures from the 1964/65 New York World's Fair. Glut had never mentioned these. What happened? So I wrote the Calgary Zoological Society, inquiring about Prehistoric Park's history and famous primeval denizens. Kevin Strange, Education Program Coordinator, kindly replied. He stated on 11/19/92:

> "Of the models constructed between 1935 and 1960 (author's note, i.e. - when, according to Glut (1980), John Kanerva retired), only 'Dinny the Brontosaurus' is still in existence. Currently, we have the following models on display.
>
> From Paul Jonas Studio, Hudson, NY - 5 models acquired in 1980; Originally from England. Purchased in bankruptcy sale & renovated by PML Exhibit Services of Calgary - 20 models acquired in 1984; Purchased from Topographics Ltd., Vancouver, British Columbia, Canada - *Centrosaurus*, acquired in 1983; Fabricated by in-house forces - *Pteranodon*, 1983; By PML Exhibit Services Ltd., Calgary - *Apatosaurus*, acquired in 1988.

> Note - PML has gone to strictly design and the fabrication division is now Matrix Ltd., in Salmon Arm, British Columbia. I hope this information is of use to you."

Now I understood. Following repainting and rearrangement of the Park models into their ordered life-through-time sequence in 1956, ground was later broken in 1978 for a new Prehistoric Park project. Then, evidently, a 'mass extinction' of the Beil/Kanerva/Koskeleinen 'fauna' happened. Despite Dr. Swinton's expert testimony as to the quality of the originals, by 1978 efforts were underway to upgrade the display, although this time with fiberglass models judged as more accurate; hence the succession of purchases as stated in Mr. Strange's letter. Two sections, east and west, were opened, respectively in 1983 and 1984.

It's a fair guess—yet lacking definitive information, *only* a guess—that the originals (through 1957), except for Dinny of course, may have been destroyed via sledgehammer to make way for the new ones. So, there's no clear record as to the disposition of the first-generation models, all of which were formed in cement and reinforcing iron. Hopefully several of the models—perhaps most likely the smaller prehistoria—were spared following their decommissioning. But if so, where could they be?

Which leads us to the most recent development. Evidently as of March 2006, a new documentary educational film was being prepared about John Kanerva's involvement in the creation of the original cast of Park prehistoria. I found out about this through an internet search, in which a rather heated 'palaeo-blog' exchange unfolded between an artist who considerately sought to commemorate John Kanerva's involvement with Prehistoric Park and his unrelenting grandson, W. Kanerva. While sparing you all the details (and I'll allow you to search on your own for this blog), it does seem as if the original concrete models were destroyed in 1983 and that one, Cal Abrahamson, "spared Dinny from the wrecking ball." Furthermore, John Kanerva's original miniature sketches, "scale-models made from plaster... and hand carved scale-models from a block of wood that he used to make his monsters," are still intact and will be featured in this new documentary by an unnamed producer. So, indeed, we look forward to this interesting production.[1]

As mentioned previously, while researching for this piece I decided to use the stamp postmark date of my 'Dinny the Dinosaur' postcard as a guide. Then I plunged into my grandparents' projector slide collection, and discovered seven photos they had taken at Prehistoric Park on that sunny August day so long ago. I can't reproduce all their photos here, but visible to me and for the first time, were 'new' shots of (1.) "Dinny" with children scrambling up the end of his very long, thick tail, (2.) a toothy *Tyrannosaurus* held within the confines of a chain-link fence, (3.) two *Dimetrodons* and an *Edaphosaurus*, (4.) a *Styracosaurus* with *Pachyrhinosaurus* standing behind, (5.) *Chasmosaurus* with sprawling front limbs and armored *Paleoscincus* in foreground with *Thescelosaurus* further down the pathway and the large concrete body of another non-descript dinosaur further along the path, (6.) *Plateosaurus* (with no signage visible), and (7.) a photo of my grandfather—smiling through all the years as if he just *knew* someday I'd find these pictures—standing next to *Corythosaurus.* Also, in a 1963 photo album, I found my grandfather's photograph of Prehistoric Park's *Stegosaurus*.

As far as visiting the Park today, well, I've seen those great Jonas statues in several locations around the country. *G-Fan* editor John D. Lees, who recently revisited Prehistoric Park, fondly remembers the old statues. "I was first at the zoo in about 1965.. so I saw all the original concrete dinos ... I was kind of disappointed to find them all gone this year ... I was quite attached to those old-fashioned guys and wanted my kids to see them. Dinny is the only one left... but he's off in a corner and hidden away..."

There's always a lead time for PT articles, so, it's possible that by the time you read this you may have already seen the projected Kanerva documentary. Alternatively, in case you haven't, maybe now you'll be inspired to learn more. It's quite a distance from Hanover Park, IL to Calgary, but if all the original Biel/Kanerva statues were still there I'd undoubtedly go.

Note (1.): In the 'palaeo-blog' exchange, W. Kanerva also criticized the Sinclair Refining Company, implying they had swiped Dinny the Dinosaur's symbol to use as their company 'mascot' to sell gasoline. However, just a quick checking of my memorabilia files led me to the original Sinclair Dinosaur Stamp Album, copyright 1935. On the 3rd page one can see a picture of Sinclair Opaline and Pennsylvania Motor Oil cans, both of which feature a sauropod on the label. My guess is that further checking would indicate that the Sinclair brontosaur logo would extend the usage of this symbol even further back. Furthermore— Sinclair had their own mechanized 'Brontosaurus' life-sized dinosaur on display at their Sinclair Dinosaur Exhibit at the Chicago 1933/34 World's Fair, predating Calgary's Dinny the Dinosaur by at least a couple of years. I also own three of Sinclair's souvenir newspapers available at the Fair, interestingly none of which clearly shows any product or advertising bearing their famous *Brontosaurus* logo. So when exactly was the Sinclair dinosaur symbol trademarked? According to Glut (1999, p.18), Messmore and Damon Inc.'s mechanized *Apatosaurus*, created in 1925 and also appearing at the 1933/34 Fair in The World A Million Years Ago exhibit, was nicknamed "Dinny."

Chapter Twenty-one—*Of Prehistoric (Sculpted) Creatures Great & Small*

I've long marveled about life-sized dinosaur models sculpted by Josef Pallenberg (1882-1946) for Carl Hagenbeck's Tierpark in Stellingen, Germany (which opened in 1907). Certainly, life-sized prehistoric animal models are no longer novelties today. In fact, increasingly, the newer sculptures are jazzed-up animatronic, interactive types (as opposed to static varieties cast in cold, heartless concrete). Unlike a century ago, paleosculptors and technicians have more materials with which to avail themselves, such as fiberglass, plastics and silicone rubber, than simply plaster or cement. Today, scientifically accurate dinosaur statues and models adorn both hallways and exterior landscapes of numerous natural history museums, whereas those of yesteryear—built with varying degrees of accuracy—were often created as roadside "tourist traps" or perhaps for special temporary exhibitions. One may find life-sized dinosaur models at many locations around North and South America, Europe, Cuba, Asian countries, Australia, the African continent, and – where this unusual practice originated – southern England. A century ago, there was only one such dinosaur "park," while now there are dozens and hundreds (perhaps thousands!) of individual life-sized prehistoric animal creations scattered around the globe.

Although largely unheralded today, Pallenberg's was the second such display and here I'll delve into how it was formerly promoted by journalists when it was viewed as more of a novelty, if not sensation. Why do Benjamin Waterhouse Hawkins' timeless yet "antediluvian" creations still exhibited at Sydenham remain celebrated today, whereas Pallenberg's seem mired in obscurity?

As an aside, Pallenberg's statuesque restorations weren't the only new life-sized, in-the-flesh prehistoric animals created between 1854 and 1909. For instance, Hawkins evidently did complete some of his "Palaeozoic Museum" sculptures at least in clay during the 1868 to 1871 period before they were demolished. During the late 1800s Henry Ward sold an impressive fuzzy Woolly Mammoth recreation, a replica of which appeared at the 1893 Columbian Exposition in Chicago. And a life-sized *Stegosaurus* restoration made from papier-mache' was sculpted for the 1904 St. Louis World's Fair Exposition, before its subsequent transfer to the Smithsonian. But indeed, Pallenberg's primeval creations constituted the world's second great outdoor dinosaur "park."

According to Dr. Claus Hagenbeck (correspondence dated 7/10/1990), "…the dinosaurs, originally designed in concrete by Josef Pallenberg in 1909 … have been reshaped and covered with fiberglass in the 1960s." A 1971 document enclosed with Hagenbeck's letter indicated there were nine life-sized prehistoric animal sculptures represented at the "Dinosaurier Tierpark." These included *Diplodocus, Allosaurus, Iguanodon, Stegosaurus, Ceratosaurus, Teleosaurus, Pareiosaurus, Plesiosaurus* and *Miolania.* Another enclosed sheet stated how Pallenberg studied under professors Rober, Spatz and Jansen at the Dusseldorf Academy and then returned to the Zoological Gardens as a sculptor. Here Pallenberg created the animal sculptures of living animals at the entrance of the zoo. And he modeled dinosaurs too. His reputation thereafter carried to South America, where, in La Plata, a Nature Park was founded for which Pallenberg labored to produce life-size primeval sculptures made of copper. It is unfortunate that the ruling government then did not provide means for completion of this work.

So where and how did the connection between extant animals and the considerably long-extinct possibly arise? The rather idiosyncratic decision to invest in life-sized dinosaur restorations rests with Carl Hagenbeck (1844 – 1913), the internationally renown wild life collector and trader who inaugurated the open-air zoological gardens at Stellingen, Germany. Carl Hagenbeck was intrigued by the pursuit of "unknown animals," and in the autobiographical *Beasts and Men* (English ed. 1909) he concisely summed up efforts to track perhaps the strangest attraction of all—yes dinosaurs then still allegedly living in Central Africa (e.g. "Rhodesia"). He financed expeditions to "the great swamps" to capture a "huge monster, half elephant, half dragon… seemingly akin to the brontosaurus." Inaugurating what has been dubbed the "Hagenbeck Dinosaur Controversy," he added, "I am almost convinced that some such reptile must still be in existence."

Although Hagenbeck's brief summary was "buried" within his book, the story caught on like wildfire, soon reaching the American Press by January 1910 where Professor W. D. Matthew and Dr. William Hornaday became commentators; the hoopla petered out by the spring of 1913. (For more on this episode see Dwight Smith and Gary S.

Mangiacopra's article "Carl Hagenbeck and the Rhodesian Dinosaurs," *Strange Magazine* no. 5, 1990.) Were the natives stoking the fires of imagination to a certain degree? Surely this sort of thing—deceiving the gullible whenever possible with prehistorical tales—has happened before.[1] What may have been the incentive? Maybe when the proposition arose of casting life-sized dinosaurs within his zoo—perhaps substituting as a 'consolation prize' for those specimens allegedly persisting in Central Africa but which defied capture, Hagenbeck simply beckoned to Pallenberg, willing him to make them "live." Or at least, 'putting two & two' conveniently together, that is my speculation. (I also wonder to what extent Arthur Conan Doyle may have known of this episode in natural history.)

Given that the dinosaurier portion of the Tierpark was the second attempt to create such a curiosity involving life-sized, in-the-flesh dinosaur restorations, and the first in which each dino-creation appears more or less "modern" (recognizing that Benjamin Waterhouse Hawkins' then 55-year old creations were once fully considered "scientifically accurate" according to contemporary standards), shouldn't Pallenberg's models have stirred quite a sensation? Why would so little generally be known about Tierpark's paleo-denizens today, say, when compared to Hawkins' monumental creations at Sydenham? Weren't these models more celebrated in their day? With some help I was able to find a sampling of early journalistic reporting (published in English) concerning the park's prehistoric denizens. Let's see what two journalists then had to say about it.

A "Mrs. C. R. Miller" composed an engaging story, "Monsters of Bygone Ages," concerning Stellingen's dinosaurs for *Leslie's Illustrated Weekly Newspaper* (Dec. 11, 1913, pp. 563, 573). Playing up the cryptozoological lost world-ish angle by way of introduction, she finally got to the point: "While these models are interesting, the average person longs to see them in the open among the trees and grass." She mentions that the "rather startling" models were sculpted from "plaster and burlap" (not concrete!), which of course wouldn't weather well in such an environment. In addition to the listing of prehistoric specimens on display cited earlier, in a published photograph one spies an *Ichthyosaurus* swimming in a lake off to the right rearward side of the bipedal *Iguanodon* statue standing on shore. She also mentions a "Tryannosaurus" (sic) displayed in the act of devouring a smaller animal. But this description may instead correspond to an *Allosaurus* posed in a hunched over-eating stance according to a Charles Knight painted restoration.

And Mrs. Miller refers to a certain publication only recently circulated in the pages of *The Strand Magazine,* claiming "The visitor may wander at will among these models, which are set up in such apparently natural positions that for a time it seems that one is standing on the great plateau described by A. Conan Doyle in his 'Lost World'." Conan Doyle's brilliant "Lost World" story had been serialized in *The Strand* from April to November, 1912. She states the models are intended to represent the Jurassic Period of 10 to 15 million years ago. (So if not a "Lost World" then perhaps the original "Jurassic Park"?) She further states the *Triceratops* family represents the Cretaceous Period (i.e. from 7 to 10 million years ago).However, in his lead-in to the story, the editor pronounces Barnum Brown's then recent discovery of *Monoclonius* as a creature thought ancestral to the *Triceratops* that lived a mere "three million years ago." Obviously the geological time scale hadn't yet been refined absolutely using radiometric data (and wouldn't be for another two decades).

However, the earliest article I've located concerning the dinosaur park appeared in a December 1910 issue of *The Strand Magazine*, titled "A Prehistoric Zoo" penned by Harold J. Shepstone. Shepstone (who carried the title of "F.R.G.S." which stands for Fellow of the Royal Geographical Society). Shepstone was a prolific journalist who wrote on a potpourri of topics for a variety of magazines. He also authored books with titles like *Wild Beasts of Today* (1931), *Polar Adventures* (1952), *Cavalcade of Ships* (1945), *A Modern Utopia* (1899), and *The Gum Lands of New Zealand* (1919), etc. And between 1910 and 1911 he also published two articles on Hagenbeck's Tierpark, the first chronologically of which was his enlightening "A Prehistoric Zoo."

Here we learn things otherwise lost in the shuffle of time (and turmoil of two world wars). For instance, Shepstone states, "In all, some thirty have been erected, but more are to follow, until we have a complete prehistoric zoo." Thirty is a greater number than Mrs. Miller accounted for in her article three years later. (Actually, she didn't directly provide a 'body count' of exhibited statues.) Shepstone specifically mentions, for instance "curious fin-backed lizards" on shoreline view (possibly *Dimetrodon* dating from the Permian Period), "prehistoric crocodiles" (probably *Teleosaurus*), a "sloth, dodo and mammoth" and 12-foot long tortoises and toads made "in stone," one of which may have been the Middle Triassic *Mastodonosaurus* a head of which was positioned breaking surface in the lake. He also states that the models have been cast in cement (not plaster and burlap).

Shepstone offers more complete documentation of how Pallenberg completed his work:

> "Fossil remains of many extinct beasts have been discovered, thus enabling man to reconstruct with marvelous accuracy the mighty beasts of bygone ages. Every care, too, has been taken to render the representations accurate. Before the sculptor commenced operations at the park he spent twelve months in preparatory work. He visited all the leading museums in Europe, including the one at South Kensington, consulted with leading naturalists, and made extensive drawings and sketches of the bones of those beasts which have been unearthed by the fossil-hunters. The American Museum of Natural History in New York rendered particularly valuable services in supplying drawings as well as measurements. Before work was actually commenced in the grounds models were built up in clay, casts taken of them and these were submitted to leading authorities for opinion. When it was found that they differed the models were remade and submitted again until they met with the desired approval. In this way the prehistoric animals at Stellingen may be regarded as scientifically correct."

So it seems as if Pallenberg went about his business most professionally and exactingly, much as did Benjamin Waterhouse Hawkins did half a century earlier.

Prominently and foremost on show around the lake were the 25-foot tall *Iguanodon*, then more correctly posed (relative to Hawkins' models at Sydenham) as a biped, and a 66-foot long *Diplodocus*, an 18-foot tall "duplicate" of a skeleton exhibited at the American Museum of Natural History (not the 84-foot long plaster specimen displayed in London). Shepstone also discussed Othniel Marsh's precocious opinions of a *Stegosaurus* with a double row of plates that was a dinosaur "trying to become a mammal." And *Plesiosaurus* was considered "half mammal and half fish." Rather than a restricted "Jurassic Park" ambiance, it seems Pallenberg intended to distinctively convey a life-through-time portrayal of prominent life forms evolving from our planetary history beginning with the Late Paleozoic into the recent Glacial Period (although emphasizing Mesozoic 'reptilian' forms). "All around, peeping out amid the shrubs and trees, standing by the water's edge and emerging from the lake itself, are wonderfully realistic and life-size models of those creatures that dwelt upon this globe from the days of the "thunder lizards" and beyond, down to the time of the mammoth, going back at least seven to ten million years." There were also representations of *Archaeopteryx* and giant dragon flies placed upon boulders outside the Insect House

Shepstone was truly impressed with the layout, concluding, "Altogether, it is a wonderful aggregation of prehistoric wild-life which Mr. Hagenbeck has constructed in his park. Then it is both valuable and instructive, as well as novel and picturesque. The animals have been so realistically conceived that one obtains, with a minimum of exercise and study, a faithful representation of those strange beasts, reptiles, and birds that dwelt upon this globe in the distant past."

Our next glimpse of the Tierpark came with Shepstone's April 8, 1911 *Scientific American* article, "Monsters of Bygone Ages: some remarkable restorations of extinct animals." This article generally repeats what was stated four months earlier in *The Strand*, yet we do learn a bit more about the park. Evidently, the *Iguanodon* was the first model sculpted by Pallenberg: *Diplodocus* came next. Also it is apparent that Shepstone conversed onsite directly with Pallenberg concerning how he designed and constructed the models. For instance, it is stated, "*He admitted to me* that he had obtained a great deal of help from the authorities of the American Museum of Natural History…" Furthermore, Shepstone "noticed in his office drawings as well as measurements. I noticed in his office, too, copies of the *Scientific American* containing articles and illustrations on the subject." Concerning those aforementioned finbacks, "These … have comparatively speaking small bodies and a curious erection down the center of their back like a frill. They are shown standing on the banks while a few are depicted swimming in the water."

And there is some clarification about the sculptures adorning a structure known as the Insect House, as "…the outside wall of this structure is composed of artificial boulders built up in the form of steps. Upon these have been placed huge flying creatures of uncouth form, and specimens of extinct birds." A photograph illustrating this article refers to a "Dactylosaurus" resting upon some boulders, although this is clearly the *Pteranodon*. It is reasonable to presume that the photos appearing in Shepstone's articles were taken by him; several are credited to him in his 1934 article, to be mentioned shortly.

Although Shepstone refers to thirty or so models displayed at the Tierpark, in photos published in the articles cited herein or in recent books, I've only seen photos of eleven genera represented as Pallenberg's sculptures. These include: *Allosaurus, Diplodocus, Pteranodon,* a *Triceratops* 'family', *Mastodonosaurus, Ceratosaurus* biting into *Stegosaurus, Ichthyosaurus, Plesiosaurus,* a 'croc-like creature' – possibly

Teleosaurus, and the *Iguanodon* statue which has a 'cameo' in the Max Fleischer 1923 silent film, *Evolution*. But I've never seen photos of his Mammoth, "fin-backs," or others rounding out the paleo-zoo 'herd' of two & a half-dozen witnessed by Shepstone.

Two decades later Shepstone wrote another article concerning paleontology in which he relied on five photos of Pallenberg's sculptures for illustration. This was a chapter titled "Big Game of Other Days" appearing in a book, *Wild Life of Our World* (Crossland and Parrish, eds., 1934). "Big Game" projects quite an ensemble of visually captivating paleosculpture! The chapter is basically about the history of ancient life with details concerning how fossils form and also how paleontologists find and reconstruct fossil vertebrates. (Unfortunately, even for its time it is fraught with technical errors.) Like many popularizers of his day, Shepstone was caught up in the "nature red in tooth and claw" aspect of prehistory, claiming "These ancient creatures were anything but a happy family and were continually at war." Shepstone seems to have been preferential toward paleosculpture, as opposed to painted restorations, for in this chapter he also relied on eight miniature, three-dimensional restorations sculpted by Vernon Edwards – whom we'll turn to now.

The *Wild Life of Our World* volume has long been in our home. It was one of my father's books, serving as my introduction to Pallenberg's artistry. And for many years I was also very intrigued by Vernon Edwards' sculptures, which were not credited to him in the book but instead to "Camerascopes-Mondiale." Decades flew by before I was able to learn more about this latter organization, eventually realizing that the sculptures were Professor Edwards' handiwork. Perhaps the best summary I've come across about these models appeared as a March 27, 1931 Press Release re-published in *The Milwaukee Sentinel* (4/5/1931, pp.14-15, 25), a story titled "Lifelike models of extinct monsters of bygone ages." Therein it states, "Scientists of the British Museum collaborate with a distinguished sculptor-scientist to reconstruct accurately many of the strange creatures which roamed the Earth in prehistoric times." These featured dioramas were created by the head of the British Museum's Geology Dept., Professor Vernon Edwards – who was the cited sculptor and a scientist. Edwards' mini-paleozo was among the first to be displayed inside a natural history museum.

Most of this article is about how Edwards went about his magnificent work. You must realize that during those pre-King Kong/Chicago World's Fair days, prehistoric dioramas (both life-sized as in Pallenberg's case, or in miniature form) remained a relatively rare commodity. Edwards instructed, "It is curious to note what a marvelous amount of information the Paleontologist is able to extract from a few ancient bones." A great number of fossil clues afforded him with suitable if not often subtle information "…which led me to attempt the reconstruction of scenes showing the animals in their natural surroundings…" So many fossils and "…other wonders beyond computation go to make the strange story written in the pages of time for the scientist to read and for the artist to record." Thereafter, as the article goes on to say, because "not everybody in Great Britain is able to make the journey to London to study (the models)… authorities have had photographic studies made of these scientific reconstructions of the huge beasts in their natural environment of prehistoric times. … photographic experts, Camerascopes, Ltd. … of London have photographed the series and mounted prints are supplied by them to schools, museums, scientific societies and individuals."

Anthony Beeson further states of that Camerascopes Ltd. "… also issued a series of twelve double stereoscopic cards to be viewed with a folding viewer. Double and joined stereoscopic cigarette cards of Edwards' models, together with commentaries by Swinton, again photographed by Camerascopes and designed for use with a viewer, had also been issued in 1930 by Cavenders Ltd. Of London as a set of twenty-four." (*Prehistoric Times* no. 117, Spring 2016, p.29) Perhaps the original means of achieving dinosaur 'virtual reality'?

Nine photos of Edwards' miniature prehistoric dioramas (five of which featured Cenozoic genera) appear in the 1931 article, a menagerie including *Arsinoitherium, Megatherium* (w/ paleo-indian), *Phenacodus, Uintatherium, Diplodocus, Stegosaurus, Machairodus, Triceratops* and *Ceratosaurus.* Edwards' dinosaur dioramas were also later highlighted in a 1934 book by William E. Swinton, *The Dinosaurs: A Short History of a Great Group of Extinct Reptiles*, including several not seen in the 1931 *Milwaukee Sentinel.* Additional genera shown in Swinton's splendid book included time-honored types such as armored "Scolosaurus," *Polacanthus, Scelidosaurus, Plateosaurus, Styracosaurus* and tree-hugging *Hypsilophodon.* In his day, Vernon Edwards was quite an accomplished and prolific sculptor.

Mike Howgate has identified ten much additional Edwards sculptures, in the 8-inch to two-foot length range, at Glasgow's Hunterian Museum (several of which seem highly reminiscent of Charles Whitney Gilmore's earlier sculptures). The menagerie includes *Diplodocus, Stegosaurus, Triceratops,* "Trachodon," *Ceratosaurus, Polacanthus, Dimetrodon, Hypsilophodon, Brachyceratops,* and *Pareisaurus.* Plaster reproductions were once available for purchase through a Thomas Murby & Co. Leaflet.

So why was Pallenberg lost in time relative to Hawkins, and the Stellingen models less known or celebrated today than those at Sydenham? I'm not providing a satisfying answer here. We are reading articles printed in English and news of the German park was probably mostly disseminated and documented in German. In fact, there are probably more such articles buried in the periodical stacks (including those printed in German or other foreign languages; I simply worked here with materials that were readily accessible). A mild cultural gap. But also, I am not certain to what extent the park was impacted by two world wars. Also, I am interested to know what became of Professor Edwards prehistoric animal sculptures. And so this story shall continue, when information and time permits.

(With thanks to Gary Williams, David Goldman and Karl Debus for facilitating copies of three of the aforementioned articles.)

Note (1.) See "The Truth *IS* Out there: On the trail of 'living' dinosaurs," by Allen A. Debus, Chapter 15, pp. 280-292, in *Dinosaur Memories* 2002.

SIX

FIG. 13.—Skeleton of Megalosaurus, restored. (After Meyer.)

(Top) An 1893 skeletal reconstruction of a sail-backed, or possibly humped Megalosaurus appearing in Hutchinson's Extinct Monsters. (Bottom) "Bird or Dinosaur?" Jack Arata's illustration exemplifying the confusing concept of what is a dinosaur versus an avian? (Original art made for hire in 1997, no copyright stated, A. Debus personal collection). See Chapters Twenty-two and Twenty-three for more.

Unsung Paleo-Monsters

What appeal could the prehistoric past possibly hold without its monsters, both factual and fictionalized – construed from 'alternative facts?' I've written quite a number of pieces through the decades concerning prehistoric "monsters" for several fanzines, and of course I've even titled one of my books *Prehistoric Monsters*. What follows is an eclectic set favoring what may be considered a 'second string' of such creatures, those perhaps less recognizable to the younger paleoenthusiasts of today. In the pop-cultural showcase world of prehistoric monster-dom, however, don't all prehistoria merit their fair share of intrigue? And so here they are again – back from a million years ago (or more), submitted for your approval.

First, a quick backstory on the third segment in Chapter Twenty-two, this piece stemming from a visit Lynne Clos (then editor of *Fossil News*) and I with her family made to the Museum of Science and Industry in Chicago. By the early 2000s I regarded myself as more or less an 'expert' on older paleo-restorations, as I thought I'd seen most or nearly all of them. Imagine my elation when I saw the 1884 'old-stego' restoration in a museum display case mentioned in this chapter! Occasionally, fresh discoveries simply fall into one's lap, and so I thought this might be significant to report on in a forthcoming issue of *Prehistoric Times*. In a follow-up issue of Frederick's magazine—letters column—a careful reader, Brad McFeeters, mentioned that this same image had been reproduced in David Norman's and Angela Milner's 1989 book, *Eyewitness Books: Dinosaur*, a volume until then unbeknownst to me. I suppose I was only the *second* person to discover the restoration, and knowing it for what it was, I wrote in the next issue of *Prehistoric Times* (no. 70, Feb/March 2005, p.5)—letters column:

"I suppose my enthusiasm just got the better of me after I tracked down the original article because of my more than casual interest in stegosaurids. Also interesting (to relate) is that after reading my article, Don Glut informed me in late Dec. 2004 how several decades ago an acquaintance (of his) mailed him a copy of the 1884 restoration, but then, without having the reference article handy, he never had any idea what that dinosaur was supposed to be. Up through today, it's not a picture I'd seen published anywhere else except in the 1884 *Scientific American* issue. Thanks for that info, Brad ... even though you took the wind out of my sails. Although I can't devote further attention to this matter, I still wonder whether my speculative reinterpretation of Marsh's 1877 paper would be supportable ... (i.e. - as stated in Chapter Twenty-two) ..."

Also thanks to linguistics expert, Carl Masthay, who confirmed for me the nearly illegible spelling as it appears within that 'old-stego' image: the identifying artist "A. Tobin" and the engraver, "Vermorekin." I had slightly misspelled their names originally. Lately, I've seen the 'old-stego' image posted on Facebook, although those downloading usually haven't identified it properly!

The *Agathaumas-Monoclonius* portion of Chapter Twenty-two was requested by the editor of *Prehistoric Times*—Mike Fredericks. He wanted an article focusing on a vintage dinosaur for issue no. 101 which was a special issue featuring Charles R. Knight. In writing this, I parlayed my interest in Knight, coupled with his magnificent 1897 restoration, with an opportunity to further advertise the resin *Agathaumas* sculpture I'd made a few years earlier. My chance (re-)perusal of an old paleontology book written by the Rev. H. N. Hutchinson led me to the 'humped megalosaur' concluding segment of Chapter Twenty-two.

Alas—there are many additional images accompanying the original fanzine articles that I could not reproduce here.

Chapter Twenty-three originally appeared in a fanzine that I once co-edited (along with Gary Williams and John Lanzendorf) named *Dinosaur World* (with nine 'meaty' issues printed between early 1997 and early 2000). This chapter's content, concerning Mesozoic birds, is no longer technically current but it's included here as a snapshot, a temporal cross-section. Readers can see how quickly and in what a state of flux was immediate furor over the 1996 discovery of Chinese feathered dinosaur specimens, announced in October 1996—just as *Dinosaur World* issue no. 1 was going to press. I began writing the piece in December 1996, although the article was printed in our second issue. Interestingly, it now seems that in the evolution of avian flight, vertebrates early on also experimented with a "tetrapteryx" (four-winged, the two 'extra' being on hindlimbs) stage of theropodous flight—exemplified, for instance, through the Early Cretaceous Chinese genus, *Microraptor*. Furthermore, the recent bird-dinosaur link is more strongly bonded through discovery and description of the emu-like *Corythoraptor*—the "most bird-like" of all dinosaurs, as reported in the popular (online) press on July 28, 2017. Yes, so much has evolved in the "dinosaur world" since the fall of 1996!

> These chapters originally appeared as articles in the following publications:
> Chapter Twenty-two—*Prehistoric Times* (nos. 68, 101, 104), respectively, Oct./Nov. 2004, p.52; Spring 2012, pp.46-47; Winter 2013, pp.41-43.
> Chapter Twenty-three—*Dinosaur World* no. 2, June 1997, pp.10-14, 42.

Chapter Twenty-two—*A Trio of Unlikely Prehistoria*

'Stego-record' breaker! The Oldest Known Stegosaur Restoration

In early August 2004 I made an interesting paleontological discovery at, of all places, Chicago's Museum of Science and Industry. Here, I came across the oldest restoration of a *Stegosaurus* (recently) identified as such. Behind a glass exhibit case I spied a quaint, old illustration published in a copy of the November 29, 1884 *Scientific American* which rather fortuitously was opened to page 343. There was little time to read the page as the museum was closing shortly. Eager to learn more about this restoration, which I didn't recall ever seeing before, I later found the article on microfilm. The article featured news about *Iguanodon* and *Brontosaurus*. A short caption to the illustration in question (Fig. 3 in the two page 1884 article) read, "American landscape of the Jurassic Epoch with reptiles and plants of the period." The artist's signed name was 'A. Tobin': the engraver was 'Vermorekin.' It was the bipedal animal at foreground which seized my attention, as it appeared to be an upright sauropodous creature, outfitted with many rows of spines. What hath nature wrought?

Further explanation was provided on page 344 of the *Scientific American* issue, which I obtained days later. (This 'next' page wasn't displayed behind the Museum of Science and Industry exhibit case.) The article had no byline. Now poised intently before the microfilm screen, I was startled to read:

"If through the admirable discoveries that have been made in recent years, we endeavor to bring to life again the fauna of the Upper Jurassic period in the United States, we shall find one that is no less rich and strange than that of the Old World. Here we have, amid araucaria and cycads, the gigantic *stegosaurus*, with a body clothed with bony plates and spines that formed a powerful armor for it, and with forelegs much shorter than the hind ones; the compsonotus (sic), with forepaws equally as well developed as the hind ones; and the strange flying reptiles, the pterodactyls (Fig. 3)."

The next paragraph in the article goes on to describe the "Brontosaurus." Which refers to a Othniel Marsh's skeletal reconstruction, labeled as Fig 4 in the article. So evidently, the upright, spined 'sauropod' in Fig. 3 was indeed a *Stegosaurus*! Well, I've already had the pleasure of identifying what seemed at the time to be the oldest *Stegosaurus* restoration of all—one by Carl Dahlgren, published in an April 1892 issue of *California Illustrated Magazine*, that I came across in 1991 and first referenced in my article "Historical Dinosaurs: Episodes in Discovery and Restoration." (*Earth Sciences History* vol.12, no.1, 1993, pp.60-69). A decade later I reproduced an image of this stegosaur—referred to as "Hypsirophus" (now known to be synonymous with *Stegosaurus*) on page 74 of my 2002 book, *Paleoimagery: The Evolution of Dinosaurs in Art*. But surprisingly, now, or so ti seemed, I'd broken my own 'record' in the category of 'oldest found stegosaur restoration,' and by seven years, no less!

The *Scientific American* article was printed only seven years after Othniel Marsh described *Stegosaurus armatus* in *American Journal of Science* (3rd series, vol. XIV, no. 84, Dec. 1877, pp.34-35). Marsh's 1877 description was not illustrated, but he speculated as to *Stegosaurus*' life appearance, claiming, "The limb bones indicate an aquatic life … The body was long, and protected by large bony plates, somewhat like those of Atlantochelys (Protostega). These plates appear to have been in part supported by the elongated neural spines of the vertebrae." Then he continued, "The present species was probably thirty feet long, and moved mainly by swimming." Marsh acknowledged that *Stegosaurus*' bones were found in association with a sauropod, "… near the locality of the gigantic Atlantosaurus montanus, and in essentially the same horizon." Immediately after his published description of the *Stegosaurus* in this *American Journal of Science* issue, Marsh went on to describe the sauropod genus, "Altantosaurus."

I wonder how and why this curious 1884 restoration resulted. The animal restored in Fig. 3 is adorned with five rows of spines from the shoulder area to the hips, where the longest vertebral row yields to 18 or 19 dermal plates. Marsh evidently was confused by *Stegosaurus*' "teeth," several of which, "… are cylindrical, and were placed in rows … are especially numerous, and may possibly turn out to be dermal spines." Although the 1884 *Stegosaurus* restoration has large 'plates' over its caudal tail vertebrae that are noted as plates in *Scientific American* text, they appear more as enlarged crocodilian 'scutes.' Marsh stated that "one of the large dermal plates was over three feet (one meter) in length." From rough proportions, it is certainly possible that several of the most anteriorly positioned plates on the 1884 restoration approached that size.

Still, there are far too many spines, all placed in the wrong position in the 1884 restoration, none at the end of the tail. Possibly those 'extra' teeth gave someone an idea that there were spines of varying sizes situated all over the body, sort of how British paleoartist Benjamin Waterhouse Hawkins restored the Crystal Palace quadrupedal *Hylaeosaurus* in 1853 under Sir Richard Owen's guidance. Then too, close proximity of sauropod bones may have resulted in the restored animal's elongated neck.

According to dinosaur bone collector Arthur Lakes (1844 -1917), when the block of *Stegosaurus armatus* was pried open, "We broke open the block in which it lay and exposed twelve long black enamelled spines … There was a pair or two sets of these spines side by side a piece with two small hour glass shaped bones close to them."[1] Because several fossils found in conjunction with the *Stegosaurus* remains turned out to be *Diplodocus* teeth and limb bones. I wonder whether some of these spines may have been sauropod vertebral spines. In 1992, Stephen Czerkas suggested that conical spines were arranged in a row along the tail, neck and body of sauropods like *Diplodocus* and *Barosaurus*. But in 1877, Marsh had apparently already noted these, although he may have misinterpreted their nature. So, Marsh may have had a mixed bag of sauropod teeth, spines and stegosaurid spikes, which he perceived collectively as 'stegosaurian armor.' (Incidentally, the 1884 restoration also ranks as one of the oldest known showing a 'sauropod' in bipedal pose. See "About the Cover," here.)

The *Scientific American Stegosaurus* restoration of 1884 seems to be a unique sauropod-stegosaur hybrid. Much of it was guesswork, and today we can only wonder who lent technical expertise.

"Agathaumas –Monoclonius"

A familiar Charles R. Knight restoration obtainable as a print from the American Museum's photographic collection (negative # 322527) bears the description "Agathaumas (Monoclonius)." This brief caption reflects a long standing state of confusion. *Agathaumas'* restored, pictorial appearance is nearly as fanciful as Godzilla's, yet the former is exalted as a "real" dinosaur owning an entry in Glut's *Dinosaurs: The Encyclopedia* (1997). *Agathaumas sylvestris* is a name assigned by Edward D. Cope on basis of fossilized ceratopsian bone discovered in Wyoming (Lance formation). Let's briefly consider how *Agathaumas* rose from the ashes of deep time only to slip into scientific backwater, along the way fulfilling an ideologically important role beyond its celebrated performance on the silver screen.

Cope was elated with his guesstimated 30-foot long, 6- ton dinosaur, a "… species …no doubt equal in dimensions to the largest known terrestrial saurians or mammals." Charles R. Knight, who in 1897 famously conferred with Cope on this genus, is responsible for boldly creating such a beautiful painting of this animal. With only fossil scrap available from outcrops in southwestern Wyoming (16 vertebrae – tail, backbone, pelvis, ilia, a few rib bones & other scrap), the beast proved mysterious, despite the "wondrous" impression it created in Cope's fertile mind, as conveyed to North America's greatest paleoartist. Why did their restoration have to be so beautiful?! But what a magnificent looking, "horn-a-plenty" face that *Agathaumas* sported! A head not based on a single shred of documented fossil evidence.

The first essay best illuminating *Agathaumas'* problematic peculiarities appeared in Glut's *The Dinosaur Scrapbook* (1980, pp. 276-281). By the early 20th century, relatively few dinosaurs had vaulted into pop-culture as science fictional, horror monsters. However, despite its dubious scientific nature, *Agathaumas* became one of our very first "true" dino-monsters, perpetuating its popularity. For many paleophiles recognized the likeness of Knight's dinosaur—nearly fanciful as a unicorn—in a sculpture made by Marcel Delgado and animated by Willis O'Brien for 1925's *The Lost World*. Arguably, if not for *The Lost World* and conflated pop-cultural morphing with *Monoclonius* - another one of Cope's named dinosaurs, once popular *Agathaumas* would have rather quickly faded from the mainstream. In fact, its name would wane long before its imposing appearance did. Rightfully relegated to questionable status by professional paleontologists, Knight's fanciful restoration generally was omitted from geology and paleo-textbooks. Today however, that is, for many sci-fi enthusiasts, *Agathaumas* remains an unforgotten, if non-celebrated curiosity.

Even before Knight's fateful meetings with Cope—resulting in Knight's small statue and the most famous painted *Agathaumas* restoration, a peculiar reconstruction of this genus appeared in a popular 1892 article by James Erwin Culver. The animal depicted in the article appears as a nondescript uintathere, although Culver clearly intended to describe a three-horned dinosaurian, not a mammal.

Following restoration of Marsh's *Triceratops*, with its imposing "three-horn" skull, Cope had reason to reconsider his earlier evaluations of material assigned to *Agathaumas* and *Monoclonius*. Glut (1980, p.277) reproduced Cope's outline drawing of a very 'triceratopsian-looking' dinosaur drawn by Cope; the name "Agathaumas" is handwritten on the original sketch. Cope also drew the first outline restoration of *Monoclonius* "*sphenoceras*" (sic), replete with some postcranial body armor scutes. Cope had described this species in 1889. This sketch, similarly lacking in detail yet more similar to Knight's subsequent restoration was reproduced as Fig. 23 in H. F. Osborn's *Cope: Master Naturalist* (1931). The *Monoclonius* head has a long, straight nasal horn, with a shorter horn shown laterally over an eye. The frill collar is smooth, not spiky – looking. (Cope's pictures were probably dated circa or post-1890.)

Under Cope's tutelage, the mental morphing process resulted in Knight's small sculpture and painting of the *Agathaumas*. Knight's painting was reproduced in an 1897 Century Magazine article by William H. Ballou, where he described two presumed species of *Agathaumas*. Ballou stated that "… one of the species *Agathaumas sylvestris* (Cope), is distinguished by its nose-horn pointing forward; another *Agathaumas sphenocerus* (Cope), by its nose-horn pointing straight upward." Furthermore, Knight's restoration of the *Agathaumas sphenocerus* (i.e. not *sylvestris*) "…is based on Professor O. C. Marsh's prior reconstruction of *Triceratops prorsus*." Cope's *Monoclonius* sketch (and therefore Knight's) restorations were fortified with dorsal armor, perhaps borrowing (however exaggerated) from J. Smit's 1893 restoration of *Triceratops prorsus* (as published in editions of H. N. Hutchinson's *Extinct Monsters*). By 1897, Cope also had reduced the size of his "elephantine" *Agathaumas* from 30 feet to 12 -14 feet long.

Previously, Cope assigned the name, *Monoclonius* (*crassus*) in 1876, and erected three more species in 1889, including "*M. sphenocerus*," which indeed possessed a straight, tall nasal horn. Essentially then, a species name attributed to *Agathaumas* in *Century Magazine* was taken from the *Monoclonius* part of the "chimera," while the generic name was as assigned to the scrappy collection of fossil bones discovered in 1872. Cope may have intended Knight's 1897 statue and painting to be a melding of *Monoclonius* species.

By the 1910s, most paleontology texts and popular works had dispensed with references to *Agathaumas*, opting instead to feature *Triceratops* as the exemplary horned dinosaur, which by then was far better known thanks to O. C. Marsh and coworker John Bell Hatcher. Furthermore, during the 1910s, Barnum Brown exhumed additional skeletal material belonging to another horned dinosaur, *Monoclonius*, a skeletal reconstruction of which was illustrated expertly by Erwin S. Christman in 1916. As Glut noted in 1980, while Hatcher and Richard Swann Lull (1907) considered *Agathaumas* to be "transitional" between *Monoclonius* and *Triceratops*, others believed *Monoclonius* was congeneric with headless *Agathaumas*.

In 1900, one of the earliest copycat restorations of Knight's *Agathaumas* appeared in an article by Rev. H. N. Hutchinson. The "clever illustration" by Lawson Wood (dated 1900), was founded upon Knight's "life-like plaster cast" as depicted in *Nature*, 8/25/1898. Except Hutchinson's description in the caption refers the animal restored by Wood to *Triceratops*. Thus began a trend where chosen aspects and features of the "appearance" of Knight's 3-horned *Agathaumas* became "drafted" or incorporated into alternative pictorial usage, yet defined in captions as "belonging" to other horned dinosaurs, usually *Monoclonius* or *Triceratops*. In a collectible series of colorful cards, "*Tiere der Urwelt*," issued in the early 1900s (e.g. circa 1902 to 1906), issued by the Kakao Company of Hamburg-Wandsbek, artist F. Long copied Knight's *Agathaumas*, although a caption refers to the beast as "*Triceratops prorsus*." Then a decade later, paleoartist Joseph Pallenberg outfitted his life-sized, fully 3-horned *Triceratops* restorations for Germany's Hagenbeck Tiergarten with spiny looking, *Agathaumas*-like frills and considerable scutey-looking, body armor.

But whenever necessary, Knight's 1897 imaginary *Agathaumas* could also magically morph into *Monoclonius*, as we see in a caption written within Barnum Brown's May, 1919 *National Geographic* article, "Hunting Big Game of Other Days." Even a decade after release of First National's *The Lost World* (1925), Knight's *Agathaumas* enjoyed further reproduction in the guise of "Monoclonius," as drawn in 1935 by V. T. Hamlin for *Dinny's Family Album*. Ray Harryhausen vivified an *Agathaumas* – like creature for his 1940 stop-motion film, *Evolution*. Besides examples shown in Glut's 1980 essay, other "Agathaumas (Monoclonius)" appearances include a 1936 chocolate trading card (see PT #67, p.26), in photos of stop-motion "puppets" intended for an aborted film - circa 1939's *The Lost Atlantis*, as a "fleshy" home movie puppet (see *Prehistoric Times* no. 78, p.7), and a 1990s toy created by Jary Lesser based on Delgado's 1925 handiwork. And then there's a miniature resin rendition of the *Agathaumas* (Monoclonius) sculpted in 2002 by yours truly, a limited edition once available for purchase.

Throughout the mid-20th century, thanks to impressive life restorations such as by Zdenek Burian and Richard Swann Lull (a late 1920s "skeletal half-restoration" mount for Yale's Peabody Museum), Lois Darling and a handful of others founded on Christman's on-paper skeletal reconstruction, *Monoclonius* became a more celebrated denizen of the Cretaceous world than the "real" *Agathaumas*. Meanwhile, in behind-the-scenes fashion, paleontologists were demonstrating that a similar yet distinct genus, *Centrosaurus*, was understood to be a far more common contemporary than *Monoclonius*. Peter Dodson expertly outlined this transitional episode in his excellent, *The Horned Dinosaurs: A Natural History* (1996).

The name "*Agathaumas*" (incorrectly spelled "Egothamus" in context of the 1925 film) was generally ignored during the latter 20th century and into the early 21st. Paleontologists now regard influential *Agathaumas* as probably *Triceratops* or *Torosaurus* instead. The former name is considered a *nomen dubium* due to a lack of suitably descriptive characters assignable to its fossils. However, once gazing upon *Agathaumas*' magnificently spiky face, many dino-aficionados will be forever smitten, favoring its distinctively 'Knightian' allure & charisma.

Over time, *Agathaumas*' popularity and scientific significance waned. However, as related in 1982 and 1994 by Brent H. Breithaupt, *Agathaumas*' fossils did indeed become instrumental in aiding Cope's eventual confirmed determination that the associated sediments (named "Lance" in 1950) were of Late Cretaceous rather than Eocene age (i.e. replacing the term, "Laramie" for these rocks). An unheralded legacy.[2]

A hump-backed, high-spined Megalosaurus

When it comes to dinosaurs, especially those we've become so familiar with like the African genus *Spinosaurus*, elongated spines forming a prominent "fin" or sail seem cool (which may have been their purpose, partially; i.e. to cool!) whereas at least in certain cases an incongruous hump would seem (with apologies to Lon Chaney Sr.) downright Quasimodo-like. Particularly, since early phases of the dinosaur renaissance, we've been accustomed to so many startling life restorations of *Spinosaurus* that we unquestionably accept how it absolutely had a prominent sail, right? I agree, although many of you may not be aware of paleontologist Jack Bowman

Bailey's researches. During the mid-1990s Bailey reconsidered consensus opinion, opting instead for a "buffalo-back" interpretation. (Jack Bowman Bailey, "Neural Spine Elongation in Dinosaurs: Sailbacks or Buffalobacks," *Journal of Paleontology*, vol. 71, no.6, 1997, pp.1124-1146.) I interviewed Dr. Bailey concerning this paper ("Dinosaur World interviews Dr. Jack Bowman Bailey of Western Illinois University," in *Dinosaur World*, no.5, Summer/Fall 1998, pp.65-71). Accordingly, I even sculpted a humped *Spinosaurus* for him (a photograph of which was published in Donald F. Glut's *Dinosaurs: The Encyclopedia – Supplement 1*, McFarland, 2000, p.332.) Other "high-spined" theropod genera have come to the fore lately as well, including *Suchomimus* now considered a second species of *Baryonyx*, as well as newcomers *Concavenator* and *Ichthyovenator* – both also adorned with curious spinous projections. But what about a real oldtimer? Yes—our very "first" theropod, *Megalosaurus*, named and described in 1822/1824.[3]

We're all familiar with those mid-19th century, Victorian restorations of *Megalosaurus* in quadrupedal pose, made famous by paleoartists such as Benjamin Waterhouse Hawkins. But, no, this article isn't about whether *Megalosaurus* had bipedal vs. quadrupedal posture. The bipedal interpretation was more or less established by the 1880s, particularly following O. C. Marsh's analyses of dinosaurs judged similar to *Megalosaurus* such as *Ceratosaurus and Allosaurus*, as well as contemporary anatomical reconsiderations of the ornithischian *Iguanodon*. By 1893, Richard Lydekker considered *Megalosaurus* and *Ceratosaurus* to be "generically inseparable." Furthermore, he opined, perhaps *Ceratosaurus* should be instead reclassified as *Megalosaurus*! Meanwhile, considerations that (bipedal) theropods, presumably including *Megalosaurus* (as established in Arthur Conan Doyle's *The Lost World* (1912), in the chapter titled "It was Dreadful in the Forest"), hopped like kangaroos persisted among some scientific circles and reigned in popular culture until the early 1930s.

But, no, this article isn't about Megalosaurus' anatomical stance nor whether it displayed a springing, kangaroo-type locomotion like Cope's "Laelaps," but rather about its presumed "hump"! Hump? Yes, hump, although I'm also not going to debate presumed physiological merits of humped vs. finback functional anatomy. I'm merely going to probe into where the humped interpretation may have begun and how it may have perpetuated. In the most recent volume of Glut's *Dinosaurs: The Encyclopedia* (Supplement 7, 2012, p. 506) a silhouetted skeletal restoration credited to R. B. Benson, shows *Megalosaurus* more or less as a "generic"-

looking theropod, without a hump or elevated spinous processes – reflecting modern opinion. But that isn't entirely how it used to be.

Cut back to 1965. By then I had already witnessed Neave Parker's celebrated *Megalosaurus* life restoration appearing on a British Museum postcard, where our subject hump-less dinosaur has a horizontally inclined posture—rather precocious for the late 1950s when the postcard was issued. This image was later reproduced in Edwin H. Colbert's *Dinosaurs: Their Discovery and Their World* (1961), as Plate 40. Gee whiz. To my eleven-year old mind, isn't that how this particular dinosaur was *supposed* to look?! Much of our personal perspective is shaped by early life experiences, based on things or events that were meaningful or enjoyable to us in youth, which made it difficult then (as well as even today) to deny the "accuracy" of Parker's restoration (which like it or not retained "canonical" status). Often, when it comes to paleo-restoration, "accuracy" is a relative term, progressing in tandem with scientific discovery and interpretation.

And so, during that impressionable time, along came a genuine curiosity, M. Jean Craig's *Dinosaurs and More Dinosaurs*, published by the Scholastic Book Services (1965), featuring restorations by George Solonevich (many of which unfortunately were printed along the middle page creases). Under Craig's two-page entry for *Megalosaurus* there appears a picture of what certainly seemed to me then as a "humped" dinosaur. Now in retrospect this seems an atavistic reproduction of much earlier humped renderings. Yes—including the large cement model still displayed on the Crystal Palace grounds sculpted by Benjamin Waterhouse Hawkins under Sir Richard Owen's supervision (1853 to 1854).

A humped *Megalosaurus*? Most of us don't think of the genus that way (anymore). How in blazes did the interpretation come about?

Today, *Megalosaurus* is considered a messy "grab bag" of fossils and names, although it certainly was an important, if not downright "stellar" dinosaur back in the early days of paleontological understanding. As Glut's 2012 published skeletal reconstruction ("after Benson") testifies, there are relatively few distinctive bones which may be confidently assigned to this dinosaur genus. In his 2012 *Princeton Field Guide to Dinosaurs*, Gregory S. Paul states for megalosaurs that the "…validity of this group is not certain; may be splittable into a larger number of divisions." A common practice has been that when you didn't know exactly what the fossil was and it's likely "theropod" – why not just call it "megalosaur."

Besides those early quadrupedal restorations, copycat versions of which persisted into the very early 20th century, *Megalosaurus* appeared in a skeletal reconstruction, I believe, for the first (or more likely, second) time as a bipedal dinosaur in an 1893 publication by Hutchinson. What's rarely noted, however, about this particular reconstruction is the spinosaur-like vertebral projection (predating *Spinosaurus'* 1915 description). How did this 1893 "spino-megalosaur" reconstruction come about? Well, six decades prior, in 1832, paleontologist Hermann von Meyer had stated that *Megalosaurus'* "…spinous process is moderately high…." Of course, this remark was published during the time when the genus was regarded as quadrupedal, not bipedal. Later, Richard Owen also examined a then new backbone segment discovered by Samuel H. Beckles in 1850, which Owen attributed to *Megalosaurus bucklandi*. The enlarged spines observed on Beckles' specimen may have granted added basis for Hawkins to build his quadrupedal *Megalosaurus* with a hump over its shoulders.

In 1922, the particular backbone fossil found by Beckles (a picture of which appears in Don Glut's massive *Dinosaurs: The Encyclopedia* (1997, p.206), sporting proportionally enlarged neural spines (or according to Glut (1997), "…with spines about five times longer than diameter of centra…" was incorporated into Friedrich von Huene's genus, *Altispinax* (meaning "tall spine"). Much later, through the smoke & mirrors of taxonomic mitosis and reassemblage, the high-spined fossil, formerly attributed first to *Megalosaurus* in 1850, and much later to *Altispinax*, was assigned by George Olshevsky to a new genus, *Becklespinax huenei* in 1991, based on the aforementioned holotype specimen (i.e. figured in Glut's *D:TE*, p.206). Sigh, unfortunately, the taxonomic history of fossils at one time or another assigned to *Megalosaurus* is fairly complex and I've probably yet (intentionally) only scratched the surface here. Olshevsky's hypothetical life restoration of a sprinting, crested-back *Altispinax* appeared on p. 52 of Glut's *The New Dinosaur Dictionary* (1982).

Alright – now going back a century of more … in two of the Rev. Henry Neville Hutchinson's popular books on prehistoric life, *Extinct Monsters* (1893 ed.) and *Extinct Monsters and Creatures of Other Days: A Popular Account of Some of the Larger Forms of Ancient Animal Life* (1910), we see the aforementioned skeletal ("spino-") *Megalosaurus* reconstruction. Figure 13 in *Extinct Monsters* was probably (re-)drawn by artist J. Smit, although credited as "(After Meyer)." Hutchinson's later *Extinct Monsters and Creatures of Other Days* establishes further anatomical ties between *Megalosaurus* and North American *Ceratosaurus*, even including Smit's life restoration of the latter. Hutchinson acknowledged that, by the 1910s, through a host of anatomical studies on dinosaur fossils and skeletal reconstructions dating from the late 1860s onward, theropods were clearly considered bipedal and rather birdlike animals.

Probably because "Meyer's" speculative megalosaur skeletal reconstruction *idea* was presumably bedecked with proportionally long neural spines, as illustrated by Smit, that skeletal reconstruction in turn led to J. Smit's life restoration of a peculiarly long-necked megalosaur—with a second restored live "specimen" seen in the distance, posed in kangaroo-like stance—Hutchinson, *op. cit.*, Plate VI. In Smit's 1893 life restoration, the longer neural spines have been translated into a humplike feature situated along its dorsal region. As Hutchinson envisioned in 1893 (p.78): "It is not very difficult to imagine a Megalosaur lying in wait for his prey (perhaps a slender, harmless little mammal of the anteater type) with his hind limbs bent under his body, so as to bring the heels to the ground, and then with one terrible bound from those long legs springing on to the prey, and holding the mammal tight in its clawed fore limbs, as a cat might hold a mouse. Then the sabre-like teeth would be brought into action by the powerful jaws and soon the flesh and bones of the victim would be gone!" This is the scene Smit captured, at foreground, in a vivid restoration published as Plate VI. Note the kangaroo-like "bounding," "springing" action of the megalosaur according to this verbal reconstruction/imagetext.

In 1910, Hutchinson reproduced the same (i.e. Smit's) spino-*Megalosaurus* skeletal reconstruction, although this time printed as Fig. 43, but the "(After Meyer)" credit was omitted. Also, Smit's life restoration of the megalosaur pair was discarded. Incidentally, Henry R. Knipe's contemporary *Evolution In the Past* (1910), takes note of the supposed ceratosaurian "horn" that "*Megalosaurus*" allegedly sported over its nose. It would seem that megalosaurs had become conflated with snout-horned ceratosaurs.

Disclaimer (of sorts): now, Hermann von Meyer died in 1869, before the date of the earliest life restoration (e.g. – dated c. 1886). I've seen claiming *Megalosaurus* as a (non-humped) biped. And I cannot vouch that Hutchinson's 1893 reference to a "Meyer" is THE Hermann von Meyer, cited previously. I also do not know when or where a "spino-megalosaur" bipedal skeletal reconstruction with elongated spines was first (if ever) published (i.e. before Hutchinson, although possibly in a lengthy 1884 German monograph by W. Dames, or maybe later in one of Richard Lydekker's 1880s publications – just a crude guess). As stated here, Hermann von Meyer did note *Megalosaurus*' "moderately high" spinous projection, although during the period when the genus was considered quadrupedal – not bipedal as in the 1893 reconstruction. So is the "Meyer" referenced in *Extinct Monsters* yet another paleontologist or an old, unheralded paleoartist who lived into the later 19th century? Sadly, I've been unable to trace any connection –if any exists - between the possibly two "Meyers."

Meyer's, or rather, Smit's spino-megalosaur skeletal reconstruction was also reproduced in E. Ray Lankester's *Extinct Animals* (1905) as Figure 146. The spirit or "mien" of Smit's 1893 life restoration of the megalosaur pair also seems to have been swiped, albeit in a different more upright pose by artist C. Whymper, as published in Henry R. Knipe's *Nebula to Man*, New York: J. M. Dent and Co., 1905, Plate XX. However, Whymper's "megalosaur," with its long neck and swollen, humpy-looking torso (also posed with legs in kangaroo posture) more closely resembles a carnivorous "plateosaur." It is shown eating a small furry mammal, in a visual inspired by Smit's and Hutchinson's 1893 imagetext. (Needless to say, this unlikely restoration was eventually extinguished, one might say, with nary a "whimper.") So, over a century ago, the high-spined megalosaur image was certainly making the rounds (first as a humped quadruped, and later borrowed over into bipedal reconstructions). For half a century, the tendency was not to create a sexy, spinosaurian (or pelycosaurian), sail-backed restoration, but instead a rather unpopular looking humped dinosaur.

Possibly because of *Megalosaurus'* incongruous humpy look, the general paucity of its fossil remains, and because Charles R, Knight never deified this genus in a popular life restoration, the "Great Lizard's" reign wavered, falling into disfavor by the mid/early 20th century. We find for example, that in Gerhard Heilmann's technically oriented 1927 book on *The Origin of Birds,* well-illustrated for its time, there are no references to *Megalosaurus*. However, therein Heilmann did publish his sketch of *Ceratosaurus* based on a 1915 sculptural restoration by Charles W. Gilmore (which may, in turn, have been influenced by several of Knight's dramatic life restorations). The hunched-over pose while feeding, however, is clearly reminiscent of Smit's 1893 restoration.

Then by 1934, William E. Swinton had very little to say about *Megalosaurus* in his popular book, *The Dinosaurs: A Short History of a Great Group of Extinct Reptiles*. Swinton acknowledged the high-spined feature, stating, "… *Megalosaurus parkeri* von Huene, from the Oxford Clay near Weymouth, Dorset, is remarkable for the size of the neural spines, which are nearly ten inches long. This unusual length is a foretaste of what we shall see later in an Egyptian form, *Spinosaurus*" (p.62). While there are no reconstructions or restorations of *Megalosaurus* printed therein, Swinton published a photograph of Vernon Edwards' sculptural *Antrodemus* diorama (Plate V). Swinton also claimed *Megalosaurus, Ceratosaurus* and *Antrodemus* (e.g. *Allosaurus*) belonged to the same family, and even described the cunning, predatory nature of the latter genus in verbiage recalling Hutchinson's aforementioned vivid 1893 description of how *Megalosaurus* hunted prey.

By the 1930s there were more complete, more fascinating and scarier looking dinosaurs to marvel at. *Megalosaurus* faded from the mainstream. That is, until Neave Parker came along with his British Museum *Megalosaurus* restoration, reproduced on collectible postcards, that proved "canonical" to several generations of paleo-philes. Meanwhile, North America's *Ceratosaurus* (formerly thought to be genetically allied to *Megalosaurus*) became increasingly and antagonistically associated with *Stegosaurus* in 20th century life restorations and museum dioramas. Today, nobody discusses *Ceratosaurus* and *Megalosaurus* under the same breath anymore.

Both *Altispinax* and a sexually repressed *Megalosaurus* merited cameos in Harry Adam Knight's 1984 thriller novel, *Carnosaur*.

Eventually, humps would yield to sails (of sorts), as in 1981 William Stout restored his impressions of a low sailback or "crested" *Megalosaurus*. However, today, consensus opinion favors that megalosaurs *resembled* the better known allosaurs, lacking humps, sails or crests. More so than any other modern paleoartist, Stout has remained on the forefront of documenting shifting views concerning the probable life appearance of *Spinosaurus*, particularly in light of more recently discovered, evolutionarily related European and African dinosaur genera.

Today, North America's finest "finback" dinosaur is arguably the *Acrocanthosaurus*. However, according to paleoartist Tracy Ford, *Acrocanthosaurus'* slightly elongated vertebral spines were sheathed in muscle. (See Tracy's "How to Draw Dinosaurs – Sails in the Mesozoic," *Prehistoric Times* no. 50, Oct./Nov. 2001, pp.14-15.) So *Acrocanthosaurus* was not equipped with a proper sail or fatty hump, but instead was a tall-spined "muscle-back."

Undoubtedly, considering the rapid pace at which new dinosaur genera keep turning up lately, even more high-spined specimens will be increasingly discovered too. With apologies to Dr. Bailey, the humped look has had its day. Whenever warranted by fossil evidence, paleoartists simply prefer sails and faster running predators. Today, in our mind's eye we envision bipedal & blood-thirsty *Megalosaurus* (lacking both hump and sail) charging after stampeding *Polacanthus*, stirring up clouds of dust along the way. But interpretations change as new evidence about paleoworlds emerges. If dino-humps ever become stylish again, remember – *Megalosaurus* was there first.

Notes: (1.) *Discovering Dinosaurs in the Old West: The Field Journals of Arthur Lakes*, eds. Michael F. Kohl and John S. McIntosh, (Smithsonian Institution Press, 1997, p.22) (2.) Osborn also noted in 1898 upon the forthcoming availability of Knight's "Agathaumas" sculpture in plaster replicas that, "The frilled dinosaur, *Agathaumas sphenoceros*, is based upon a prior restoration published by Professor O. C. Marsh of his *Triceratops prorsus* … The tubercular character also given by Mr. Knight to the epidermis is conjectural." (Osborn, "Models of Extinct Vertebrates, *Science*, vol. 7, June 24, 1898, pp.841-845.) (3.) That is, dismissing Robert Plot's much earlier "Scrotum humanum."

Chapter Twenty-three—*New Wings on the Paleo-Perch - (a 1996 perspective)*

For over a century, while a consensus of paleontologists maintained that dinosaurs were cold-blooded and *not* feathered, in contrast to birds, it was easy to make distinctions between birds and dinosaurs. Then it was proposed that dinosaurs were warm-blooded, like birds, which blurred the distinction. Meanwhile, for the past decade a host of new Mesozoic avian species has descended upon the paleontological community as if out of some Hitchcock-ian nightmare! The new fossil fowl and other recent discoveries have clouded our picture of what is decidedly 'dinosaur' versus what is rightfully 'bird.' Has vertebrate paleontology gone … 'to the birds'? Most certainly! Let's see what new ideas and theories scientists are flocking to.

First, a precautionary statement. Some of the following discussion may seem complex as we shall delve into matters of avian and theropod classification. But I promise not to leave you … (ahem!)… 'out on a limb.'

Mesozoic birds have undeservedly suffered in modern times. Until the 1970s there was only undisputed evidence for four genera, all discovered in the 19th century. The most famous and controversial has been *Archaeopteryx*. Poor *Archaeopteryx* barely had time to stretch its stiffened, lithified wings after being freed from its 150-million year-long slumber in the Solnhofen lithographic limestone, only to become ensnared in the nasty Victorian evolutionary debates. A decade ago (i.e. mid-1980s), an astronomer even had the audacity to suggest that *Archaeopteryx's* feather impressions were forgeries. One specimen, the third then known—discovered in 1956—was thought to have been stolen a few years ago. And today the battle over the nature of its flying efficiency and true relationship to other organisms both living and extinct is far from over.

On metaphysically founded, functionist grounds, Richard Owen placed (hot-blooded) *Archaeopteryx* midway between modern birds and (cold-blooded) pterosaurs. Thomas H. Huxley, Darwin's "bulldog," regarded it as a warm-blooded evolutionary link between dinosaurs and modern birds. On the basis of comparative skeletal anatomy one possibility was that close relationships existed between as yet undiscovered ornithischian dinosaur species similar yet ancestral to the *Hypsilophodon.* Alternatively, saurischian dinosaurs such as the *Compsognathus* may have been most closely related.

By the mid-20th century, based on the observation that the wishbone in certain dinosaurs appeared to be absent, little *Archaeopteryx* was thought to have descended from a Triassic pseudosuchian ancestor. Compounded by a dearth of other Mesozoic avian fossils, *Archaeopteryx* was long heralded as the direct ancestor of all modern birds ... until recently. During the 1970s, paleontologists began reconsidering the nature of bird/dinosaur skeletal anatomy in light of the warm-blooded dinosaur theory.

A Triassic Ancestor?

Huxley and others speculated that the ancestors of *Archaeopteryx* would be ultimately discovered in Paleozoic deposits. Danish ornithologist Gerhard Heilmann postulated and even skillfully depicted a fanciful creature, the "Proavis," with rudimentary flight adaptations (hypothetically) emerging in the lower Triassic. And Triassic "bird" footprints had apparently already been discovered in the Connecticut Valley. (As we shall see, based on the latest evidence, Alan Feduccia estimated that the ancestral first bird may have lived during the Upper Triassic, about 211-million years ago.) If a more ancient common bird/dinosaur ancestor had existed, wouldn't dinosaurs have been feathered like birds?

A decade ago, the announcement of Sankar Chatterjee's *Protoavis* offered tempting glimpses into the evolutionary origins of birds. Although Chatterjee's claims were hotly debated from the outset it seemed possible that the 225-million-year-old specimens referred to *Protoavis* were avian in aspect. If so, then winged, feathered bird-like dinosaurs arose independently a multiple number of times during the Mesozoic era. Some speculated that *Protoavis* had descended from more ancient paleo-dinosaur stock (e.g. *Herrerasaurus*), while *Archaeopteryx* and other later avians evolved from more advanced theropod ancestry (e.g. coelurosaurs).

Excitement over Chatterjee's discovery prompted a restoration by artist John Sibbick, which I first saw printed in the *Chicago Tribune* (Monday, June 12, 1989). The caption to the restoration read, "Artist's reconstruction of the smallest dinosaur, 'Protoavis.'")

Protoavis appeared to be more bird-like in certain respects than *Archaeopteryx* yet sported more primitive adaptations such as four-toed feet. It was debated whether the remains represented an early "theropod bird-mimic" variety or possibly a direct ancestor leading to modern birds. Alas, following a 1991 description of the fragmentary cranial remains, John Ostrom soundly rejected Chatterjee's claim for avian affinity. Major disappointments stemmed from Chatterjee's unpersuasive style of presentation, which failed to account for fortifying evidence that *Protoavis* possessed avian qualities and the fact that the fossils were generally unavailable to researchers.

Although a host of new Mesozoic avian or near avian species have been identified during the past decade, uncertainties in establishing true relationships between them, as well as to possibly ancestral dinosaur species have only escalated. Neglecting doubtful *Protoavis*, more than twenty new fossil bird genera (depending on your classification scheme of choice) have been discovered, all of which seem to be geologically younger than *Archaeopteryx*. Let's review some of the better known or more significant varieties.

Generally speaking, some of the most remarkable discoveries have recently been made in five geographic settings: Late Jurassic/Early Cretaceous deposits of northeastern China; the Cretaceous of Mongolia; the southern half of South America (i.e. Patagonia); and Late Jurassic/Early Cretaceous deposits in central Spain.

Las Hoyas Bird:

There are possibly five genera of fossil birds known from Late Jurassic/Early Cretaceous 115-million-year-old deposits from Spain, although presently two are of doubtful or fragmentary nature. Fossils known from Las Hoyas, Spain have been most widely publicized yielding specimens referred to three genera—*Iberomesornis, Concornis,* and *Eoalulavis*.

Of this trio, the most recent discovery is *Eoalulavis*, The little bird with a 17-centimeter wingspan was found in a remarkable state of preservation. Scientists could readily discern feather impressions arranged in a natural position and a special wing apparatus evolved to enhance low speed flight maneuverability (known as the alula, or 'bastard wing'), as well as the fossilized contents of its final meal. Apparently *Eoalulavis* fed on aquatic organisms and could expertly control its landings and take-offs due to the presence of the alula on the wings' leading edge. This is the oldest known species for which there is positive evidence of such a sophisticated structure.

Despite the fact that *Eoalulavis's* hand structure differed from that of modern avians (i.e. hand and wrist bones rigidly fused in modern species into a structure known as a carpometacarpus), the fossil species had evolved a means of controlled flight far beyond the capacity of *Archaeopteryx*. While nature can provide alternative adaptational solutions to address similar ecological challenges, evolutionary forces must build on existing structures in ancestral lineages.

Next, for its geologic age (only about 115-million years), sparrow-sized *Iberomesornis* displays remarkably advanced skeletal features suggestive of a perching adaptation. Enhanced flying capabilities are suggested on basis of a strut-like coracoid, as well as an ulna bone, longer than its humerus. *Iberomesornis* also is known to have had a bird-like pygostyle (rather than a bony dinosaur-like tail as in the case of *Archaeopteryx*), and lacked gastralia (mass-adding, belly or abdominal ribs). However, other primitive characteristics indicate *Iberomesornis* was 'intermediate' between *Archaeopteryx* and other contemporary birds. Another Las Hoyas specimen referred to *Concornis* also apparently evolved arboreal perching and enhanced flight adaptations.

Concornis belongs to an avian category known as the "opposite birds" (i.e. "enantiornithes," a category established in 1981). Although examples of 'opposite birds' had been discovered over a century ago, they had been misidentified as theropod dinosaurs (e.g. *Ornithomimus*). Although *Iberomesornis* is closely related to the enantiornithes, if has been placed in its own clade.

Concornis and relatives differed from modern birds (i.e. "neornithes") in their skeletal structure. For instance, primitive enantiornithes had toothy jaws and most had wing talons. Also, bones in their ankles fused in reverse (opposite) fashion compared to modern birds. How? Modern birds have the foot bones fused from the bottom up, whereas in 'opposite' birds, the fusion of from top to bottom. Most intriguingly, when viewed cross-sectionally, their hind limbs reveal a reptilian growth ring pattern. Also, enantiornithine sternal features generally are not as developed as in modern birds. However, their breastbones were keeled, unlike the condition in *Archaeopteryx*. These signs indicate that, metabolically, opposite birds may not have been fully warm-blooded. Furthermore, here is likely indication that feathers evolved *before* endothermy (warm-blooded metabolism).

Some authorities have claimed *Archaeopteryx* was a possible direct ancestor of the enantiornithes. Alternatively, it is possible that the ancestor of modern and opposite birds lived at about the same time as did the little saurian bird—*Archaeopteryx*, which, by then, may have been an evolutionary 'side line,' a surviving (then) "living fossil."

Enantiornithine species rapidly diversified about 125-million years ago. By the Late Cretaceous, they apparently enjoyed a worldwide distribution, only to become extinct along with their dinosaurian brethren at the end of the Cretaceous.

A Mongolian Enigma—Mononkyus:

Next to the South American *Patagonykus*, perhaps the strangest of all the new avian fauns is the flightless and apparently non-feathered genus, *Mononkyus*, which definitely must be regarded as one *strange* old bird! *Mononkyus* created a sensation after its announcement in early 1993, even reaching the cover of *Time* (April 26, 1993). Bones of this highly unusual 75-million-year-old turkey-sized variety were first collected in the Gobi Desert in 1922, only to languish, catalogued as a "bird-like dinosaur" in a museum drawer. Significance of this discovery went unrecognized, until recently.

It has been debated whether *Mononkyus*, which had toothy jaws, should be considered a bird, versus a dinosaur. This is because *Mononkyus* is 'dinosaur' from its trunk through the end of its tail. However, its most unusual skeletal features, an avian-keeled sternum and stubby arms ending in knobby claws, may have evolved from theropod ancestors. The supposedly avian nature of *Mononkyus's* hand and breastbone has been challenged. Were the arms used for mole-digging, an adaptation which is difficult to comprehend in birds? If this is what *Mononkyus* used its arms for, then can analogies also be made to those puny *T. rex* arms? Or were *Mononkyus's* forelimbs and hands derived from an already specialized avian structure? On the basis of its specialized forelimb structure, *Mononkyus* cannot be a direct ancestor of modern birds even though it shares more traits with modern birds than the geologically older *Archaeopteryx*.

Much as paleontologists now must consider a variety of possible origins for the different categories of winged, flying Mesozoic avians, so must equally perplexing matters concerning the flightless nature of other Mesozoic avians be settled. Did *Patagopteryx* and the mononkysaurs evolve directly from flightless 'protobird' dinosaurs closely related to "ostrich dinosaurs"? Or, just as winged avian flight may have developed a multiple number of times during the Mesozoic from different theropod lineages, in turn, some branches derived from these winged descendants may have split off and converged independently toward a secondary flightless condition.

More Asian Avians:

Besides *Mononkyus*, at least two additional birds merit attention. According to some paleontologists, the early Cretaceous *Ambiortus*, known from a relatively complete specimen found with associated feather impressions, may represent one of the oldest ornithurine birds possibly on the lineage of modern birds. The claim has been disputed. Evidently *Ambiortus* was a capable flyer. E. N. Kurochkin, who described the genus in 1985, claimed it was a likely possibility that, "typical carinate birds were already in existence by late Jurassic, so that at that time, *Archaeopteryx* would already have been a primitive *relict* form … If so we may expect the origin of birds to be pushed further back in time than late Jurassic." (*Cretaceous Res.*, 1985, vol. 6, p.276)

Yet another bird, *Gobipteryx*, known from the Gobi Desert, represents a clade closely related to the enantiornithes. Known merely on basis of two crushed skulls and some fossil embryos, it appears *Gobipteryx* was an active flyer.

An entire avifauna is flocking from fossiliferous beds in seven localities of northeastern China. Visions of soaring pterosaurs along with the occasional *Archaeopteryx* typically frame our picture of Mesozoic skies. However, tangible evidence for an abundance of birds coexisting with pterosaurs and dinosaurs is no more readily apparent than in these Late Jurassic/Early Cretaceous deposits.

The 135-million-year-old Early Cretaceous, Chinese, sparrow-sized *Sinornis* ranks among Paul Sereno's well-publicized successes. Described in 1992, this well preserved enantiornithine displays a remarkable admixture of primitive and derived skeletal features. *Sinornis* had a pygostyle, a wrist joint allowing folding of the wings anticipating the condition in modern birds, and a reversed first toe indicating a perching adaptation. From its relatively well-developed sternum, *Sinornis* was an adept flyer that retained teeth and belly ribs (gastralia). Because it was recovered from lacustrine (lake) deposits, it was concluded that *Sinornis* may have evolved in inland, freshwater wooded habitats.

Confuciusornis and *Cathayornis* are sparrow-sized Chinese enantiornithines fully equipped with arboreal perching adaptations. A skeletal reconstruction of the *Confuciusornis* was published in *Science*; the specimen was found with associated plumage traces revealing notable differences when compared to *Archaeopteryx*. Its bony tail is reduced to a long pygostyle comprising eight vertebrae. Teeth are absent and bones associated with flight musculature are broader. However, as in the case of *Archaeopteryx*, *Confuciusornis* had an unkeeled sternum and gastralia. The abundance of remains attributed to *Confuciusornis* could imply that the species socially congregated in flocks.

Most intriguingly, modern ornithurine species are also known from these Chinese deposits. Here is a rare circumstance where enantiornithine precursors of modern birds are known to have cohabited, sharing in song, while competing for resources. Birds on the modern limb so far identified include *Liaoningornis, Gansus, and Chaoyangia*. The absolute age of these deposits has been estimated to range from 142 to 121 million years. If the latter, younger age is correct, the fauna would have been nearer contemporaries of Mongolian *Ambiortus*.

Sparrow-sized, arboreal *Liaoningornis*, known from a partial, headless skeleton, had a keeled sternum, possibly the earliest evidence for such a structure. Sufficient material attributed to *Chaoyangia* composited from remains of several individuals warrants skeletal reconstruction. *Chaoyangia* was clearly built for powered flight, having a modern aspect to its keeled sternum and shoulder girdle. The new wading genus had teeth and also bore skeletal similarities to *Ambiortus*. *Gansus* is thought by some to represent the oldest record of true modern birds, although alive and well so long ago in the Early Cretaceous!

The relative age of the Chinese fauna has implications for our understanding of early avian evolution. For if these birds evolved shortly after the age of *Archaeopteryx*, a possible interpretation is that two subclasses of birds defined as 'sauriurae' (including *Archaeopteryx* and the 'opposite birds') split, diverging from the ornithurae (ornithurines) in the Early or Middle Jurassic. However, an alternative interpretation that modern birds descended from enanthiornithines—which were in turn derived from *Archaeopteryx*—would be fortified if the absolute age of the Chinese fauna happens to be much younger (i.e. 121 million years).

South American Discoveries:

A gaggle of South American avian specimens is also roosting in the minds of paleontologists. Three genera, two flightless birds—hen-sized *Patagopteryx* and newly discovered *Patagonykus*—are known from 85-

million-year-old Argentine deposits of the Patagonian steppe. *Patagonykus* (possibly resembling a diminutive *Phororhacus* from the Tertiary period) had vestigial wings, yet *Patagonykus's* forelimbs highly resembled those of the Mongolian *Mononkyus*, a controversial creature to be discussed shortly.

Of the three Argentine varieties, *Patagopteryx* is thought to be most closely related to modern birds, and is believed to be closely related to the (ornithurine) taxon including the more familiar Cretaceous toothed diving birds, *Hersperornis* (known for over a century from deposits in Kansas) and *Enaliornis*, known from England. The recent Argentine discoveries provide convincing evidence for the occurrence of faunal interchange between Laurasia and South America during the Late Cretaceous.

Besides *Archaeopteryx*, rarely have restorations been attempted for Mesozoic avi-fauna. Skeletal reconstructions ordinarily illustrate scientific descriptions, except in cases where preservation is poor. Ed Heck of the American Museum of Natural History, has proven to be one of the few unsung heroes involved in restoration of unheralded Mesozoic birds. Another is John O'Neill who has painted *Iberomesornis*. (*Science*, vol. 267, 2/3/95, p.638)

In Heck's painting of the Argentine birds, published in Luis Chiappe's contribution to the June 1995 issue of *Natural History* (p.53), three genera were depicted—*Patagopteryx*, an unnamed 'relative' of *Mononkyus,* and a splendidly feathered falcon-sized enantiornithine, *Neuquenornis.* The latter closely resembles the possibly ancestral *Archaeopteryx,* which lived 65 million years *earlier*. In so doing, Heck provided a rare opportunity to sense how these nearly forgotten creatures may have appeared in life.

Malagasy Discoveries:

Recently, an avifaunal wealth has come to light in Late Cretaceous deposits of Madagascar. Although receiving less publicity than associated dinosaur discoveries, the new bird fossils are at least as remarkable. One species possesses a miniature replica of the slashing claw known from the feet of *Velociraptor*! To considerable degree, this hawk-sized (evidently) flying griffin fortifies relationships between birds and theropod dinosaurs. None of the fossil species so far known from this region are on the modern lineage, placing doubt on a hypothesis that more recent (Holocene) birds evolved on Madagascar from Late Cretaceous Malagasy ancestors.

So What is a Dinosaur?

So what does all this have to do with dinosaurs? After all, with exception of *Archaeopteryx*, which may have actually been a feathered flying dinosaur, aren't birds just simply ... birds? *Triceratops* doesn't even look like a bird ... well, okay, except for that parrot-like beak, and it also had teeth... but so did Mesozoic birds. Unfortunately, the answer to the question posed above isn't all that simple, and in fact, things may become quite a bit murkier still.

One wonders why, years ago, a consensus of paleontologists remained so reluctant to accept the now rather obvious bird-dinosaur connection, especially in light of new fossil discoveries, which clearly unify the two great clades. Most recently, fossils from Asia dramatically favor the view that birds and certain theropod dinosaurs are as closely related as paleontologists dared to dream was the case two decades ago, yet in complex ways that are difficult to unravel. The fabric of evolution is mirrored in that complexity.

In the heralded 'Age of Man' vertebrates conventionally classified as 'dinosaurs' no longer exist except in our hearts and imaginations. Yet birds do in the present and for the most part, based on what we know, modern birds differ in numerous ways from dinosaurs. That is, until one probes the nitty-gritty details and considers a particularly thorny question: 'In the Mesozoic, what was a bird and what was dinosaur'? We find that our conceptualized categories quickly crumble, as there are many transitional species that cannot be readily accommodated in our 20th century categories.

Why is this so alarming, especially since we take pride in our grasp of evolutionary principles (even though the actual, contingent historical process of evolution remains poorly understood even in this day)?

A century ago, the very idea of evolution itself proved far less controversial than the concept of humans having evolved through randomized processes from apelike ancestors. By analogy, is it possible that so many of us who have grown up yearning for those mythical dinosaurs cannot accept the ludicrous-seeming possibility that our favorite Mesozoic monsters are more closely related to Hanna Barbara's "Tweety Bird" than they might be to ferocious modern crocs, alligators and Komodo Dragons? Strike a vibrant chord and outcry ensues!

In their recent studies, paleo-ornithologists and paleoanthropologists have traversed similar terrain. For the past two decades, Paleontology is benefiting more from the discovery of 'links' between birds and dinosaurs (or apes and humans) than were discovered in the entire preceding (e.g. 19th) century. Furthermore, hominid and Mesozoic fossil avians are so rarely discovered that a single new fossil can have significant bearing on the "evolutionary trees" of apes and dino-birds, respectively. For instance, the new avian discoveries of the 1990s alone have profoundly altered the phylogenetic sequences leading to modern birds. The current framework (of which there were different versions in 1996, depending on authorities chosen) is really a 'moving target' that surely will be further modified as more evidence accumulates.

It is astounding to consider the weighty evolutionary implications posed by such scanty evidence … attesting for nearly 100 million years of avian evolution. New theories are certain to 'hatch,' yet without overturning Darwinian philosophy. More fossils are desperately needed for study because when it comes to avian paleontology it certainly will be better to claim a fossil bird in the hand than to have it obscured in Darwin's "branching (evolutionary) bush."

So when it comes to dinosaurs, from preliminary reports, we now have a new Chinese compsognathid (informally named 'Sinosauropteryx' but now (i.e. then) known as *Compsognathus prima*), which may have sported feathers in life. (Although it is not yet concluded as yet that impressions found on the fossil can be regarded as feathers.) This fossil comes from the same beds as have other undisputed Chinese avians. Feathers on a basal (i.e. primitive) coelurosaur would suggest that all descendants would have also sported feathers, including the mighty *T. rex* (unless it lost them secondarily much as in the case of how modern elephants may have lost most of their hair), and any avian 'descendants' thereof.

For many years, in his artistry and writing, dinosaur expert Gregory Paul has championed the cause for feathered dinosaurs. The physiological reasons for his theory are sound, yet, to date, the fossil record has not supported his ideas. (Note—as of 2017, circumstances have remarkably changed!) No evidence for feathered dinosaurs has ever been found. (Note—no longer a true remark, as since 1996 numerous feathered dino-specimens have turned up!) Beyond lack of feathered fossil remains, there are convincing arguments that can be made on basis of skeletal similarities in support of close postulated relationships between birds and certain (derived) dinosaurs.

Perhaps the most controversial aspect to this debate is whether some genera conventionally classified as dinosaurs descended from genera which would be regarded as birds or 'proto-birds.' Paul has constructed a phylogenetic tree (appearing on pp. 223-224 of his 1988 volume, *Predatory Dinosaurs of the World*), indicating that several (theropod) dinosaur lineages such as those leading separately to *Velociraptor, Oviraptor,* and *Dromaeosaurus* may have been derived from proto-bird stock that gave rise to *Archaeopteryx*.

In an *Omni* contribution (June 1994), dinosaur expert George Olshevsky stated that this should come as little surprise since birds came before dinosaurs anyway. (One third of Olshevsky's reasoning has been crippled, however. This is because University of Chicago paleontologists have falsified the long-revered Cope's Law, which forms an important premise for Olshevsky.) Dinosaur genera such as Upper Cretaceous, Mongolian *Oviraptor, Avimimus* and Argentine *Alvarezsaurus* have all been cited for their decidedly avian skeletal features. The nesting behavior of *Oviraptor* has been compared to that of modern altricial birds. (See Michael Novacek's *Dinosaurs From the Flaming Cliffs* for example.) A century ago, to an ardent evolutionist—Huxley perhaps—these ideas would not have seemed so outlandish!

Understandably, these ideas have created some squawking because the idea of dinosaurs having descended from even more avian-like creatures is not universally accepted. However, evidence is mounting in favor of a revised approach. For in a recently published *Science* article (the one in which new discoveries of *Confuciusornis, Chaoyangia* and *Liaoningornis* were reported—vol. 274, 11/15/96, pp.1164-67), the authors prophetically conclude:

> "The new information on the early geographic and temporal distribution of birds may also indicate a long avian history in the Jurassic. We would expect that the common ancestor of the Sauriurae and the Ornithurae would predate *Archaeopteryx* and that we may reasonably search for birds in Middle Jurassic and older beds. This exacerbates one of the most obvious conundrums facing the theory of a dinosaurian origin of birds. The dinosaurs thought to be most like birds are primarily Late Cretaceous in age and are younger than *Archaeopteryx* by more than 76 million years. This temporal paradox has led some dinosaur experts to argue that birds gave rise to certain Late Cretaceous theropods."

Without endorsing the idea, the authors admit that the problem of bird origins and theropod evolution is paradoxical. However, much of their argument rests on their reported age (circa 140 m.y.a.) of the deposits that have produced the Chinese avifauna. A new dinosaur textbook, *The*

Evolution and Extinction of Dinosaurs (by David Fastovsky & David Weishampel... i.e. 1st ed., 1996) embraces the formerly controversial idea that all birds *are* dinosaurs. The real trick will be in resolving which evolutionary lines produced everyone's favorite bird?—as some would have it—*Tyrannosaurus rex*.

Notes: Accordingly, restorations of Mesozoic birds and *Archaeopteryx* have noticeably 'evolved' during the past 130 years. During the 1860s, Riou was the first to introduce an outline of the flying bird then known from a headless specimen, into an idealized landscape scene of the "Upper Oolitic Period" for Louis Figuier's *Earth Before the Deluge* (1865). By the early 20th century, *Archaeopteryx* had assumed a conventional avian image in restorations completed by Charles R. Knight, Joseph Smit, Othenio Abel, and Gerhard Heilmann. For many years, Maurice Wilson's three-dimensional *Archaeopteryx* sculpture at London's British Museum was a standard of excellence.

During the early 1970s, John Ostrom suggested that birds learned to fly from the ground up, and that *Archaeopteryx*'s skeletal anatomy closely resembles that of theropod dinosaurs naturally because birds are direct evolutionary descendants of theropod dinosaurs. Gregory S. Paul's 1977 restoration featuring scaly-looking, land-roving *Archaeopteryx* illustrated a shift in understanding. Paul's *Archaeopteryx* casts a blended image that is partly avian, yet remarkably dinosaurian. (However, Paul believes that birds learned to fly after adapting an arboreal existence.)

The theoretical framework for avian evolution was superbly captured in a painting by Mark Hallett published in the June 1985 issue of *Zoobooks*. In this diagram, Hallett ingeniously symbolized the origin of flight and evolution of birds from small running dinosaurs through *Archaeopteryx* and its hypothetical ancestors, directly to a modern form. All figures are depicted chasing dragonflies, implying that flight arose in terrestrial, not arboreal, creatures.

More recently, however, John O'Neill (19193) introduced new images of the ancient bird. O'Neill's restoration, depicting an airborne *Archaeopteryx*, sporting a jay-like head crest of feathers, was published in *Science* (Feb. 5, 1993).

In a fascinating illustration illustrating the chapter *when it was originally published* in the June 1997 *Dinosaur World*, Jack Arata's contributed image (completed under my direction) was spied on page 42, not reproduced here. Here we see a number of ideas metaphorically conveyed by the image of evolutionary branching bush (although unintended to represent a 'ladder of progress). Although quite impossible to definitively sort out true relationships between represented organisms, there were a number of significant questions raised therein. At the top, one must wonder, to what extent survival of modern birds' common Mesozoic ancestor depended on flight capabilities. Were there other adaptational factors equally important or even more significant? Images at the base (root) of the 'tree' symbolize paleontologists' concerns for origin of birds. Were they dinosaurian (coelurosaurian), or crocodilian? Referring to 'bubbles' in Jack's figure, ... sequentially moving from the base in a counter-clockwise direction, there are other interesting questions to consider regarding dinosaurs and birds. (1) To what extent was the apparent success of ancient birds related to the ultimate extinctions of pterosaurs? (2) What relationships will be drawn between modern birds and dinosaurs, based on the new downy dinosaur fossils discovered in China? (3) Were mononkysaurids birds or dinosaurs? (4) Did large flightless birds coexist with dinosaurs in the Upper Cretaceous? (5) Just how modern in aspect were early Cretaceous birds of Spain and China? Would they appear substantially different from the modern variety? (6) Does *Archaeopteryx* represent an anachronistic variety of avians? (7) Just when did the first non-flying protoavid evolve? Did it survive because of climbing abilities? Was it a feathered organism? (8) Did dino-birds that learned to fly inherit their remarkable abilities from a tree-climbing ancestor? To what extent (and purpose) did the evolution of feathers contribute to the success of the true avian ancestor?

SEVEN

A fantasy portrayal showing a 'what-if' brooding atmospheric, apocalyptic merging of two Lovecraftian pseudo dino-monsters: Pacific Rim's Otashi *'versus' a cyclopean otherworldly creature spawned from the pages of Stephen King's* The Mist *(film, 2005). (drawing made for hire by Jason Croghan, copyright 2017)*

Mythic Monstrosities

It's become increasingly common these days to read about the most tangential forms of dino-monsters, or of what such creatures may represent in an apocalyptical setting. The subgenre of *kaiju* fiction is clearly allied to Godzilla's omnipresent popularity, a cool-looking metaphorical monster which in turn is dino-derivative. Take this as a very brief 'outline' of the flowering of dino-*daikaiju* lore and fiction—emphasizing examples where prehistoric animal-aspect creatures so often fight mysteriously among themselves, or materialize out of gray mists in our plane of existence.

And so, without further ado…

> These chapters originally appeared as articles in the following publications:
> Chapter Twenty-four—*Monster!* no.27, March 2016, pp.95-99.
> Chapter Twenty-five—*G-Fan* no.108, Jan/Feb 2015, pp.88-90.
> Chapter Twenty-six—*G-Fan* no.111, Winter 2015, pp.50-51.

Chapter Twenty-four—*Reflections of Doomsday: When did Dinosaurs become Apocalyptical?*

An immediate off the cuff response from readers would simply be, "the 1950s," which is largely correct ... for the first attack wave. Certainly monsters like Godzilla, Rhedosaurus, Anguirus, Rodan, and (later) Gorgo and the Paleosaurus paved the way in the apocalyptical dino-monster filmic arena. In a 2016 interview actress Gillian Anderson of *The X-Files* fame opined, "Sci-fi is apocalyptic." Certainly that strange assortment of 1950s giant dino-monsters may further be categorized, accordingly, as "apocalyptic" too – although their prophetic messages and dark symbolism have evolved considerably through the decades. Collectively, they help paint a trending vision of a dystopian future. But let's examine the matter and the question at hand in more detailed fashion, placing matters in historical context.

By the early 20th century, dinosaurs and other prehistoric vertebrates became central objects of evolutionary ideology and even publicized scientific controversy, long before they took on more dire metaphorical meaning (as in fantastic fiction). Some genera known to science—such as the "*American* Mastodon" and *Diplodocus*—had already become rather 'politicized.' And dinosaurians were certainly viewed by scientists and laymen alike as chief 'spokesman' in our general understanding of how life evolved through the dim corridors of geological time. In this vein, they were clearly 'natural' (albeit extinct) creatures, although often described by popular writers as "monsters."

Decades prior to invention of nuclear weapons, societies understood that world war was a projected natural geopolitical outcome. Think of H. G. Wells' 1898 novel, *The War of the Worlds*, for example, or George Chesney's *The Battle of Dorking* (1871), both metaphorically presaging Germany's then futuristic European invasions. Dinosaurs had not yet been cast into the fray though, partly because their gestation in human consciousness had only just begun. Later, in the wake of World War II, because mankind's industry transforms Earth's surface, oceans and atmosphere at an increasingly alarming rate, it became recognized that Man is a veritable geological force. Yes—comparable to major forces rocking the planet in primeval times.

So perhaps, by mid-20th century, it became fitting for fossil creatures out of our geological past—dinosaurs and pseudo-dinosaurians alike—to symbolically 'warn' mankind of our polluting "human volcano" - our current and ever present, reckless folly. And as our planetary trespasses seemingly multiplied, there were increasingly new creative avenues for fiction writers and imagineers to explore and exploit in evoking images of condemning dinosauria. It's sorta like if … well, since 'we' (our species as an array of separate cultures arranged in independent political strata) won't logically listen or react to the cause of our own silent screams, perhaps the haunting image of an instructive yet haunting dinosaurian persona – presenting an offer we simply won't be able to refuse ultimately—will arrest our attention instead. Just maybe?

And so while an early fictional dinosaur of the movies plundered London (in 1925's *The Lost World*), such fantasy occurrences were rare and certainly not so apocalyptic. However, movie themes dramatically changed during Cold War times, particularly during early 1950s invention and refinement of both USA's and the Soviet Union's hydrogen bombs, as noted in contemporary angst-ridden fare, projected in *The Beast From 20,000 Fathoms* (1953) and *Gojira* (1954). A slightly earlier film, 1951's *The Lost Continent* (with its compelling ad tagline – "Thrills of the atomic-powered future! Adventures of the prehistoric past!"), suggested a tepid relationship linking dinos to then looming fears of radioactively charged missiles. This is when dinosaurs (a term used loosely throughout here) began to be seen by the public as 'apocalyptical.' But, before exploring the literary side of the question, how have their filmic roles since conveyed reflections of doomsday. And which kind of "doomsday" do we mean?

For the most part, dinosaur films of the late 1960s and 1970s, a period corresponding to the early years of the 'dinosaur renaissance,' essentially served as cool movie 'props.' Well, of course, the Godzilla series continued with titles like *Godzilla vs. the Smog Monster* and others blatantly decrying man's ecological and environmental excesses (which had been of increasing concern since the 1950s). Or as in *Godzilla 1985*, the heightened 1980s prospect of nuclear exchange with the Soviet Union was portrayed. By process of 'extrapolation,' many variations of *kaiju* subsequently evolved, several of which may be counted among the ranks of pseudo-dinosauria. Yet arguably it wasn't until 1993's release of Steven Spielberg's *Jurassic Park* that dinosaurs arrived with another kind of (i.e. bio-tech) 'warning,' even though essence of the intended dire message was more poignantly conveyed in Michael Crichton's "*JP*" novels. Apart from such 'messaging,' most of the movies' substantial hype surrounded the uncanny realism with which technical staff resurrected their dinosaurs on camera.

Crichton wasn't the first to explore ramifications of artificial reproduction, or 'cloning' of prehistoric animals – especially vicious dinosaurs. Nor was he the last. A spate of less memorable science fiction films followed on the heels of *Jurassic Park's* filmic success. Some of these dealt with the more obvious Frankensteinian threat of biotech and bioengineering – (likened, by Crichton, to the potential of nuclear power in scope). But metaphorical messages of uncontrolled reproduction (not unlike what happened in Karel Capek's 1937 novel, *War With the Newts*) mirroring man's tendency to overpopulate with concomitant exploitation of vital resources were more successfully conveyed in sci-fi literature.

It is clear that by the mid-1980s, a bifurcation in cultural meaning of 'renaissance' dinosauria had taken place, if sensed only on a subliminal level. For besides the unwise spawning of more accurate-looking, yet artificial dinosaurs using biotech, writers and scientists alike increasingly speculated over the prospect of dinosaurs cast in humanoid guise. Reptiloid creatures, spawned in sci-fi literature that in today's lingo might or could loosely be considered today as "dinosauroids," go back decades – even to H.G. Wells heyday. But THE "Dinosauroid" was a figurative, early 1980s scientific 'thought experiment' invented by Dale Russell and sculptor Ron Seguin. In their hypothetical published scenario, what if the Cretaceous/Paleogene (K-Pg) asteroid or comet had missed Earth instead?

The underlying rationale for the Dinosauroid's significance stemmed from an early 1980s discovery that the K-Pg mass extinction event was caused by collision of a ~ 6-mile diameter comet or asteroid with Earth. This event took on a symbolic aura during a contemporary heightening of Cold War angst, versus the Soviet Union during the early years of the Reagan administration, with the prospect of prolonged "nuclear winter." Scientists contemplated how intelligent dinosauroids hypothetically might have evolved by now, had Earth not been so dramatically unlucky then. So 'yes,' scientists agreed, likewise, if we aren't careful, man's civilization and species could also be catastrophically extinguished (somehow), not unlike how dinosaurs suffered extinction 66 million years ago. However unlike circumstances of the dinosaurian demise (of course agreeing that birds *are* modern dinosaurs), as an intelligent species we *should* command sufficient rationality and common sense to prevent a self-imposed kind of (analogous) catastrophe (but will we?).

So returning to the original question when did dinosaurs become *apocalyptical* dino-monsters both in film and literature?

Certainly, despite prior, brief filmic incursions toward city centers as in the 1925 silent "Lost World," giant dino-monsters became highly familiar and condemning visages during the early to mid-1950s. Although humanoid/reptiloid half breeds occasionally haunted the silver screen during this period as well (e.g. 1957's *The Hideous Sun Demon*), through the magic of suitmation they increasingly transformed into Dinosauroid-inspired creatures during the 1980s and early 1990s, as humans metaphorically melded with dinosaurian cousins into a variety of half-dino-half-man monstrosities appearing on camera (televised programs especially). Arguably, awful dinosauroids were the most horrifying pseudo-dinosaurs of all simply because they were recognizably human! Then, thought provoking ideology concerning contingency of the K-Pg extinction transformed into further reflections on chaos theory (basically, life finds a way, but also, $h*t happens!) with futuristic fears of an overarching, overweening biotech industry at large, in Spielberg's *Jurassic Park* (1993) (based on Crichton's 1990 bestseller). From those bloodthirsty "raptors" to mighty "Indominus rex," we all cowered at their cunning!

In English fantastic literature, parallels to the film industry may be reckoned. The only prehistoric monster to survive the journey back from the Amazonian plateau in Arthur Conan Doyle's 1912 *Lost World* novel was a small pterodactyl (not that stop-motion animated brontosaur of the 1925 film!) which escapes into the night sky. However, in a 1906 April Fool's Day elaboration, *Chicago Tribune* readers were beguiled with breaking 'news' that giant dino-monsters had launched a massive attack from an Atlantis reemerging in the Arctic. After storming through Canadian forests and swimming southward through the great lakes, they ultimately wage a desperate 5-weeks long battle upon our Windy City, versus the military. But that early "Trib" entry was farcical – tongue in cheek, not so much 'apocalyptical' in tone.

Decades later, however, Carson Bingham's 1960 pulp novel, *Gorgo*, signified entry of literature's apocalyptical giant dino-monster – far more effectively conveying a sense of pending doomsday than via the 1961 film. Not long before, James Blish's 1958 sci-fi novel, *A Case of Conscience*, essayed the instance of a world-destroying dinosauroid race – the "Lithians." (Those aforementioned warlike Newts in Capek's novel—while also implying Germany's threat—were marine amphiboid creatures, not so much 'reptiloid' in nature.) And then following the 'Alvarez' asteroid impact extinction theory of 1980, Russell/Seguin-inspired Dinosauroids went on to populate short stories and novels, especially from the mid-1980s onward.

Thereafter, DNA-bioengineered dinosaurs or other *kaiju* of doom ran amuck in numerous sci-fi stories when Crichton's novels scored commercial success. Or, recently, reflecting the latest biogenetic twist & trend, we've enjoyed novels like Jeremy Robinson's (eXtreme *X-Files*-ish) *Project Nemesis* (2012) and David Sakmyster's and Rick Chester's *Jurassic Dead* (2014), partly inspired by AMC's hit show, *The Walking Dead*. Both tales meld Crichton's literary concept of genetically altered dino-monsters with that of humans transforming (respectively via laboratory experiment or infection) into 'dinosauroidal' *daikaiju*.

So, since the early 1950s, due to dread stemming from the hydrogen bomb's potential to annihilate mankind, with later scientific ties made three decades later to a primeval Armageddon caused by an impacting asteroid, then with the unnaturalness of mankind's exploitation and uncontrollable ruination of the planet reflected in a blight of artificial dinosaurs recreated by mad scientists, "Gaia" – Mother Earth may be telling us through our symbolic, yet evolving perception of dino-monsters that our time is nigh. By sending us monsters out of our geological past to threaten us, you can tell she's pissed!

And we've got no place to hide.

Chapter Twenty-five—*Toward a Unified Theory of Dinosaurian Kaiju: A Pacific Rim-inspired, 'what if?*

Giant dino-monster daijaiku ordinarily emerge from mysterious sea depths. Often there's an oceanic paroxysm signaling or causing their arrival. Even though they *resemble* dinosaurs, they're amphibious and clearly unlike any dinosaurs known to traditional science. Plus, they're impossibly huge. Some are winged varieties. Okay now—ever see 2013's *Pacific Rim*, contemplating *kaiju* featured therein? By comparison we find that, generally, they're, "Like a dinosaur, kind of, only an order of magnitude larger than any dinosaur that ever lived." See where I'm going here? Bear with me on this.

The *Pacific Rim* scientist character who cracks the code for us is Dr. Newt Geiszler (actor Charlie Day), a psychologist and *kaiju* 'groupie' who neurally mind-melds with a *kaiju* brain matter in said film (as well as in the novelization from the screenplay). He learns far more beyond what only several Rangers fighting the robotic-looking *kaiju* may already suspect, as we shall soon recap here. Except, *kaiju* belching forth from the Pacific Ocean "breach" (leading into another dimension known as the "anteverse") aren't robots at all. They're organisms with silicate founded DNA, rather than carbon-based. And they evolve rapidly: those passing through the breach trend ever larger in form with each successive attack. Some of them manifest organs and components which even vaguely resemble more familiar creatures both from our present and prehistory.

What does Newt deduce? In the movie genetic dino-connections to *kaiju* are accounted for, but rather downplayed. There's nary an indication of dinosaurian *kaiju*-ness during the neural drifting Newt conducts. However, upon returning to consciousness after his first experimental drift, Newt is able to conclude that they've (i.e. *kaiju*) have been here before and that the dinosaurs were a "trial run." He also realizes that when the giant *kaiju* "Otachi" is killed by the human-controlled Jaeger Gypsy Danger in Hong Kong, that he may be able to harvest its fresher brain to perform another drift with, although this time taken from its more accessible hindquarters – its dinosaurian "secondary brain." (He ends up utilizing the brain matter of a just-born juvenile instead.) Newt possesses and wields this much knowledge in the film, but doesn't go all the way with a theory of *kaiju* origins therein—as I intend to here.

Although what Newt learns is a little murky and debatable as conveyed in the film, there's more elaboration and explanation presented within passages of Alex Irvine's 2013 *Pacific Rim* novelization from the screenplay. So what may we glean from Newt's cerebral brain-drifting *kaiju* experiences and interpretations thereof?

The fact that *kaiju* emerging from the breach during the 21st century vaguely resemble dinosaurs turns out *not* to be a mere coincidence (nor resultant of convergent evolution). Through *kaiju*-enhanced perspective offered via two mind drifts and consequent flashbacks, Newt observes that *kaiju* had breached Earth experimentally before from their dying world, first gaining a beachhead on the former continent Pangaea. But at that time they realized our atmosphere wasn't optimal for them. They would abide until the atmosphere was sufficiently polluted and thus favorable for their full force arrival. And they came later to retest the waters, so to speak 100 million years ago during the Mesozoic Era. However, conditions still weren't right for their kind.

As writer Irvine explains in the novelization, "The dinosaurs had been a dry run. Whoever sent them hadn't liked what they found. So they waited for the climate to change, and while they waited they did a carbon-to-silicate upgrade. Bam! You got *kaiju*. The silicate molecular basis gave you the additional strength you needed to get bigger and carry more mass as well as carry more information on a genetic level." Furthermore, respective organs within their massive bodies contain DNA that is identical. Despite each *kaiju*'s outright morphological uniqueness, they're not unlike clones created in assembly line fashion. In a later passage it's stated plain as (Charlie) day: "(Newt) almost told Chau that he thought the dinosaurs were an earlier, cruder version of the *kaiju*, but he just barely held himself back."

In short, the dinosaurs and other grotesque creatures of the Permian and Mesozoic ages were earlier carbon-based *kaiju* prototypes. Yes, there's an inherent anti-pollution message. The alien species spawning the *kaiju* are waiting for an Earth – inevitably polluted, just the way humans have altered our planet with its acidic oceans and ozone-depleted atmosphere. *Now* is the right time and as Newt muses, "…hell, we terraformed it for them."

The plated and spike-tailed dinosaur we've come to know as *Stegosaurus* is of particular note in the novelization, being mentioned several times by Newt following his first *kaiju* Drift. (Readers will recall that this dinosaur is of special significance when it comes to comprehending

Godzilla's suitmation design and supposed evolutionary ancestry). After noting that *Stegosaurus* had a "second brain" in its pelvic area (which is far from entirely true), Newt is able to hypothesize that recently felled *kaiju* must also, analogously, have had a secondary brain, allowing them to move incredibly fast and with remarkable precision. Also one of the *kaiju* attacking Hong Kong, known as "Otachi," has a pseudo-stegosaurian "three-pronged tail (with) serrated thorns" which, once severed, seems to have a deadly mind of its own. Anguirus, as divulged in *Gigantis the Fire Monster*, is believed to have had three brains, one situated in its pelvic region – evidently a *kaiju* trait.

Newt marvels at the awesome sight of Otachi, "The hundred – million plus years between dinosaurs and Trespasser (i.e. the very first *kaiju* witnessed from the Pacific breach) in 2013, well, that was a lot of time to refine your prototypes and get them field-ready." As Newt concludes, Otachi is *Stegosaurus'* "second draft."

But Otachi also has immense wings, grafted from that assembly line in the anteverse! As we know some of the dinosaurian (carbon-based, first draft) *kaiju* like Rodan are winged creatures. And, curiously, adult Rodans—long regarded as prehistoric anachronisms—were first seen emerging from a volcanic shaft or cavernous vent in the earth. Not from the sea, but still from a cavern *underground*. Now it's time for my 'extrapolation.'

Besides, Rodan, there's Godzilla, Anguirus, the Paleosaurus Behemoth, Gorgo and Gargantua, to ponder for example, which have all risen from the sea. Scientists offer reasons for why and how they've come to plague man … nuclear bombs, human-spread radioactive and chemical pollution, our hubris and arrogance But have we misinterpreted these alleged "warnings" all along? Paleontologists and geoscientists have certainly had things wrong before. Aren't these dino-monsters more than allegorical manifestations of Mother Earth's vengeance for wreaking planetary havoc? Aren't some, like Godzilla, sent to 'restore' a proper balance in Nature? Possibly, but …...

But maybe, just maybe, instead, they're all reawakened *kaiju* from that anteverse, having invaded Earth millions of years ago across *former* breaches that have since been rendered impenetrable through violent subterranean plutonic workings of plate tectonics and deep sea sedimentation. Maybe as a result of planetary convulsions they were left trapped within Earth's interior, retained in a dormant state of suspended animation, waiting to be set free. After all, a quarter-billion years of plate tectonics, sufficient time for our solar system to whirl around the Milky

Way galaxy through one revolution, may have sealed off former test "breaches" from the anteverse which Newt wasn't aware of.

And while we're at it, think about all those stories and legends concerning the 'hollow' Earth and its entrapped prehistoric fauna—*kaiju*, popularized by notable authors such as Jules Verne, Edgar Rice Burroughs, Lin Carter (and others). Scientists, alchemists, romantics and crackpots alike, have long dreamed of an interior Earth where peculiar organisms may yet thrive. In 'myths' and fictionalizations such as in *At the Earth's Core*'s "Pellucidar," well, we know human explorers could have never actually reached the true iron-nickel, super dense and hot "center" or very core of the earth. (Fossilized skeletons of dinosaurs were, for example, recovered from deeply dug mine shafts during the later 19th century, but *never* so deep as in "Earth's Core.")

So, instead, our intrepid 'hollow Earth' explorers must have inadvertently penetrated into much shallower, tumultuously pinched-off passages and chambers that once bled *kaiju* from the anteverse into our own. This would of course explain the exotic, prehistoric-aspect organisms found therein. It would seem that the anteverse left behind numerous *kaiju* that were either preserved in suspended animation undersea or underground, or which otherwise thrived in artificial, subterranean environments, evolving for millions of years until the present. Being closer to Earth's mantle they're more capable of withstanding higher doses of radioactivity. Every so often one or more of these dangerous creatures is 'reawakened' from entombment into modernity through various means, escaping to the surface of our planet, for a while surviving in 'lost' exotic places – then challenging civilization's armies.

One way to ascertain or falsify my unified theory would be to compare DNA of the recent silicon-based *kaiju* with carbon-based dinosaurian DNA. However, with exception of *Jurassic Park* lore, to date and at best scientists only have extracted a few kinds of biomolecules from dinosaur bone; no DNA yet. Plus, there's no guarantee they'd be able to match elements of a silicon-based to a carbon founded genome. So I guess you can't prove me wrong, huh.

Meaning it must be true?

Chapter Twenty-six—"*Things in the Mist*": *Stephen King's foray into creepy giant dino-monsters*

Yes—that is, although in the published version of "The Mist" (1980), but included as part of his collection, *Skeleton Crew* (1985). One doesn't ordinarily regard Stephen King as a writer of giant monsters, but it is clear such things festered in his mind during the writing of "The Mist." For he recalled experiencing writer's block in coming up with the story.

> "I couldn't think of a thing. The harder I thought, the more easily nothing came." Then after visiting a local supermarket, his muse struck. "…looking for hotdog buns … I imagined a big prehistoric bird flapping its way toward the meat counter at the back, knocking over cans of pineapple chunks and bottles of tomato juice. By the time my son Joe and I were in the checkout lane, I was amusing myself with a story about all these people trapped in a supermarket surrounded by prehistoric animals. I thought it was wildly funny – what the Alamo would have been like if directed by Bert I. Gordon. I wrote half the story that night …" (p.568)

Thank goodness for those elusive hotdog buns!

Now probably most of you are familiar with the 2007 movie version of his story, which doesn't offer much in the context of his original idea invoking prehistoric monsters, scaled up to gigantism via "Mr. BIG." I love the movie, but its filmic cgi-monsters are mostly 'buggy' or 'tentacular' horrors. Nothing wrong with that fearful formula for the big screen, but just for fun let's examine the original published material and see what might have been. While King evidently doesn't revere Japanese monster movies or science fiction, as we'll see, there's even a reference to 'Ghidrah' in "The Mist."

The basic 'Alamo' concept of this gripping novella, with its quasi-science fictional component, is well conveyed in the film. As in the film, protagonist David Drayton with his son get holed up in a supermarket, just when things – no!, the entire universe – … *shifts*. Out of nowhere, a milky white, slow-moving, thin acrid-smelling fog permeates over their New England town, smothering, enveloping everything around them. Those who exit the supermarket usually are soon heard screaming at the horror of unseen (or only partially visible) things in the mist. The pseudo-scientific cause of this mess seems to be an alternate dimension accessed in consequence to a mad experiment gone awry conducted nearby by military scientists—the top secret "Arrowhead Project" which locals talk about only in hushed tones.

The titular "mist," recalls opening scenes in *The Incredible Shrinking Man* (both film and novel by Richard Matheson), presaging the beginning of awful changes. But the major problem isn't the fog, or the mist itself, but rather those things that dwell within it, a horrific ecology which has entirely overwhelmed and superseded our own. Unable to see clearly within the mist, these essentially 'alien' organisms seem to thrive on a sense of smell and acoustics, using blind scent to stalk their human victims. Melded with certain phrases and passages in the novella, some of the described creatures – "Lovecraftian horrors" (p.141) - testify to Stephen King's imaginatively contrived prehistoric derivation and inspiration (as cited above).

The enveloping Mist environment is referred to in geological terms. First an earthquake shakes the vicinity, thus demarcating the new space-time order from Earth's former existence. Later, Drayton considers the as yet unseen giant things out in the parking lot as, unlike elephants, but instead "…something from the primordial ooze." (p.122) When King's winged, prehistoric flying creature invades the supermarket, "It looked a bit like the paintings of pterodactyls you may have seen in the dinosaur books…" (p.107) This monstrosity, later referred to as a "dinosaur bird" (p.124), is extinguished by the terrified group. But later, Drayton, dozing fitfully, establishes a subliminal link between the Arrowhead Project and a *gigantic* 'Rodan-sized' pterodactyl that 'emerges' from a waterspout over a lake. The nightmarish pterodactyl's "…. prehistoric wingspan darkened the entire lake from west to east. Its beak opened, revealing a maw the size of the Holland Tunnel. And as the bird came to gobble up my wife and son, a low sinister voice began to whisper … *the Arrowhead Project … the Arrowhead Project.*" (p.111) Yes – a most peculiar association between the prehistorical and the mad science project that conceivably "… had pulled this entire region into another dimension as easily as you or I would turn a sock inside out." (p.139)

Drayton, his son and a handful of others conclude that escaping the supermarket and its increasingly crazed mob represents the best course of action, and so they do manage to drive out onto the local highway in a van. After escaping in his van, Drayton drives on slowly, carefully on limited gasoline, envisioning encountering *anything* that could come out of the Mist, including *Ghidrah the Three-headed Monster*. (p.150) Well, this bit of foreshadowing leads not to Ghidrah itself. Instead something else shows up, something comparable in size! In the movie version at this point of the story, we see a huge striding creature with tentacles waving atop. But in the novella, Stephen King places the dimly witnessed creature in an alternate context:

> "It may have been the fact that the mist only allowed us to glimpse things briefly, but I think it just as likely that there are certain things that your brain simply disallows. There are things of such darkness and horror ... that they will not fit through the puny human doors of perception. It was six-legged. I know that its skin was slaty gray ...clinging to it were scores, hundreds, of those pinkish 'bugs' with the stalk-eyes. I don't know how big it actually was, but it passed directly over us. One of its gray, wrinkled legs smashed down right beside my window ... only two Cyclopean legs going up and up into the mist like living towers until they were lost to sight. ... It left tracks in the cement of the Interstate, tracks so deep that I could not see the bottoms." (p.151)

Drayton's son Billy asks if what they've just partially seen is a dinosaur. Well, certainly not one ever seen on Earth, or known to science.

True – not all of the monsters in the published story are even remotely 'dinosaurian' as opposed to simply terrifying per Mrs. Carmody's livid apocalyptical ranting. (p.141). Besides the more dinosaurian-aspect, or 'prehistoric' giant monster and "bird," there are also tentacled creatures, super web-spiders, winged bugs that fly into the supermarket, and a huge lobster-handed thing. And those parasitic bugs clinging to that immense 6-legged 'dinosaur' presage those witnessed in 2008's *Cloverfield*. Yet it is the *prehistoric* element—unusual for King—that catalyzed this story.

Despite the references to Ghidrah and a Rodan-sized pterosaur, giant monsters aren't exactly Stephen King's cup of terror tea. For instance, he doesn't really address that genre in his 1981 book *Danse Macabre*, outlining the influential history of horror and sci-fi writings and film. Therein, however, he does mention the 1961 movie, *Gorgo*, but nary a word of the

Japanese film industry. Toho fans, don't feel 'dissed' here though; RKO's 1933 *King Kong* was also omitted.

One might say, *King* was no follower of the industry pioneered by *Kong*.

Note: Page citations refer to Stephen King's, *Skeleton Crew*, Signet Books, 1985 ed.

EIGHT

(Top) Jack Arata's 2005 illustration showing Ape Gigans battling the Shark-crocodile. Original art made for hire, no copyright stated, A. Debus personal collection. See Chapter Twenty-seven for more. (Bottom) A scene from Film Classics Inc.'s Unknown Island (1948). See Chapter Twenty-eight for more.

Planet of Apes versus Dino-Monsters

The primal theme of mammal-versus-reptile seems universal. Does it have anything to do with the Cretaceous-Paleogene mass extinction event 66 million years ago, when—as the story is usually told, dinosaurs lost in the great evolutionary sweepstakes, whereas we, or rather our mammalian ancestors, 'triumphed?' Perhaps (as of this mid-2017 writing) no film has ever captured such brute forces in combat on center stage more effectively for those young at heart than did 1962's *King Kong vs. Godzilla* (U.S. release 1963). Sources indicate that Merian C. Cooper envisioned his Skull Island ape as a "prehistoric" being, as did Toho Studios with their titanic challenger, Godzilla. This movie formed a crucial part of my psyche back in 1963 when my father took me and my brother Rick to see it at a Chicago Heights matinee; it was my first Godzilla film witnessed in a theater. And for several impressionable years it became my favorite movie!

The theme and spirit of this primal battle endures on DVDs and in memory. In early 2017, *kaiju* expert Justin Mullis noted that a 1933 story he came across ("The Menace of the Monsters") related how a giant ape battled a colossal theropod-like dinosaur in a metropolitan center, thus predating the Toho epic by three decades! As mentioned in Chapter Nineteen, Don Glut tackled the theme in his amateur stop-motion 1962 flick "Tor, King of Beasts," while a handful of years later Toho did a far less convincing job with their scene of Kong fighting "Gorosaurus" in *King Kong Escapes* (1967).

But first, I must note that Chapter Twenty-nine is not a 'story,' but a tall-tale 'spoof' capped with an awful pun, inspired by a number of publications available in 2005, plus the 2005 *King Kong* film. I included it not only in memory of those 'fossil foolers' appearing my *Dinosaur Memories*

(2002)—to lighten things up a little—but also because of an interaction at one G-Fest event not too long ago. An enlightened fellow "g-fan" approached me several years after the item was published—meaning he actually *remembered* it after a handful of years—just to tell me how much he'd enjoyed that two-part *G-Fan* article. Because the concluding 'cliff-hanger' in Chapter Thirty echoes much of what has already been covered in the preceding chapter, a "part two" to this chapter, here, was omitted.

Let's now indulge in the theme of apes versus pseudo-dinosaurs for just a bit, before winding toward an inevitable conclusion. For those who might be interested, I extended this particular theme in a pairing of articles appearing in the pages of Martin Arlt's *Mad Scientist* magazine. These were titled, "From Kong to Godzilla: A Survey of Giant Dino-Monster Novelizations" (*MS* no.29, Winter 2014), and "Battle of the Giant Monster Novels: King Kong vs. Godzilla" (*MS* no.30, Winter 2015).

> These chapters originally appeared as articles in the following publications:
> Chapter Twenty-seven—*Scary Monsters Magazine* no.57, Jan. 2006, pp.46-49.
> Chapter Twenty-eight—*G-Fan* no.76, Summer 2006, pp.38-41.
> Chapter Twenty-nine—*G-Fan* nos. 78-79; Fall, 2006, pp.10-14; Spring 2006, pp. 23-26.
> Chapter Thirty—*G-Fan* no. 117, Fall 2017 (in press).

Chapter Twenty-seven—*Two Mysterious Monsters of Early Science Fiction Literature*

Okay—quick! Which pair of monsters were created 64 years after Frankenstein's Monster, yet decades *before* literary invention of other classic creatures such as Count Dracula, Phantom of the Opera, the Morlocks, and Mr. Hyde? Furthermore, *who* created this largely forgotten pairing of abominations? Give up?

While the answer to the first question is relatively obscure, we can only speculate about the latter. The two neglected monsters in question are (1.) Ape Gigans, and (2.) the Shark-Crocodile, both of which appeared in 1872 and 1873 editions I've read of *Journey to the Center of the Earth*. (According to Arthur Evans—a Verne scholar, the earliest known English translation dates from 1871.) But before you all cry out 'Jules Verne's' hallowed name in unison as their alleged creator, I must hold you at bay, waving my lit torch of trivia.

No—it was decidedly not Jules Verne who created Ape Gigans and Shark-Crocodile, as Jeff Rovin, author of *The Encyclopedia of Monsters* would otherwise have you believe. Rovin mentions the beasts in two entries in his enjoyable book, while uncharacteristically providing erroneous background information. So it's time to describe these monsters, while outlining how they ever got into chapter 40 of Verne's *Journey to the Center of the Earth.*

Ape Gigans is described as an "antediluvian gorilla," or the progenitor of the hideous monster of Africa. "Fourteen feet high, covered with coarse hair, or a blackish brown, the hair on the arms, from the shoulder to the elbow joints pointing downwards, while that from the wrist to the elbow pointed upwards … Its arms were as long as its body, while its legs were prodigious. It had thick, long, and sharply pointed teeth—like a mammoth saw." When approaching the frightened explorers, Ape Gigans groaned sounding like "fifty bears in a fight." Its opponent, Shark-Crocodile, is "…about the size of an ordinary whale, with hideous jaws and two gigantic eyes … his jaws were at least seven feet apart and his distended mouth looked large enough to have swallowed a boatful of men." While,

"… much of his body resembled that of a crocodile, his mouth was wholly that of a shark." The carnivorous Shark-Crocodile has relatively puny-sized legs. To me, Ape Gigans sounds more or less like Kong, while the Shark-Crocodile resembles and, in part, may have been derived from descriptions of fossils of a real Mesozoic aquatic saurian known since the late 18th century—*Mosasaurus*.

Interestingly, Ape Gigans and Shark-Crocodile may rank among the fifteen oldest monsters of (modern) fantastic fiction. Ape Gigans and Shark-Crocodile made their only appearances in a short sequence added to English editions (e.g.1871(?), 1872, 1873) of Verne's *Journey*. (However, I haven't checked all comics adaptations of Verne's novel. See Note at end.) But Verne didn't write these particular passages of the story and neither paleo-monster appeared in previous (1864, 1867) French editions of "Journey." So an unknown author, presumably the same individual who translated Verne's novel into English, may have been the culprit. Perhaps creativity temptation got a little out of hand and he just couldn't help himself from crafting a bit of fiction himself.

Details of Verne's *Journey* novel differ significantly from the best filmed version (20th Century-Fox, 1959, directed by Henry Levin). Yes, surely the basic idea is quite the same. After deciphering a runic cryptogram written by alchemist Arne Saknussemm, a geology professor leads an expedition into a volcanic vent hoping to arrive at the 'center' of the Earth. Tracing Saknussemm's footsteps, the explorers cast off for Mount Sneffels in Iceland, and after penetrating eighty-seven miles inside the crust, they encounter a variety of 'prehistoric' organisms (including flora, fauna and fossils). Prehistorical elements of their life through geological time journey are much more pronounced in the novel (especially the 1867 French edition) than in the 1959 movie, and this is where the Ape Gigans and Shark-Crocodile fit into my tale. (For more background on inspirations leading to Verne's novel, read William Butcher's Introduction to the 1998 Oxford University Press edition.) Consequently, Verne's Central Sea (or more correctly, the "Lidenbrock Sea") is enlivened with representations of prehistoric life.

When I was a young boy I avidly listened to an old 33-1/3 RPM LP recording titled *Space Stories and Sounds* narrated by Bill Stern (produced by MGM—Full Fidelity Lion; L-70086), which on side 2 contained a quick synopsis of Verne's "Journey." Stern dramatized a monster scene from the 1864 novel in which two colossal aquatic lizards—the *Plesiosaurus* and *Ichthyosaurus*—battle to the death, while the terrified exploring team is tossed to and fro in the mighty waves onboard a raft made of fossilized wood (which wouldn't be expected to float anyway). This cataclysmic episode was illustrated by Edouard Riou for "Journey's" 1867 French edition.

So, understandably, when I saw the 20th Century-Fox film shortly thereafter, I was disappointed to see that this exciting scene had been omitted. In fact, there is really only one significant example of prehistoric life which made its way from the novel into the 1959 now-classic film—a boring forest of giant mushrooms, which in the novel dates from an ancient period (the Silurian). And our favorite prehistoric monsters from the 1959 movie—that marvelous herd of sail-backed *Dimetrodon*, was a genus that hadn't been scientifically described until the early 1870s, far too late for Verne to have known of them at the writing of his novel. Just 'for the record,' Stern's narration for my old favorite *Space Stories and Sounds* recording didn't include references to Ape Gigans and Shark-Crocodile; following his terrific marine saurian combat description, perhaps it would've sounded too repetitive.

In his novel, Verne included scenes with prehistoric fish, pterodactyls, paleo-mammals, and man. However, the Ape Gigans and Shark-Crocodile were not his literary handiwork. One subtle clue can be spotted in that phrase where the anonymous author described the Ape Gigans as the "progenitor" (an evolutionary term?) of the African gorilla. Although Verne wrote about prehistoria, he was decidedly not an evolutionist, disagreeing with Charles Darwin's theories. Verne didn't read English; conveniently, the first French edition of Darwin's *The Origin of Species* appeared in 1863.

No, Verne's prehistoria, situated inside Earth's caverns, were 'antediluvians' that got there when flooding took place through fissures and faults within Earth's overlying crust. Besides Darwin's contentious book, *Journey* was highly influenced by ideas and theories contained in a popular and far less controversial geology book available in France—Louis Figuier's *Earth Before the Deluge* (1863, 1st ed.). While Figuier's book was translated into English during the mid-1880s by British geologist Henry Bristow, there's no clue as to whether Bristow also may have been employed in translating Verne's classic novel. Until 2005, when Jack Arata undertook the assignment from me for a *Scary Monsters* magazine article, no one (including Riou) had illustrated the ensuing battle between Ape Gigans and Shark Crocodile.

I close with passages excerpted from an 1873 edition reproduced by Signet Classic in 1986 (an edition described by Verne 'purists' as bowdlerized), describing frightening combat between these two early mysterious paleo-monsters of sci-fi literature. Perhaps someday the author of this passage—as 'narrated' by 'Axel,' (not 'Alex' as played by Pat Boone in the 1959 film) will be identified! This scene seems oddly influenced by Verne's dramatic *Plesiosaurus* vs. *Ichthyosaurus* scene, staged earlier in the novel:

> "The gorilla placed himself on the defensive, and clutching a bone some seven or eight feet in length, a perfect club, aimed a deadly blow at the hideous beast, which reared upwards and fell with all its weight upon its adversary. A terrible combat, the details of which it is impossible to give, now ensued. The struggle was awful and ferocious. I, however, did not wait to witness the result. Regarding myself as the object of contention, I determined to remove from the presence of the victor. I slid down into my hiding place, reached the ground, and gliding against the wall, strove to gain the open mouth of the cavern. But I had not taken many steps when the fearful clamor ceased, to be followed by a mumbling and groaning which appeared to be indicative of victory. I looked back and saw the huge ape, gory with blood, coming after me with glaring eyes, with dilated nostrils that gave forth two columns of heated vapor. I could feel his hot and fetid breath on my neck; and with a horrid jump—awoke from my nightmare sleep."

What!? Nightmare *sleep*? Yes, it was all a dream (akin to another much more significant 'waking dream' Axel has already experienced aboard the raft made of fossilized wood). And (boo! Hiss!) Axel never even witnessed the battle while it was raging at is full tilt and intensity. No question that, overall, Verne was the much better writer. But notice how uncannily this scene presages another, the Kong—*T. rex* stop-motion animated battle staged 60 years later by RKO! While the moniker of the mosasaur-like Shark-Crocodile seems imaginatively inspired by the real *Ichthyosaurus* (whose name means "Fish-lizard"), Ape Gigans precedes more famous apes of monster moviedom.

Note—Interestingly, Ape Gigans and Shark-Crocodile may rank among the fifteen 'oldest monsters' of (modern) fantastic fiction. That is, if besides Frankenstein's monster we count Victor Hugo's Quasimodo or the Hunchback of Notre Dame created in his 1831 novel, John Polidori's 'Vampyre' (1919), Thomas Prest's 'Varney the Vampire' (1847), Sheridan Le Fanu's 'Mysterious Stranger (1860), and 'Carmilla' (1872), three additional early werewolf stories, and several prehistoric creatures—particularly a long-necked plesiosaur and ape-men described and illustrated by Pierre Boitard in his 1861 novel *Paris Before Man*. Also, the Classics Illustrated 1957 "Journey" comic book allied to *Space Stories and Sounds* did not incorporate Ape Gigans or Shark-Crocodile. For more on this, see my article, "A Record of Space & Time," *Mad Scientist* no.22, Fall 2010, pp.25-32.

Chapter Twenty-eight—*From "Incognitum" to Odo: Prehistoric Roar*

In the midst of late 2005 *King Kong* fervor, it's relatively easy to neglect the strangest 'Kong'-related film of them all. While the matter isn't completely unknown, no writer has yet explored this tangential movie oddity (to be unveiled shortly) stemming from the art of curious association.

Toho films and the Japanese didn't invent the idea of featuring giant dino-monster battles in movies, let alone published sci-fi stories. For instance, long before Toho's first such entry, *Godzilla no Gyakushu* (1955), National Pictures released a 1925 silent film *The Lost World* with its dinosaur fights (based on Arthur Conan Doyle's novel). *The Lost World* featured several stop-motion animated scenes pitting fearsome *Tyrannosaurus rex* versus horned dinosaurs, no doubt inspired by Charles R. Knight's contemporary museum paintings. Several years earlier, similar scenes highlighted World Cinema's fifteen-minute picture—*The Ghost of Slumber Mountain* (1919). Nearly half a century before, French writer Jules Verne frightened readers with a terrific aquatic saurian combat, waged between two 100-foot long sea-monsters, *Ichthyosaurus* versus *Plesiosaurus* in a chapter of his 1864 novel, *Journey to the Center of the Earth*.

In 1955, Toho took matters to an unprecedented extreme in *Godzilla no Gyakushu* (U.S. released title—*Gigantis the Fire Monster*, 1959). Here, a new revived colossus—pseudo-ankylosaurian, Anguirus (a.k.a. Angilas) fought Gigantis, introduced as another individual of the same Godzillean stock which had been defeated before with the oxygen destroyer. Their titanic struggles spilled over to Japan's Osaka, where buildings toppled, the military was defused and innocent people were burned and crushed to death.

And for film trivia buffs, there's an additional three second scene of two fighting stop-motion sauropods in *Godzilla no Gyakushu*. While film experts haven't been able to trace the true origin of this short scene, it *may*(?) be footage chopped from *The Lost World*. Interestingly, *Gigantis* also incorporated a short sequence spliced in from an American production,

Unknown Island (Film Classics Inc., 1948), a title to be discussed momentarily.

A pairing of dino-*daikaiju* fighting their way through civilized areas of the globe certainly became all the rage in Japanese science fiction/horror film following *Godzilla no Gyakushu*'s release. In fact, no production company had ever or has since tackled this particular theme more brilliantly than Toho. Hollywood may have initiated the trend toward battling monsters with Universal's *Frankenstein Meets the Wolfman* (1943), although both monsters were human-sized and non-prehistoric. In the two decades leading to *Godzilla no Gyakushu*, Hollywood released at least two fantastic films—RKO's *King Kong* (1933) and United Artist's *One Million B.C.* (1940), featuring dinosaurians which battled other 'prehistoria' in unknown, misty, prehistoric places instead of in urbanized settings.

In all the better Kong movies, the colossal, prehistoric ape eventually battles a theropod-type dinosaurian … to the death, as he did in RKO's production and also Delos W. Lovelace's 1932 novelization of the original screenplay. Plus, against all seeming odds, Kong never loses to his worthy theropod adversaries, as once again demonstrated in Peter Jackson's 2005 Universal Pictures version. The best scenes of an earlier Kong entry, *King Kong Escapes* (Toho, 1967), highlight a suit-mation battle between Kong and a gigantic tyrannosaurid referred to as Gorosaurus. Kong also struggled mightily against a mutated raptor-like dino-monster named Gaw in Joe DeVito's and Brad Strickland's illustrated novel, *Kong of Skull Island* (2004).

Arguably not the best, but certainly the most titanic, anticipated and (for me) inspirational battle ever waged between Kong and a pseudo-dinosaurian was produced on film—*King Kong vs. Godzilla* (U.S. release 1963). This was THE giant monster movie that really hooked me at a young, impressionable age. Instead of finishing the battle in a primitive island setting, Kong fought Godzilla in urbanized Japan. Only Kong surfaces from the waves at the end of the movie after both titans flop into the sea, suggesting that, as in his prior RKO performance, another pseudo-theropod dinosaurian—i.e. this time Godzilla—has been vanquished. (Gigantis and Anguirus had themselves tumbled into the ocean in an early *Gigantis* scene.)

Okay—probably so far none of this is new to Godzilla fans and experts. But where did this trend of battling dinosaurians toppling buildings and brushing humanity aside really begin? That is, who originally conceived, as documented in writing, the earliest 'proto-battle' between prehistorians, anticipating all the famed painted and filmed dramatic portrayals of fighting paleo-combatants, static and flowing restorations mirroring man's subjective impressions of prehistoric violence and savagery, witnessed both in primitive settings and civilized areas of the globe, toward which the human mind experiences such visceral reaction?

Nope, it wasn't Conan Doyle's invention; nor was it Verne's. You'll never guess the answer to this question.

As a clue, note the strange gorilla-suited monster battling a horned tyrannosaur in the 1948 movie, *Unknown Island*. From glimpsing the suit-mation scenes alone, it would almost seem as if *Unknown Island's* producers anticipated *King Kong vs. Godzilla* by fifteen years. But, hey, that's not Kong, nor is it even intended to be an ape, although suit-mation sizes of Toho's Kong and *Unknown Island's* 'ape' were similar. Donald F. Glut discusses the gorilla suit worn by actor Ray 'Crash' Corrigan for *Unknown Island* in his 1978 book, *Classic Movie Monsters*. In fact, Glut explains, this 'ape' was intended to be a *giant ground sloth*! 'No way,' I hear you saying. Yes, "WAY"! But aren't giant ground sloths considered the most dullard of all the ranks of prehistoric vertebrates one could choose from to make a loosely-termed 'dino-monster,' let alone early form of *kaiju*? While it may seem so today, in the beginning, giant ground sloths were all the rage.

It was especially through study of giant ground sloth remains that paleontology, in a larger sense, matured to its present state. For a century, scientists' discoveries of, particularly, *Megalonyx*, *Megatherium*, *Eremotherium*, and *Mylodon* cultivated important scientific themes prevalent during the 1790s, through the late 19th century. While in America, dinosaurs, horned Tertiary mammals, toothed birds and even faked fossil monsters inspired dramatic headlines, for a century, the less publicized, venerable giant ground-sloths remained on vertebrate paleontology's ideological forefront. For vested within their fossil traces was an invaluable microcosm or proving-ground of theoretical ideas.

While it has been suggested that the American Mastodon (i.e. a genus of prehistoric proboscidean, or elephant) represented the New World's first culturally and scientifically significant 'monster' shown to be prehistoric, such a claim diminishes the importance of the giant ground-sloths, creatures attracting attentions of stalwart paleontological pioneers such as Georges Cuvier, Charles Darwin, William Buckland, Charles Lyell, Joseph Leidy, Richard Owen and, yes! … even former president Thomas Jefferson.

Which is why I found *Unknown Island* so provocatively peculiar. And that is your final clue.

Besides introduction of a pseudo-sloth and the fact that this was the first (non-animated, i.e. besides Disney's *Fantasia*, 1940) dinosaur movie filmed in color, there's nothing all that original about *Unknown Island's* plot, based on Robert T. Shannon's decent story. (According to Mark Berry, in 1970, a giant ground-sloth was also planned for an unfinished production "Zeppelin vs. Pterodactyls.")

Unknown Island was recently released on DVD so you can readily enjoy your own purchased copy. Basically, the story is a Skull Island/Lost World-ish morph, with explorers and animal collectors returning to a taboo island one of the party glimpsed from the air in a war plane years earlier. After their ill-advised arrival, they're attacked by a herd of savage 'cerato-tyrannosaurids,' a fin-backed *Dimetrodon* and the gigantic sloth, prop designs inspired by paleoartist Charles R. Knight's paintings published in the February 1942 *National Geographic Magazine* issue.

Of the sloth, Berry remarks, "The 'giant sloth' is a puzzling creation, but veteran movie ape-man, 1930s cowboy-actor Ray 'Crash' Corrigan somehow breathes some menace into it." Rather like Kong, the sloth eventually engages one of the rubbery suit-mation theropods in a climactic struggle, with the latter falling over a cliff, hardly an original conclusion to an epic dinosaurian battle even then.

Corrigan owned his own ape suit which he customized for the film by adding extra fur, head and facial enhancements. However, his sloth suit lacks the characteristic, enormously powerful tail which these animals used to anchor themselves on the ground while tearing down tree limbs or uprooting tree trunks (as was popularly thought), and has far, far too many teeth. Like the dinosaur-suited actors suffering in 100-degree heat during filming (one of them fainted while on camera—a scene which made it into the final cut), Corrigan must have also been rapidly dehydrating in his sloth-ape costume (even though he was not 'aping' a sloth so convincingly).

While it's tempting to snicker at Corrigan's sloth, so ineffectually reprising Kong's RKO performance, now try to regard these scenes from a different, perhaps more enlightened perspective. Instead of another bad instance of Hollywood-gone-ape, what we're being treated to here is a sell-'suited' appearance of an original, charter member prehistoric monster—*Megatherium* in a 'fitting' long overdue sci-fi role. You see, because of their 19th century paleontological prominence, from the time of Verne's novel through the 1920s, giant ground sloths made repeated appearances in published science fiction novels and stories. However, they've been curiously neglected in science fiction film ... that is, until *Unknown Island* came along. Would it be too magnanimous to proclaim that it's almost as if

Unknown Island's producers realized the historical significance and essentiality of their (relatively short-tailed, fanged) paleo-monster sloth? So, in resurrecting the giant ground sloth onto this particular film, someone in the film industry finally spoke up for them? Probably, but still, bloody good show!

Now for the final answer. Where did the time-honored tradition of fighting prehistoric monsters all begin? All right, here it is.

Incognitum. That means "unknown." As in perhaps "Unknown" Island's title, like where a giant ground sloth attacks cerato-tyrannosaurids? Like King Kong's Skull Island too. But much, much earlier, during the late 18th century, the term *incognitum* was formerly and rather loosely applied to "lost" animals, those no longer appearing extant, such as the tusked and presumably carnivorous Mastodon, or its hairier elephantine cousin, the Woolly Mammoth. These were termed unknown 'incognita' by early naturalists because until the 19th century, their natural history remained so poorly understood; for a time scientists didn't even realize these mysterious species were prehistoric.

Eventually, in a 1797 presentation, fragmentary and incomplete fossilized remains of the first known North American giant ground sloth were named *Megalonyx* (or the "Great Claw") by Thomas Jefferson. For our purposes here, Jefferson also envisioned the first epical battle between such species which turned out to be prehistoric! Just imagine the potential billing on this paleo-fight!

"Great Claw vs. *Incognitum*"! (Or, perhaps in modern parlance, *Megalonyx* vs. Mammoth!)

That's what Jefferson imagined, as documented in his seminal American Philosophical Society paper published in 1799 where he described *Megalonyx* (i.e. *Megatherium's* giant ground sloth evolutionary cousin) suggesting further "… that he was more than three times as large as the lion; that he stood as pre-eminently at the head of the column of clawed animals as the mammoth stood at that of the elephant, rhinoceros, and hippopotamus: and that he *may have been as formidable an antagonist to the mammoth* as the lion to the elephant." (My italics.) In another passage, Jefferson stated what a formidable "destroyer" was his (megalonyx) Great Claw!

Well, there you have it. That's the earliest documentation we have of battling prehistoria as envisioned by modern, intellectual man … and by founding father Thomas Jefferson, no less. It should be stated that Jefferson doubted whether his conceivably one-ton weight 'super-lion' *Megalonyx* was extinct, as he was a 'crypto-zoologist of his time. But weren't Gigantis,

Anguirus, *Unknown Island's* cerato-tyrannosaurids and Corrigan's sloth (fictionalized) *cryptozoological* monsters too, and therefore, In a sense, creatures new to science, or 'living fossils,' like Jefferson's perceived *Megalonyx*? Jefferson and Conan Doyle had similar fertile imaginations.

The germ of the incorrect idea of *Megalonyx* as 'super-lion' was apparently planted in Jefferson's mind through powers of suggestion by one Col. John Stuart, in 1796. Stuart also convinced Jefferson that hunters had experienced close encounters with 'super-lion' Great Claw—although it wasn't *that* 'lion'—around their campfires at night when they heard awful roaring sounds.

So, yes, long before Toho's Godzilla successes and invention of a steady stream of *kaiju* eiga battling through downtown city streets, or Conan Doyle's "Lost World" monsters ... still before Skull Island's Kong and Verne's battling sea saurians, the dramatic and appealing idea of the (primeval) monster fight—although staged in modernity—was conceived and documented for the first time by Thomas Jefferson.

Now as twisted as it may seem, try to imagine a gathering of evidently like-minded souls sitting round a campfire, tellling tales of natural mystery and bestial horror ... an occasion shared by Eiji Tsuburaya as well as Jefferson (who are all then strangely linked through giant ground sloths and peculiar gorilla suits), as well as Conan Doyle, Verne, Charles Knight and Kong's creator Merian C. Cooper who would fit right in too. (You can supply a Japanese-English translator.) And they're all having a grand ol' time until, disturbingly, something out there in the woods—something *unknown* begins to roar.

Note—In 1942, Jefferson's paleontological views were heavily critiqued by prominent biologist George Gaylord Simpson who stated that Jefferson's *Megalonyx* paper was "regretfully poor." Indeed, Jefferson's imaginative paper owed much to outright speculation: it certainly wouldn't pass muster today. For considerably more concerning "Incognitum" and Great Claw, see Chapters Two through Four in my 2010 *Prehistoric Monsters* volume.

Chapter Twenty-nine—*Godzilla vs. King Kong: Skull Island's 'True' (alternative) Tale*

What have writers of the past documented about these two old giant monsters, and, quite apart from the movies, is it possible to piece together their un-natural histories from dusty pages?

Quite a bit has been written about the 1962 Toho movie, *King Kong vs. Godzilla* that, for many impressionable American G-fans of the time, represented first exposure to Japanese *kaiju*. According to J. D. Lees, Godzilla was the only *kaiju* featured in this film though, because "... *kaiju* eiga means 'mysterious beast,' and there's little mystery about Kong. He's an oversized gorilla, and he doesn't possess any mysterious powers. Kong definitely lacks a sci-fi element, which is an essential part of the *kaiju* genre." (*G-Fan* no.74, p.8) Perhaps Kong could loosely be envisioned as a surviving variant, or enlarged quasi-evolutionary descendant of the real 10-foot tall Plio-Pleistocene genus, *Gigantopithecus*. And can it be that the Kong of celebrated movie-dom is somehow less 'mysterious' than the Kong of fantastic literature?

Fortunately (again apart from those familiar films), several writers have delved into Kong's murky past, facilitating our ability to judge just how mysterious Kong really is in fantastic literature. Be mindful that besides their works, which shall be summarized and reinterpreted here, there are a number of other manuscripts and testimonies lurking about whose authenticity on the matter may be considered, at best, doubtful. Perhaps chief among them is a comical 1976 book purportedly narrated by Kong himself. In this extreme oddity, *My Story* (as divulged to writer Walter Wager), Kong claims to have been born on February 29, 1912, and tells of his capture, ultimately resulting in his 1933 acting exertions for RKO. In discussing his later film career, Kong had this to say concerning a certain famous co-star: "I gave the silver screen another chance in 1963 when I starred in 'King Kong vs. Godzilla,' a harmless Japanese romp that co-starred a young kid with horrendous breath. Fire shot from his mouth, and his posture was atrocious. I taught the kid plenty…" Yeah, right! As if the real Kong could talk (to humans). Well, I'd wager that Wager's Kong was one and the same as that twisted ape-actor Carmen Nigro who died at 84 in 1990, no doubt wearing his infamous gorilla suit while telling Kong's 'true' story to gullible listeners. Anyway, as you'll see, other more reliable documentation supports that the real Kong was born prior to 1912.

Donald F. Glut has listed a number of other publications (and actors) allegedly offering ('true') insights as to Kong's life history in Chapter VIII of his *Classic Movie Monsters* (Scarecrow Press, 1978, pp.311-312, 358-359). Movie producer Glut's own novel *Frankenstein in the Lost World* (1976, 2002), featured a giant ape named Tor that defeats a tyrannosaur in bloody combat in a mysterious primeval setting which can only be accessed through a chronal (time) eddy. But as we know, time warps don't exist anywhere on Earth; Glut's tale of Tor is clearly a yarn. Total fiction, albeit inspired by Kong—although decidedly not a record of Kong himself.

In a 2003 article, another Kong scholar, M. J. Simpson, made readers Kong-nizant of Kong's many tales in print, listing obscure entries such as Walter F. Ripperger's 1933 *Mystery Magazine* serial, on through a spectrum of 2005 publications. Be apprised, however, that many of these were Kong ripoffs, designed to cash in on Kong's genuine celebrity, often merely borrowing from second-hand information. Accordingly, these doubtful gems shall all be weeded from further consideration here as we delve into Kong's real tale of mystery and his illustrious past. Therefore, herein, in attempting to splice together the true natural histories of this mighty pairing of pseudo-prehistoric beasts, we shall only rely upon authentic records pertaining to Kong and Godzilla.

In Godzilla's case, we have a lesser selection of trustworthy documentation to rely on (disappointingly, the key Japanese 1950s novel, *Godzilla*, has never been commercially translated into English). So, rather than anticipating how *King Kong vs. Godzilla* might have been novelized from a screenplay, instead let's reconstruct what may have really happened so many decades ago by focusing on what writers and journalists, chiefly Delos W. Lovelace, Marc Cerasini, and Brad Strickland with John Michlig, and of course, 'Carl Denham' have said about these beasts. The strange tale which unfolds will take us from Skull Island's confines and eventually to Odo Island in the Pacific.

While the earliest published novel of concern to us here would be Lovelace's *King Kong* (1932), for glimpses of Kong's younger life and rearing we must instead first turn to the masterful *Kong-King of Skull Island* (2004), both a sequel and 'pre-quel' to Lovelace's treatment, written by Strickland and Michlig (conceived, created and illustrated by Joe DeVito). In tackling King of Skull Island, however, not only are we struck with the fact that Kong was menaced by some decidedly bad dinosaurian dudes, but that, apparently, differing 'gospels' have been written about Kong's early life as well as Skull Island's horrific fauna and natural history. We'll attempt to weave the elements of these apparent 'parallel universes' together into a sensible whole. Shall we begin?

Lovelace's Narrative (circa 1932):

Delos W. Lovelace incorporated prior written and conceptual elements by (in order of significance) Merian C. Cooper, Ruth Rose (and James Creelman), and Edgar Wallace, who all seemed to know something about the *real* goings-on surrounding the notorious figure, Kong's promoter—Mr. Carl Denham. With resources like that to draw upon, perhaps it's no wonder that Lovelace's novel is exceptionally well done. From the outset, Cooper was influenced by American Museum of Natural History zoologist W. Douglas Burden's discovery and celebrated capture of Komodo Dragons (from Komodo Island in Indonesia), and a book, *Explorations and Adventures in Equatorial Africa*, by explorer Paul Du Chailu, who had written about the capture of a wild gorilla. In an RKO company memo written on December 18, 1931, Cooper wrote about a "prehistoric giant gorilla, fifty times as strong as a man—a creature of nightmare horror and drama" who would become the star of his next film whose working title was "Giant Terror Gorilla." Was the creature based on fact? Certainly! After Ernest Schroedsack's wife Ruth Rose refined a screenplay, Cooper entrusted his friend, journalist Lovelace, with the task of novelizing the script for publisher Grosset and Dunlap.

Kong's story is a 'beauty and the beast' tale, and Kong himself is referred to throughout the novel as a mysterious "beast-god." As Carl Denham predicts, however, to know actress Ann Darrow is to go 'soft' for the beautiful girl—that is, losing ability to maintain one's composure and dominance. It is Ann who suggests Kong's 'prehistoric' nature after spying Skull Mountain's Great Wall. When movie producer Carl Denham deduces that Kong may be an enormous gorilla, Ann exclaims (in the novel, although probably verbatim), "But there never was such a beast ... At least not since prehistoric times." And soon she is cooroborated, as Captain Englehorn's, i.e. Captain of the ship *Wanderer*, crew encounter living dinosaurs and other scaly prehistoric reptilians inside the jungle behind the Great Wall. (They also find huge insects with possible 'prehistoric' connections corresponding to those real gigantic bugs of the Carboniferous Coal Age, and an "asphalt morass" clearly inspired by southern California's La Brea Tar Pits, although this dating from the recent Ice Age.)

Denham, with the most scientific mind of all the characters in Lovelace's narrative, professes "Prehistoric life! ... (Ann) was right last night on the ship ...But she only had the start of it. She guessed the beast-god was some primitive survival. But if this thing we've killed means anything, the plateau is alive with all sorts of creatures that have survived with Kong." And they're absolutely right. Later, Denham suggests that Kong is "... something more than beast. He's one of nature's errors, like all the others, but he was ... not an error. And in that huge head of his is a spark. Ann means something to him."

Besides the giant bugs, the fearsome reptilian fauna—'Nature's mistakes"—seems hell-bent on destroying every last human intruder. As we'll see, this element of Lovelace's tale seems entirely consistent with that of the two Denhams (father and son). First gun and gas-toting sailors and crew are attacked by a spiky-tailed quadrupedal dinosaur which is dispatched with a gas bomb and bullets. Next a brontosaur chases them out of the river. Then after Kong battles a herd of three-horned *Triceratops*, Ann attracts a "...grotesque, hopping creature of very little less than Kong's own bulk." This is the carnivorous "tyrannosaur" which would like to make a snack out of Ann; she is spared only because of Kong's heroic exertions. But no, as we'll soon learn, this witnessed 1932 struggle wasn't battle-hardened Kong's first tussle with a theropod dinosaur nor would it be the last. And although this carnivorous 'King Dinosaur' isn't a *kaiju* eiga-saur, Lovelace's combat scene is perhaps one of the best written in all of dino-monster literature. (In Lovelace's manuscript Kong battles what many interpret from his description to be a tyrannosaur immediately after defeating *Triceratops*, linking "beast-god," Rex, and 'Tops'—the primeval triumvirate.)

There's more to the prehistoric theme as well. Kong shivers after killing an aquatic Elasmosaur, as he was "... of the loathing his species has had of reptilian things since the dawn of time..." Finally, before his capture and voyage to New York where he becomes symbolic of (prehistoric) savage nature exploited, Kong easily dispatches a winged pterodactyl.

Much of Kong's misfortune in New York has already been played out metaphorically or analogously during the sea voyage to Skull Mountain leg of the story. For instance, Ann has befriended a sailor's pet monkey, Ignatz, and climbed the *Wanderer's* Crow's nest with Driscoll (foreshadowing their ascent to the top of the Empire State Building later). The islanders, like Kong himself in New York, represent racial 'inferiority' to be exploited, and a giant-toothed variety of *Pteranodon* (of the same genus, yet smaller species than the Japanese Rodan) foreshadows biplanes which will fell Kong from New York City's highest point. Evidently, modern civilization is removed from a prehistoric society only by degree. Just how evolved and 'civilized' are we, after all?

Many more of us have probably seen the 1932 RKO film than those of you who have read Lovelace's novel. So please be aware that our opinions of Kong and Skull Mountain's primeval bestiary are fully biased by that never-to-be-forgotten, disbelief-suspending trick photography, so deftly employed by those marvelous magicians—Willis O'Brien and Marcel Delgado. So I submit to you that their film was merely an elaborate illustration of journalist Lovelace's written account ('work' with me on this, okay?), and that for the true representations of the animals present on Skull Mountain/Island (i.e. "Mountain" in the 1932 novel), we must instead reinterpret Lovelace's notes, obtained secondhand from Denham's and Driscoll's verbal, first-hand account and, to a considerably lesser degree, from the *Wanderer's* other survivors as well.

Lovelace (who may be forgiven since he was a journalist, not a scientist!) presumes that dinosaurs found on the island are exactly those species already known to science, without acknowledging that such island survivors would have evolved in over 66-million years! (We'll return to this point shortly.) Also, bear in mind that Lovelace did not participate on the *Wanderer's* expedition (despite those unsavory rumors connecting him to Ann Darrow).

First, that 'spiky-tailed' quadruped was interpreted by the movie masters of illusion as a *Stegosaurus*, although without real footage or a photo of the animal on which to base their filmic restoration, there's no guarantee at all of accuracy. Lovelace's description doesn't even mention a row of plates along the vertebral column which is what stegosaurs are famous for, so that particular creature may have not even been dinosaurian!

However, Carl Denham's Project Legacy diary of 1937—which we'll come to shortly—does record existence of a 20-foot long, plated variety informally named "Atercurisaurus," equipped with numerous sharp defensive spines along its hip and tail in addition to stegosaurian-sized plates over the backbone. So maybe that's the animal actually confronted by the *Wanderer's* crew, although mistaken for a genuine, though long-extinct *Stegosaurus*.

Next, Lovelace describes three dinosaurian encounters in swift succession. An aquatic brontosaur chases the men from a raft onto the shore; then they're attacked by a trio of belligerent, charging horned dinosaurs attributed by Lovelace to genus *Triceratops* (which Kong dispatches using slabs of hardened asphalt as hurled missiles), followed by Kong's mighty clash with the aforementioned 'hopping' carnivore that was transformed through the magic of cinema into a roaring, stomping *Tyrannosaurus* for the 1933 RKO feature film. But was this energetic crypto-zoological creature truly a *T. rex* as known to science, or was it perhaps instead an enlarged form of raptor-like creature, or something more akin to the kangaroo-leaping "Laelaps," depicted in 1897 by Charles R. Knight as fierce, bounding beasts? And were those horned dinos really genus *Triceratops*? We'll need further information, which we'll come to shortly, to help us decide.

Note that none of the dinosaurs were described by Lovelace as 'tail-dragging' creatures (an indication of cold-bloodedness), and they all displayed remarkable energies in their motion and agility, especially during the carnivore's climactic struggle with Kong. Therefore, I suggest that the dinosaurs Lovelace described were in many ways consistent with others' descriptions. And this is my 'segue' to the fabulous Project Legacy notebooks maintained by Carl Denham himself, who did indeed return to Skull Island in the mid-1930s.

Carl Denham's Project Legacy Diaries:

Perhaps he fled seeking anonymity following Kong's New York City escapade. Then again, Carl Denham may have been strangely enchanted by horrors that lived on Skull Island. Perhaps dreaded memory of its deadly fauna summoned him, a 'death wish.' Or was Denham simply seeking soul-maddening confirmation that he ultimately was indeed responsible for the death of the Kong race's last mighty member? Perhaps it was partly for solace or in penance for his cruel crimes—the exploitation of nature, that he felt compelled to return and atone. No, Kong never had a son despite what RKO's executives may have led you to believe. Carl Denham did return with Kong's remains, however, which were interred beyond Skull Island's immense wall. But unlike Kong, Carl Denham did have a son!

In Denham's notebooks, which were polished and published finally in 2005 by the Weta Workshop under Peter Jackson's direction, we learn of Project Legacy, in which university scientists returned with Carl Denham during the mid-1930s to record the prehistoric fauna and document the island's strange and aberrant natural history. While this book suggests that team members perished during the explorations, the fact that none of us had ever heard of this information until recently indicates that the notebooks were not known to the western world until following Carl's son's—paleontologist Vincent Denham's—trip to the island in 1957, when he met his father and was presented with all the precious Project Legacy records, so meticulously compiled and under such dangerous conditions through two decades.

It seems as if Carl may have been Project Legacy's sole survivor. Several top-secret expeditions were sent to Skull Island, but after the early reconnaissance attempts, all the latter primarily became rescue missions. Whatever dated records were obtained from each of the subsequent missions passed into survivalist Carl's hands, who received and, importantly and conscientiously, warned members of each respective mission as they waded ashore, unwittingly, to their doom. Back in New York, there must have been a major cover-up over the grievous Project Legacy affair, given that so many lives were lost, reputations sullied and so many financial resources hopelessly drained. By the mid-1950s, scientific research had been directed toward study of outer space, rockets, nuclear weaponry and Cold War spy technology. In hushed quarters, scarce mention of Skull Island and the doubtful fauna must have been relegated to the same stagnant breath as 'Piltdown Man.'

Furthermore, although Vincent tried to keep the island's location secret throughout his lifetime, the notebooks eventually fell into the hands of fellow American Museum of Natural History colleagues. To keep true to his word about never divulging the island's location, by 1957, Vincent had altered records about the island's having disappeared into the waves circa 1945, as indicated in the Weta Workshop publication. But now, as we know, the island must have still been intact over a decade later, when Vincent traveled there with Jack Driscoll (as documented by Brad Strickland's and artist Joe DeVito's secondhand, 2004 account). The island had long been seismically unstable and was on the verge of tectonic destruction. But nobody could have predicted the time (nor the rapidity) of its submergence beneath the waves. Skull Island most likely submerged by 1959 or 1960, as no island corresponding to those sketchy map coordinates has ever been identified either at sea or via satellite images.

The monsters of every conceivable phyla that lived on Skull Island were of untold ferocity and cunning. I have room but to mention a few. And here understand that, clearly, Carl Denham was more exact and scientifically accurate in his reporting of what he saw there than as conveyed by Mr. Lovelace. Savage and scrapping for survival, nature supremely 'red-in-tooth-and-claw,' beyond anything Darwin would have imagined, were Skull Island's denizens. Let's consider just a few of the paleo-varieties.

First, those 'Triceratops' mentioned by Lovelace were actually one of several other species mentioned by Carl Denham during the second bloody expedition. Probably, the horned variety which the *Wanderer's* crew observed during the 1932 mission were those named by Project Legacy as "Ferrucutus," a 30-foot long horned *Triceratops* descendant. An easy mistake to make. Ferrucutus suffered mighty struggles on the battlefield with *T. rex's* gigantic evolutionary descendant—"Vastatosaurus," or simply "V. rex."

But the island was also horribly infested with far more cunning and secretive dinosaurian forms, cousins to the Late Cretaceous raptor dinosaurs of Asia and North America. These were not observed during the abbreviated 1932 *Wanderer* exploration, but their existence became painfully obvious to Project Legacy's team in all of the deadly, subsequent missions. In fact it is now quite clear that Project Legacy's members fell victim to dino-raptor attack. Shortly, through Vincent Denham's more scientifically correct commentary, we'll describe in more detail the significance of these vile and insidious creatures. (And here we must equate the strange "Venatosaurs" with the creatures known to the natives as "deathrunners." It all makes me cringe, wondering whether Jack Driscoll's delirious deathbed confessions may have inspired Michael Crichton's writing of *Jurassic Park*! Driscoll died in 1989.)

But before delving further into the natural history of Skull Island's dino-raptor fauna as we will shortly, we must entertain Project Legacy's speculations about Kong himself, as documented through Carl Denham. Denham wasn't a scientist but a showman and film producer. Yet as Lovelace astutely realized, he understood scientists and the ways and means of paleontology.

A notebook Vincent found in the collection of aging manuscripts tagged the species King Kong belonged to as "Megaprimus kong," meaning "Big primate." Without formal published description, however, this isn't a name recognized by taxonomists. Mention it today in the hallowed halls of academia and you'll hear twitters of Sasquatch or even, tsk, tsk, the Abominable Snowman. Nor is there any remaining biological record of the kongs' existence to work from. From apparent archaeological evidence though, it does seem that the original, ancestral 'proto-kongs' may have been brought to Skull Island from mainland Asia by Stone Age people whose descendants eventually built the Wall (and relics comprising an Old City within the Wall), millennia ago.

Due to the ferocity of the local saurian fauna (V. rexes, dino-raptors, etc.)—in a sense late-surviving ancient stock 'equivalents' to Moa birds, dwarf mammoths, and the famous Dodo birds, which all perished as humans encroached into island settings during the Pleistocene—natural selection bias soon found kongs adapting to the terrors of Skull Island by acquiring enormous size. Soon a race of 25-foot tall apes developed, characterized by tremendous strength, and sharp talons; these were also the brainiest jungle genus, even smarter than "deathrunners." Otherwise, lacking such key adaptations, ancestral kongs wouldn't have survived, including *the* Kong, who became more human than animal and, yes, *the* ape that fought Godzilla's predecessor in a colossal struggle over a century ago, as recounted by Vincent Denham!

Recollections of Dr. Vincent Denham:

In 1957, Carl Denham's son, Vincent—a Ph.D. paleontologist with the American Museum of Natural History, discovered a map which led him to Skull Island, accompanied by Jack Driscoll, the same heroic gentleman from the 1932 voyage. Due to the top-secret nature of Project Legacy, Driscoll never returned to Skull Island until 1957, nor was he fully aware of Carl's whereabouts during the intervening time.

Incidentally, the Driscolls, Ann (nee' Darrow) and Jack, died over a decade ago (c. late 1990s) without leaving any *written* record as to their Skull Island adventures. But it is known that Vincent Denham encountered an aged native storyteller, Ishara, who related the archaeological history of the island, its tribalistic struggles, the indigenous fauna, and told of the ancient "kong" race. Ishara, owner of a pet *Archaeopteryx*-like, reptilian bird which mimicked human voices, had witnessed an intelligent kind of mega dino-raptor dinosaur, which formerly infested the island interior, terrorizing the natives. That is, one in particular. A mutant theropodous creature we'll come to shortly named "Gaw" by the islanders.

Ishara also spoke of the young Kong's experiences on the island during the mid to late 19th century, and explained how the practice of offering human sacrifices to Kong originated from an even older tradition involving Gaw.

Vincent also encountered his father, Carl, who as previously mentioned, had disappeared from society after secretively returning to Skull Island with Kong's body. He was amazed to find that for over twenty years, Carl had maintained sketch books and notebooks—Project Legacy's, well, "legacy"—outlining the island faunal natural history. For instance, in a

January 1955 entry, Carl wrote "The dinosaur life of the island is gradually dying out, as it died out millions of years ago in the rest of the world…"

In one book adjacent to Carl's sketch of a raptor skull, he ponders whether the raptor breed still exists. But that obviously must have been one of Carl's earliest entries, because other (later) Project Legacy notebooks recorded observations of late-surviving members of the dino-raptor breed. Vincent, who had become cynical of mankind's plundering tendencies, yet sensitized to the islander's plight after learning their history from Ishara, vowed never to divulge Skull Island's location to the civilized world, nor to crate back and exhibit any of the *remarkably fresh* bones of recent dinosaurs in his great museum, which of course would have stirred a sensation.

As opposed to how they were restored onto film by Delgado and O'Brien in 1933, Skull Island's dinosaurian fauna were active and very likely warm-blooded, a finding consistent with paleontologist Vincent Denham's observations made during his 1957 exploration of Skull Island, confirming his father's, as documented in his notebooks years prior to the "dinosaur renaissance." Although the exact date is uncertain, by 1960, Vincent safely returned to New York, following a harrowing escape from Skull Island at the time of its fiery destruction.

Legend of 'Gaw-d-zilla':

As you now know, during their 1957 adventure, Driscoll and Vincent learned of and found evidence for a highly intelligent and vicious species of 'raptor' dinosaurian breed, known to the natives as "deathrunners." In fact, Jack Driscoll was 'herded' by a small pack of deathrunners, "like a goat led to the slaughter," during his 1957 trip, a peril which he barely survived, and a tale which decades later may have pricked Michael Crichton's eager ears!

Deathrunners are a sinister theropod variety, highly derived from the more familiar Upper Cretaceous genus—*Velociraptor*. Lovelace doesn't mention them in the 1932 book, but then this may be because during the abbreviated initial island visit, such beasts weren't ever encountered by the *Wanderer's* crew. Deathrunners communicate verbally and even can wield simple weapons! (Surely they could open doors too—a *Jurassic Park* reference.) During the late 19th century, Kong himself had numerous encounters with these awful creatures, killing them by the dozens when their blood-thirsty packs spilled over him. Over untold centuries, Kong's ancestors exterminated most of their ranks, nearly causing extinction of the deathrunners.'

'King Dinosaurs,' akin to that famous North American species of apex tyrannosaur, as featured during the RKO film, abound on Skull Island too. However, years after release of the famous 1933 RKO 'documentary' film *King Kong*, paleontologist Vincent Denham noted that King Dinosaurs are not identical to *T. rex*. In fact, Vincent astutely realized the unlikelihood of *T. rex* existing as a 'living fossil' species through an incredible 66 million year span, that is, *without* evolving. King Dinosaurs—or if you will—"V. rexes," as previously named by Carl Denham, are evolutionary descendants of tyrannosaurids, primarily scavengers feasting off corpses left behind by deathrunners.

There is another more nondescript form of (far less intelligent) raptor dinosaurs lurking about named "slashers," which function like grim soldiers commanded by deathrunners. Quite a host of carnivory, eh! This place is a 'Dinotopia' gone wild!! But the greatest of all the indigenous sauria has yet to be described, and this awful dino-monster is Gaw!

Every so often (as reported by DeVito, Strickland and Michig— based on Vincent Denham's records, of course), over the course of many decades, one member of the deathrunner pack would grow continually into a "super-deathrunner," a queen that kills its rivals and controls the pack, as is "necessary for the continuation of their species." During the late 19th century, in Kong's youth, such a creature did indeed rule the island, and the natives named this monster, Gaw. The 'Gaw' encountered by Kong is at least some 70-feet long, adorned with five-foot long spikes along its vertebral column and tail, and had proportionally long arms and clawed fingers and razor-sharp teeth.

Gaw commanded her attendant deathrunner packs to swarm over the kongs on numerous occasions, nearly wiping them out, until only one—he who was destined to be known as "King Kong"—survived. This is how Kong's parents were savagely murdered, and the bitter memory of bloodshed led to a brutal showdown between Kong, the last of his kind, and the immense Gaw (not the last of its kind, however). In one prior skirmish, Kong had observed how Gaw retreated in the face of a charging horned dinosaur, and so, yes, in the end, Kong dramatically rams a handy, broken 'Tops' skull horn through Gaw's scaly neck. (On Skull Island those 'Tops' – e.g. Ferrucutus – horns are just lying around everywhere; they make handy weapons.)

Gaw wasn't the only historical member of its lineage! Bear in mind that the genetic constitution of a 'Gaw' individual is already highly predisposed to mutation. And with its vertebral spines, Gaw in its natural state distantly resembles another famous creature that we all know and sometimes revere. So, get this, could the legend of Gaw be allied to that of another paleo-monster? Could another subsequent Gaw have escaped Skull Island's volcanic subsidence before the early 1950s, migrating stealthily to a Pacific island, where it experienced enormous radiation dose, triggering its regulatory genes to expand growth into a true saurian abnormality, a freak of nature—a veritable dino-*daikaiju*? Could a hydrogen bomb test blast have caused an ordinary Gaw dinosaur to transform, mutate into a 'Gaw-d-zilla"? (Folks, this is far, far more than just a bad pun; it's my thesis!) And if you doubt the veracity of my account, bear in mind that, unlike Lovelace, I'm both a writer and, more importantly an investigative scientist!) I think the evidence may be more than circumstantial!

Consider! From Gaw's skeletal remains witnessed in 1957, Vincent Denham noted the enormous body, the tall vertebral spines which could have mushroomed under powerful doses of radiation, its long arms, voluminous brain capacity and that remarkable skull shape—all verging on that of a more famous amphibious 'dinosaur.'

Then there's the native tradition of virgin sacrifices often tied to King Kong, but according to Vincent originating with Gaw on Skull Island during the 19th century. As documented by John J. Pierce (see p. 17 of 1998's authoritative *Official Godzilla Compendium*), much later, we learn of a beast which "… the fisherman of Odo Island identify … with a legendary monster of the past that could be appeased only by virgin sacrifices…" Hmmmm.

Yes, it's plausible that another 'Gaw' queen had resumed its terrible, instinctive ways, although this time at Odo Island, near Japan. Then, following a hydrogen bomb test, the scarred, horribly mutated monster became enraged, hell-bent on a course of destruction throughout Asia.

Okay you doubtful souls, remember, Godzillasaurus in Toho's 1991 film was 'merely' a movie like RKO's *King Kong* or Toho's *King Kong vs. Godzilla*. And here I'm merely trying to sort out the written records, making relative 'sense' out of apparent contradiction and confusion … creating harmony from evident disorder.

And so?

So my premise is that while Kong never fought a Godzilla ('Gigantis') fire-breathing type creature as sensationalized in the 1962 Toho film, Kong indeed defeated the *progenitor*-type of a Gaw-d-zilla dino-monster in bloody combat (although without any hype or fanfare), but over a century ago! Now at last you know the true (alternative) tale of 'King Kong versus Gaw-d-zilla.'

Kong and Godzilla as Kaiju in Monster Literature?

While Kong may be heralded as America's own iconic monster, Godzilla is Japan's definitive icon of '*kaiju*-ness.' Just how inexplicable is Kong, say, 'versus' a monster like Godzilla in fantastic fiction? If Godz is a *kaiju* in the movies, is it also such in books, and how does Kong's sense of *kaiju*-ness compare? While Kong's past and origin is relatively well documented, there's relatively little of an analogous nature available to the western world, in English, concerning Godzilla's background. Marc Cerasini's informative 1996 published manuscript, *Godzilla Returns* (Random House) provides essentials, as documented by 'real' journalists situated in Japan covering Godzilla's second terrible coming.

Relying on field notes and observations, rather cryptically Cerasini distills much of what had only been previously surmised by scientists. "Godzilla is a radioactive creature. He's not really a dinosaur, although he probably was a dinosaur once. But now, after the exposure to radiation that created him, Godzilla is more like a living, breathing nuclear fusion reactor."

In other respects, Godzilla is an ordinary animal whose physiology and organs are ruled by biological principles. Godzilla also has the ability to control the radiation it emits. "When Godzilla is angry or threatened, he gives off increasing amounts of intense radiation. This energy surge is climaxed by the ray that he fires through his mouth. The creature (also) has a heart, lungs, a stomach ..." And according to herpetologist Bill Holmstrum of the Bronx Zoo, "The popular misconception is that Godzilla breathes fire. Actually, he breathes a stream of compressed radioactive air that bursts into flame on impact."

Beyond Godzilla's radioactive breath, Cerasini provides a rational explanation for the iconic monster's apparent invulnerability to conventional military weapons. For instance, cannon shells which strike Godzilla's leathery hide actually do penetrate, causing him to bleed ... but only momentarily. Then within micro-seconds, the wound is healed, "... as if it had never been inflicted."

"Godzilla's greatest defensive weapon is his amazing regenerative powers ... The radiation that created Godzilla also mutated the creature's molecular structure ... Godzilla is capable of complete regeneration of damaged tissue instantaneously. It seems ... Godzilla cannot be killed unless he is completely disintegrated." Kind of like Frankenstein's Monster.

Cerasini never suggested Godzilla's probable ties to Skull Island's Gaw. Nor has anyone previously recognized the potential for a vertebrate animal known to mutate so aggressively, naturally, such as a 'Gaw'—to be so drastically effected when bathed in massive radiation doses. Whereas other vertebrates of lesser genetic fortitude and viability would have simply perished from exposure to the nuclear fireball, only a genetically superior Gaw-like creature could have withstood and absorbed the life-nurturing, cell-transforming gamma and x-rays, albeit from some 'safe' relative distance (i.e. not situated at 'ground zero.').

Both King Kong's and Godzilla's sense of 'mysteriousness' would seem diminished at the level of a novel, relative to movie portrayals. In books, scientists are grappling for technical explanations for form and function of the encountered monsters; in movies, this sense of real scientific origin is often sacrificed instead for metaphor and symbolism as to what these strange creatures ('giant mysterious beasts') represent, now that they're on the loose. So, their '*kaiju*-ness' comes off better somehow in film. Conversely, in film the rule of thumb is 'spare me the boring scientific details,' and so we are left with many lingering questions as to how and why, exactly, crypto-creatures like Kong and Godzilla came to be.

We may conclude that due to the relative body of literature, neither Kong nor his prehistoric Skull Island dinosaurian foes are to be regarded as *kaiju* of fantastic fiction, nor are they filmic *kaiju*. Conversely, even though Cerasini provides plausible 'scientific answers concerning its biology in his 1996 novel, Godzilla remains a valid *kaiju* icon of film.

And as far as my 'Gaw-d-zilla' thesis of origin is concerned ... well. If you can accept and comprehend the likelihood of mutations evolving toward gigantism in aftermath of a hydrogen bomb blast (which is eerily close to proven scientific fact, as opposed to unproven 'fact'), then you can readily accept my non-*kaiju* (dinosaurian) theory of origin for the mightiest *kaiju* of all—as documented and derived from the incredible library of fantastic fiction.

With sincere apologies for unraveling the mystery of it all.

Notes—You may 'take' this chapter, perhaps analogously, to Don Glut's treatment of Universal's and Hammer's seemingly conflicting plots in a succession of 'Frankenstein' films produced between 1931 and 1970—as recounted in his *The Frankenstein Archives*, 2002. Although bounded by prevailing biogeochemical controls, throughout planetary history, atmospheric oxygen content has fluctuated. During the 1840s, Richard Owen, for instance, believed dinosaur extinctions were tied to shifting oxygen levels in Earth's atmosphere. In a 2005 *Prehistoric Times* article (no.74, p.28), Kong author and illustrator Joe DeVito explained to interviewer Mark Berry how he deliberated whether he should suggest in his 2004 book that tail-dragging dinosaurs seen in the 1933 RKO film *King Kong* were resultant of atmospheric oxygen levels that have declined since the Mesozoic Era, based on a controversial 1988 study regarding bubbles of amber—as suggested by one of my old college professors (R. M. Garrels). That old report, however, has been superseded & rather falsified by more recent scientific information (*Science*, vol. 309, Sept. 30, 2005, pp.2202-2204). Therefore, despite the disbelief-suspending power of stop-motion animation, the prospect of whether tail-dragging dinosaurs were ever actually witnessed on Skull Island either in 1933 or 1957 must be respectfully challenged and, here, doubted.

There is one more pivotal reason behind the Skull Island cover-up of the 1950s, which was deemed necessary to redirect further blame from the American shore. Perhaps its most sinister saurian and Japan's most pivotal nemesis monster, deemed one & the same by paleontologists and geneticists 'in-the-know' was indeed spawned on Skull Island, after all! Furthermore, after Godzilla's first romp through Tokyo in 1954, and following American and British scientists' comprehension of what this new 'fire-breathing' monster most likely was—incredibly, a mutation of the sort of creature documented by Project Legacy, as unveiled by Vincent Denham after 1957 through the restored notebooks, the U.S. government decided to erase any connections whatsoever to the new Japanese monster, which the U.S.A. had not only 'created' through early 1950s Pacific hydrogen bomb testing, but which they also had prior evidence of regarding its existence and probable genetic tendency. It could be argued that the U.S. was entirely to blame for the second Gaw's migration from Skull to Odo Island; did Project Legacy team members perhaps encounter a living specimen which, in order to survive, was chased from Skull Island's shores with superior weapons? At this juncture, such a scenario could never be proven. However, diplomats can be harsh geopolitical critics and America, by then preoccupied with a war in Viet Nam, which was becoming unpopular on the global scene.

Chapter Thirty—*Modes of Survival: Godzilla 'versus' Kong*

Our two foremost gigantic monsters, Godzilla and King Kong, are of prehistoric ilk. However, their hallmark species have managed to cheat throes of extinction and ravages of geological time by surviving into modernity. How? By mainly appealing to the authorities—Toho Studios (4 Godzilla films) and author Joe DeVito's 2017 novel *Kong of Skull Island*—let's compare evidence to see how Kong and paleo-*kaiju* Godzilla both managed such a seemingly insurmountable trick. But first a warning—some of this evidence regarding Godzilla may appear contradictory.

First let's examine four of Godzilla's notable appearances, bearing on the monster's most notable, yet contrasting origination(s):

In Godzilla's case we resolve whether the monster's existence is preferentially due to any of the following circumstances (or combination thereof): longterm crypto-survival in a lost world setting; "resurrection" = hibernation stasis then "awakening"; natural speciation; synthetic evolution due to exposure to anthropogenic wastes and toxins; radioactive mutation. Fortifying the keloid-scarred, 'mutated' appearance of four godzillas (i.e. in films *Gojira, Godzilla Raids Again, Godzilla vs. King Ghidorah,* and *Shin Godzilla*), some level of mutation is clearly a common denominator—as manifested via the monsters' orally issued 'nuclear ray,' generated within its biophysical atomic furnace, then amplified or 'pulsed' by flickering of an array of jagged vertebral fins (osteoderms). Normal species do not exhale nuclear rays on cue or willfully, enhancing each of the aforementioned Godzillas' status as *daikaiju*.

But in both 1954's *Gojira* and 1955's *GRA*, Godzilla's general appearance is hypothesized to be a result of 'cross' evolution in the late Mesozoic Era of terrestrial and marine creatures into an 'intermediate,' semi-amphibious creature that survived—adapted to life in oceanic deeps, followed by its resurrection from the (Mesozoic) depths due to hydrogen bomb experimentation. Although perhaps implied (and intended by the special effects staff), those jagged fins and other bodily disfigurements and gigantic size are never directly described as resultant of radiation exposure mutations in either of these movies, although Dr. Yamane recognizes that the two godzillas (and presumably Anguirus) have each absorbed massive amounts of radiation. The fact that the first two godzillas are essentially identical—disregarding several obvious suitmation differences—is further indication that their general appearances (e.g. "morphologies") are somehow 'right' for their species. Furthermore, a fossil record for the Anguirus exists, as this huge 'ankylosaur' even appears in a dinosaur book restoration; so evidently its species must have been previously known to science on the basis of fossils.

Despite Yamane's hypotheses expressed in the 1954 and 1955 films, Godzilla clearly isn't a true "dinosaur," although qualifying as a kind of semi-amphibious 'dino-monster,' evolving in the Mesozoic age. But dinosaurian ties are more evident in 1991's *Godzilla vs. King Ghidorah*, as the Godzillasaurus is dubbed a terrestrial sort of 'dinosaur.'

In *Godzilla vs. King Ghidorah*, Godzilla's origination story takes quite a different turn. This time mutation resultant of exposure to hydrogen bomb radiation *is* the key ingredient leading to Godzilla's 'creation' and generally familiar appearance (although this 1991 Godzilla—when it finally appears—looks markedly different in detail from the first two). This 'second' Godzilla type supersedes the 1954/55 specimens due to developments insidiously stemming from a time warp. And the story introduces another creature, a mysterious Godzilla precursor known as the "Godzillasaurus," a tall 'tyrant king'-ish-looking fellow whose species has survived presumably in relative evolutionary stasis, in a lost world Lagos Island setting, since the Mesozoic. Like so many sci-fi crypto-'survivors,' Godzillasaurus seems to be the last of its kind though. This time, newly introduced Godzillasaurus suffers not one but *two* bodily mutating exposures to high level radiation.

The first occurrence ("erased" in a time warp sequence) is the 'original' 1954 Bikini Atoll bomb blast, while the second occurs in 1992, in an oceanic Bering Sea depositional setting. The *second* time, an abandoned sunken nuclear submarine has released nuclear wastes causing mutation of the Godzillasaurus.[1] Although unclear, it seems as if both times Godzilla's mutations may (possibly) have resulted in identical morphological changes. So the mutation process on a Godzillasaurus specimen would seem consistently 'linear,' determinable, not random, even if the nature of cellular-mutating nuclear isotopes may (possibly) differ. (Oddly, despite the time warp occurrence, individuals have retained knowledge of the original Godzilla that attacked Tokyo in 1954—even though that event and its avenging dino-monster were erased from this parallel universe.)

Another means of preserving a crypto-prehistoric species is via hibernation in ice. While this disbelief suspending gimmick has been used (paleontologically) in the case of freezing two famous monsters—the Deadly Mantis and Superman's foe, the Arctic Giant—Godzilla/"Gigantis" was only subjected to much shorter term preservation in glacial ice (e.g. in between the end of *Godzilla Raids Again* presumably through the beginning of *King Kong vs. Godzilla*). Although I don't buy it, it's also possible that the dino-monster in *KKvG* was yet a *third* Godzilla that somehow also became exposed to H-bomb radiation while drifting in an ice floe from a polar region, thus precipitating its nuclear ray. The 1953 Rhedosaurus was preserved long-term in Arctic ice only to be freed in an experimental H-bomb inferno, without apparent mutation from radiation exposures, otherwise.

In utter contrast to evolutionary circumstances in its earlier films, *Shin Godzilla*, doesn't appear to have any clear ancestral ties to real dino-monsters—even though the product of its progressive 'phylosynthesis' does generally appear pseudo-dinosaurian. This new Godzilla creature presumably transforms through four—according to Robert Scott Field in *G-Fan* no. 114 (Winter 2016, p.15)—life stages, while being frozen during a fifth transition. The three most prominent, witnessed forms are an amphibious, 'salamander-like' or 'eel' stage in which the creature's gills are evident. Then it metamorphoses into a 'stegosaurian—young adult' variety, followed by its growth into a colossal form resembling the monster g-fans are fond of, developing some of its classic characteristics. Yet even this version evidently has even further potential to continue to transform … into what? Something with wings? (Also see "Shin Gojira: some observations on the new Gojira," by Stephen R. Bissette, *Monster!* No.31, Nov. 2016, pp.97-99)

I used two terms in the paragraph above: metamorphosis and phylosynthesis. The former term is usually applied to insect biology—which Godzilla decidedly is not. The second term was coined by Charles De Paolo, author of *Human Prehistory in Fiction* (2003). In his chapter pertaining to Edgar Rice Burroughs' 1918 novel, *The Land That Time Forgot*, De Paolo defines phylosynthesis as an original construct from terms, *phylogeny* and *synthesis*. He states further, applying "unilinear biology,"…Precisely what Burroughs attempts to do in his fiction is to unify the human phylum, conceptually, and to account for its origin and evolution." (p.37-38) Burroughs penned a trilogy of novels in 1918 (*LTF's* two 'sequels' are *The People That Time Forgot* and *Out of Time's Abyss*) in which cellular aquatic creatures progressively yet mysteriously transform 'upward' into modern humans during the course of each creature's individual lifetime—passing through many known organisms known to paleontologists (ordered sequentially in context of geological time) along the way—such as dinosaurs. Aided, or perhaps tainted by toxic radiation doses, isn't this more or less what happens to the precursor that evolves into the Shin Godzilla?

Shin Godzilla's possible origins as a "relic dinosaur" are pooh-poohed by scientists. Instead they come to realize that it is a "perfect organism" stemming from an ancient marine life form—"a god incarnate." In this case, however, the rapid-driving evolutionary mechanism is man's nuclear pollution of the seas. While we may learn the true nature of those 'humanoid' creatures witnessed on its tail tip—replete with forming jagged osteoderms in a Shin Godzilla sequel (proposed for 2020), it's interesting how their vaguely reminiscent human shapes suggest an awful genetic tie between the monster and ourselves.

Before moving on to discussion of Kong, there were also two American Godzillas—the first one known as "GINO" (TriStar 1998), and the other filmed by Legendary, released in 2014. The godzillas in these films differ from each other, in terms of their origins. In case of the Legendary Godzilla, the titular dino-monster is a survivor from prehistoric times, evidently unaltered morphologically by the "atomic bomb" that was used to attack it in 1954. Perhaps that nuclear 'ray' issued from its maw was a mutated manifestation of that event though, but then again maybe not, as indicated below. For as explained in Greg Cox's 2014 *Godzilla* novelization (p.126), Godzilla is "… millions of years older than mankind… from a time when the Earth was ten times more radioactive … The animal—and others like it—*consumed* that radiation as a food source. But as radiation levels on the surface naturally subsided, these creatures adapted to live deeper in the oceans, further underground, absorbing radiation from the planet's core. … We call him *Godzilla*." So the Godzilla in the 2014 production is a lone survivor, reawaked from the sea depths by a nuclear submarine in 1954, as clarified in the novelization. In utter contrast, however, TriStar's Godzilla is a recent H-bomb mutation, an "incipient species," a "mutated aberration." TriStar's Godzilla is rather not unlike the outlandish godzilla character in Mark Jacobson's 1991 *Gojiro* novel—a little reptile once happily basking on his Pacific "lavarock" before being exposed to H-bomb radiation.

Like his anatagonist Godzilla, King Kong is also of prehistoric provenance. (A quick aside: Is Kong also a *kaiju*? I would say, without elaboration here, both the original 1933 Klassic Kong and the one starring in the 2005 filmic remake are not true '*kaiju*' per J. D. Lees' definition) In Kong's case, we're immediately faced with a definitional dilemma; why do we suspect that Kong is "prehistoric"? There is nothing in the known fossil record quite like him (except the much more diminutive *Gigantopithecus*). In fact, it's been rather taken for granted that Kong *is* prehistoric. Why?

Apart from Kong's fanged and fisticuffs association with Mesozoic monsters in the 1933 film, how did such a conclusion originate? Author and sci-fi movie-maven Bill Warren has aptly remarked: "The survival of prehistoric animals other than dinosaurs is rarely treated in films, or anywhere else for that matter. Dinosaurs are inherently more colorful, and most other prehistoric animals (at least the mammals) resemble creatures alive today." (*Keep Watching the Skies-The 21st Century Edition*, Vol.1, 2010, p.470) Fortifying his legendary tussles with several prehistoric animals, 1933-Kong's anachronistic "prehistoric-ness" stems from two key references: For one thing, Merian C. Cooper perceived Kong this way—as prehistoric; for another, novelist Delos W. Lovelace designated Kong as 'prehistoric' in his 1932 novelization of the RKO *King Kong* movie script in two clearly written, suggestive passages.

In a 1933 interview, Cooper remarked, "… the thought struck me—what would happen to this highest representative of *prehistoric* animal life [i.e. Kong] in our materialistic, mechanical civilization? Why not place him at the pinnacle of the tallest building {the Empire State Building}—symbol in steel, stone, and glass of modern man's achievement and aspiration—and pit him against modern man?" (My italics …. Cooper quoted from p.48 of Ulrich Merkl's 2015 book, *Dinomaina: The Lost Art of Winsor McCay…..*)

Although the expedition to Skull Mountain/Island lacks a degreed scientist type of any kind, story character/actress Ann Darrow provides the necessary interpretation concerning Kong in Lovelace's 1932 *King Kong* novel, a tie-in to the film:

"But surely there never was such a beast!" Ann laughed uncertainly. "At least not since prehistoric times." Then later, Carl Denham corroborates Darrow's insightful suggestion, declaring, "Prehistoric life!" he ejaculated, and slowly turning to Driscoll he cried out. "Jack! She was right last night on the ship. Ann, I mean. But she only had the start of it. She guessed the beast-god was some primitive survival. But if this thing we've killed (i.e. a *Stegosaurus*) means anything, the plateau is alive with all sorts of creatures that have survived along with Kong."

A similar recent 'take' on Kong's paleo-antiquity/anachronism is cast in Will Murray's 2016 novel, *King Kong vs. Tarzan*. Here, Murray sees

Kong as a "...prehistoric survival doomed to extinction" (p.244), or the "last of its kind," an evolutionary dead-end. Therefore, we should regard Kong and his filmic and novelistic successors as "prehistoric." Okay – so how did he (or his species) survive out of prehistoric times? Murray also briefly grapples with an explanation of the forces that 'preserved' the species on Skull Island. So was Kong "... perhaps some evolutionary atavism," that survived in a place where, mysteriously, "... shrouded in perpetual fogs—evolution continues to run wild"? Will Murray's apt novel is more concerned with the African battle between Tarzan (of the apes), and the gigantic prehistoric ape (as opposed to other prehistoric animals abounding on Skull Island).

But another new entry tackles the latter matter fully ... next time.

Note: (1.) Which leads to questions as to whether Mothra's gigantic size is mainly attributed to radiation exposures (most likely the case), or instead resultant of evolutionary descent from prehistoric survivors—as those from the Paleozoic 'Carboniferous' age when giant insects did rule the day, such as *Meganeura*, the 'giant dragonfly' known from fossils. I shall refrain here from entertaining whether the absurd notion that Godzilla is a hybrid 'cross' between two dinosaurs—*Stegosaurus* and a carnosaur—as implied in *King Kong vs. Godzilla* (1963).

NINE

A meteorite crashes in ancient Arizona, wiping out the dinosaurs. From Fantastic Adventures *(April 1948). See Chapter Thirty-one for more. (A. Debus collection. copyright 1948, Ziff-Davis Publishing Co.)*

Astro-Paleontology

A rampant theme of 1980s dinosaur renaissance pop-culture was the emphasis on the startling dinosaur-killing asteroid of doom. It is likely that more publications concerning mass extinctions were printed during that interval than others having to do with *paleobiology* of prehistoric organisms. While this has settled by now—the pace and plethora of new mass extinctions theories and books (but not the hypothesis) falling by the wayside – for me, such concepts remain a cauldron of utter fascination. One of the high points during this period was getting to meet both Stephen J. Gould and David Raup—both staunch proponents of the asteroid extinction theory—in the mid-1990s at the University of Chicago on an evening occasion that my Dad facilitated.

I've written about the fabled "K-T" (or now, the "K-Pg", or the Cretaceous-Paleogene mass extinction) asteroid extinction theory elsewhere, and you won't find too much on the concept here. But this section does examine theoretical aspects that I found captivating. Regardless, few paleontological debates can approach the scale of 1980s furor concerning to what extent starry heavens wreak havoc on Earth, inadvertently directing the course of future planetary evolution.

And so here we go again: except, this time, more on the 'fringe.'

> These chapters originally appeared as articles in the following publications:
> Chapter Thirty-one—*Prehistoric Times* no.75, Dec/Jan 2006, pp.38-39.
> Chapter Thirty-two—*Fossil News: Journal of Avocational Paleontology*, vol. 19.3, Fall 2016, pp.26-31.

Chapter Thirty-one—*Meteor Crater's "Impossible but True"*

Just before my first quarter in college, in August of 1972, my grandfather and I visited Arizona's famous Barringer Crater—perhaps the best preserved and most recent of all the great meteorite craters on our planet. The crater situated in the brilliant Arizona desert was formed by an iron-nickel bolide vaporizing upon impact some 25,000 years ago. Energy dissipated by the fireball entering Earth's atmosphere between 10 and 20-kilometers per second has been likened to that of the famous 1908 Tunguska Siberia event. Arizona's crater, roughly 4,000-feet in diameter and about 570-feet deep, was first examined by scientists in 1891, triggering a wave of controversy which I won't delve into here. Intriguing as they are, it's not my intention to blather on about meteorite craters here, but if you're interested in the history and science of Arizona's famous crater, read chapter 3 in Kathleen Mark's absorbing book, *Meteorite Craters* (University of Arizona Press, 1987).

So far this is intended as brief background for a minor discovery of my own—a peculiar 'lost' reference to Arizona's crater found in a 1948 issue of a pulp sci-fi mag—*Fantastic Adventures,* where dinosaurs are viewed as (of course, unlikely) victims of the energetic visitor from space!

During the 1980s and early 1990s, I was spellbound by the mass extinctions debates, fueled by a certain paper published in *Science* in 1980 unmasking the 66-million year old impact event that doomed the dinosaurs (vol. 208, pp. 1095-1108, "Extraterrestrial causes for the Cretaceous-Tertiary extinction," by Luis Alvarez, et. al.). To my knowledge, however, no one had ever connected dinosaur mortality with the Barringer Crater blast. Dating impact craters is fraught with difficulties, although geologists have known for decades that the Arizona crater formed in a relatively recent event—whereas we've also known for decades that no dinosaurs (except birds) survived beyond about 66-million years ago. But science fiction magazines occasionally dabble in outright speculation just for the hell of it—because they can.

And so I was delighted to see on the back cover of the April 1948 issue of *Fantastic Adventures* (vol.10, no.4) a reproduction of a painting by James B. Settles, illustrating how doomed dinosaurs (possibly a *Protoceratops* and three sauropods are visible) witnessed that meteorite impact in ancient Arizona, something which they most certainly did not do. (*Protoceratops* wasn't native to North America. *Leptoceratops* was.) The caption to Settles' painting appearing over the back cover illustration reads, "Impossible ... but true. Ten million tons of flaming, meteoritic matter sank their blazing mass into Crater Mountain, Arizona. We think it happened a hundred million years ago, when the Earth still housed the ancient reptiles."

An accompanying description on page 177 underscored the fantastic visual, which some of you may be interested in reading. So, without further ado, here's "A bomb out of space," by A. Morris and James B. Settles, as it appeared in the December 1948 issue of *Fantastic Adventures*.

> "In the Western part of the United States, in the state of Arizona, there is a phenomenon of nature known as Crater Mountain. For years, scientists have been speculating on the origin of this famous landmark. The cavity in the mountain is about 3,800-feet across, reaching a depth of almost 600-feet. To form a crater of this size was, scientists agree, a herculean feat of old Mother Nature. But just how the trick was accomplished the same scientists have not entirely agreed upon.
>
> The largest consensus of opinion, and probably the more likely, is that at some time in the distant past, a huge mass of meteoritic material hurtled out of space, much like a super-interstellar bomb, and crashed into the mountain in what we now know as the state of Arizona. The resulting collision was so intense that the mountain was literally torn asunder, leaving as an aftermath, the huge jagged crater that our eyes can witness today. The force of collision was so intense that tons of molten rock were thrown as far as one mile, and possibly more, from the cataclysm. This is borne out from recent scientific research in the area where blocks of limestone, corresponding to the same material in the immediate vicinity of Crater Mountain were found a mile and more away. All of this data has been used to determine the probability of a meteoritic crash.
>
> As to the exact time of this collision from outer space, there is even more speculation. About the safest estimate made so far is that it occurred at least 'a great many' centuries ago. Other estimates have run as far as a hundred million years ago. If this last estimate is true, then the incident occurred at a time when

giant reptiles ruled the Earth. It is indeed likely that many of them perished in the cataclysm, and ever more likely that the sight was witnessed by many more of them. One can only speculate as to what thoughts, if any, traversed their embryonic minds when the flaming meteor screamed down from the sky in a howling torrent of flame and fury. An impossible fact, but true."

In retrospect, while their scenario was entirely "possible," the authors' 'facts' about dinosaur deaths being associated with this particular meteor strike are decidedly untrue. Note that Morris and Settles never suggested that the impact caused extinction of all dinosaurs, as, even then, their 100-million year old date would have been much too long to agree with the fossil record of (what is now known as the) Cretaceous-Paleogene dinosaur extinctions. Yet, Settles' imagery presaged what was to follow three decades hence when science went 'mad,' in the aftermath of that infamous 1980 *Science* paper. A few years following publication of the April 1948 *Fantastic Adventures* issue, a paleontologist named De Laubenfels placed the hypothetical asteroid-dinosaur death link on firmer, although still speculative scientific footing. (See chapter 32 in my *Dinosaur Memories* (2002) book, for more on De Laubenfels' theory.)

The fall of 1972 passed quickly as I battled college courses, and somehow I graduated: 'Impossible ... but true!' some of my professors were heard to exclaim. Yet I recall my now long ago visit to Meteor Crater as being a pivotal event in shaping later interests in the mass extinctions controversy, dawning shortly—only a few years thereafter.

Chapter Thirty-two—*Going for the Cycle: Dark Matter, Mass Extinctions, and the Resurgence of Paleontological Periodicity*

Dark matter! That's the fresh new ingredient supporting theories of periodic mass extinctions, an idea that may be as foreign to non-scientists as anyone could imagine. And yet physicists proclaim that dark matter is the essence of the universe, comprising a whopping 85% of all matter.

Dark matter is mysterious, unseen stuff, uncannily different from the ordinary matter (known as "baryonic matter") that makes up our bodies, the Earth, and the stars. But the influence of dark matter can still be measured, and its effects can be observed, however indirectly. The gravity of dark matter gives rise to the familiar shapes of galaxies and prevents their "billions and billions" of stars (as the late cosmologist and astrophysicist Carl Sagan is remembered—apocryphally—for saying) from dissipating, spewing outward into intergalactic space.

Dark matter is also implicated in paleontology's most apocalyptical episodes. In a 2014 issue of *Physical Review Letters*, Harvard University physicists Lisa Randall and Matthew Reece outlined their theory that periodic cometary bombardment was responsible for mass extinctions, marrying the idea of dark matter to the oscillatory motion of the solar system, which bobs up and down through the plane of the Milky Way.[1] The theory may have seemed alarming, in part because paleontologists have traditionally taken a dim view of starry-eyed theories of the Earth.

A Bit of Background

A couple of decades ago, paleontologist Jack Horner rather famously stated that he didn't really care how dinosaurs became extinct; he was more concerned with how they lived. This remark reinforced the idea that paleontologists should focus more on the study of the aspects of geohistory for which they were trained and avoid considerations of phenomena for which little evidence existed. As chief spokesman for this popular-science notion, Horner relieved tensions within the academic community where controversy continued to rage over the great Cretaceous-Paleogene (K-Pg) dinosaur die-out and mass extinctions in general. Many colleagues eventually wearied of the debate and simply parroted Horner's mantra.

Despite Horner's sobering remark, no geological event could seem more dramatically relevant to human existence on Earth than does the K-Pg event of sixty-six million years ago. Without that apocalyptic event, humans likely wouldn't exist. And if that world-altering impact in the Yucatan doesn't seem relevant today, consider the meteorite (sixty feet in diameter) that exploded over Chelyabinsk, Russia, on February 15, 2013 or the comet/asteroid that struck the planet Jupiter on March 17, 2016, both reminders that grander devastation may come again.

The now traditional "dinosaur and the asteroid" story has been retold through many venues and formats, often captured in dramatic paleo-imagery.[2] Beyond that singular tale, however, unprecedented scrutiny of the K-Pg event in the mid-1980s led to heightened interest in the possibility that *periodic* mass extinctions were somehow triggered astronomically—for example, through "comet showers" that occurred on the order of every twenty-six to thirty-three million years. Such fascinating theories proved inescapable as "dinomania" swept the nation.

Ironically, identifying astronomical cycles as the cause of terrestrial events has historically proven dangerous to scientific reputations. If the then-reigning paleontological establishment considered it blasphemous to suggest that a large asteroid slammed into Earth sixty-six million years ago, exterminating most dinosaurs and many other species along with them, the idea that something out in the universe triggered such catastrophes *repeatedly*, every few million years, was condemned as crackpot. In the mid-1980s, science seemed to have gone mad ... or at least to have become entirely science *fictional*.

The Question of Periodicity

Escalating such world shattering furor, in 1984 University of Chicago paleontologists David M. Raup and J. John Sepkoski, Jr., published their statistically fortified case for a twenty-six million year periodicity for twelve extinction events over the last quarter billion years.[3] In their article, they stated:

"Although the causes of the periodicity are unknown, it is possible that they are related to extraterrestrial forces (solar, solar system, or galactic).... [W]e favor extraterrestrial causes for the reason that purely biological or earthbound physical cycles seem incredible, where the cycles are of fixed length and measured on a time scale of tens of millions of years."

Fascinated by the astrophysical implications, their colleagues didn't take long to devise prospective causes in support of Raup and Sepkoski's statistical data.[4]

Barely skipping a beat, the April 19, 1984 issue of *Nature* included several papers that attempted to explain and fortify conceptual periodicity in mass extinctions despite the absence of direct evidence. The most famous of these elaborate ideas involved the hypothetical existence of a small binary star dubbed "Nemesis." According to the theory, in the course of Nemesis's eccentric orbit, its gravity nudged swarms of comets as they closely approached the Sun every twenty-six million years, dislodging them from the Oort Cloud toward the inner planets.[5]

Two additional papers in that same 1984 *Nature* issue dealt with the Sun's motion perpendicular to the galactic plane. The Sun's bobbing motion crosses the mid-plane of the Milky Way galaxy on the order of every thirty-one million years, becoming a "pace-maker" for mass extinctions. According to a paper by Michael Rampino and Richard Stothers,[6] greater mass density, in the form of giant interstellar hydrogen gas clouds, would presumably be encountered within this middle-most part of the plane, and this greater mass density would perturb icy debris in the Oort Cloud, dislodging swarms of long-period comets toward the inner solar system where some would eventually collide with planets, Earth included.

In retrospect, a thirty-one-million-year astrophysical period fits geological data (e.g., mass extinctions and impact cratering) better than does a twenty-six-million-year cycle. Randall and Reece reconsidered twenty-six terrestrial impact craters with geological ages of less than 250 million years and a diameter of no less than twenty kilometers. Their revised calculation showed that the impact-cratering rate intensified every thirty-two to thirty-five million years. Contrary to Rampino and Stothers, moreover, they argued there was insufficient mass (*ordinary* baryonic type mass) within the galactic mid-plane to launch a sufficient flux of comets Earthward.

Those controversial 1984 *Nature* publications (and another published in the January 3, 1985, issue that concerned a "Planet X" cometary "trigger") generated a rash of rhetoric, perhaps culminating in Richard A. Muller and Walter Alvarez's infamous rebuttal to a *New York Times* editorial ("'Miscasting the Dinosaur's Horoscope"). Objecting to a suggestion that evidence for periodic extinctions had "faded," Muller and Alvarez defiantly stated, "You suggest, 'Astronomers should leave to astrologers the task of seeking the cause of earthly events in the stars.' May we suggest it might be best if editors left to scientists the task of adjudicating scientific questions?"[7]

Three decades ago, in other words, the discovery of apparent periodicity in mass extinctions fueled the public imagination, set the media on fire, and had scientists drawing battle lines. By the time Natalie Angier's investigation of periodic mass extinctions appeared as a cover story in the May 6, 1985 issue of *Time* magazine ("Did Comets Kill the Dinosaurs?"), the nation was buzzing with fantasies of mass extinctions, killer asteroids, and Nemesis the Death Star. And yet the search for grand "megacycles" in the last 500 million years of geological phenomena had been a closet obsession of geologists for quite some time.

Still, by the mid-1990s, a consensus was eventually reached that the periodicity rage had indeed "faded." Volcanologists, for example, were considering an alternative theory that the timing of mass extinctions coincided with the ages of volcanic basaltic "traps" that recurred approximately every thirty million years. Such terrestrial catastrophes took place when magmatic plumes migrated from Earth's mantle, poisoning the ocean-atmosphere system and contributing to global "anoxia" (deoxygenation of the ocean) and greenhouse conditions in the atmosphere.

In his 1998 book, *Night Comes to the Cretaceous: Dinosaur Extinction and the Transformation of Modern Geology*, geologist James Lawrence Powell concurred that, "in spite of the evidence for extinction periodicity, and its importance if true, interest seems to have waned," adding that:

> "One of the reasons [for this] is that periodicity falls deep in the cracks between disciplines—it is not really the province of geologists, or paleontologists, or astronomers, or statisticians, or anyone—it is a scientific orphan in a world of limited time, scarce resources, and orthodoxy. Even the pro-impactors need not endorse it—impact could be of great importance in earth history and not be periodic."

Enter Dark Matter

When I interviewed the then-retired Dr. Raup in late 1999 to ask him about the status of the "periodicity theory" in the scientific community, he agreed that interest in (astronomical) periodicity had decreased. Regardless, Raup remained hopeful that his original 1984 claim would someday be vindicated—the theory that an astronomical mechanism existed that was capable of slinging comets out of the Oort Cloud periodically, producing a regular series of era-ending mass extinctions.

Possibly, it has been.

Instead of presuming that our galactic plane is composed of more highly concentrated ordinary "baryonic" matter that causes perturbation of comets during the solar system's wavy, vertical, bobbing oscillation, Randall and Reece (aided by a suggestion from fellow physicist Paul Davies) proposed that a thin, concentrated disk of dark matter existed within the galactic mid-plane, rather like a slice of ham in a sandwich. In other words, Earth's solar system periodically passes through a "double disc" in which two kinds of matter are sandwiched. It is the Oort Cloud's tidal interaction with the "ham" that periodically nudges numerous comets into eccentric, highly elliptical orbits toward the inner part of Earth's solar system approximately every 32 to 35 million years. The resulting "comet showers," which occasion large impacts and mass extinctions on Earth, may persist for up to two million years—a sliver of geological time.

Of course, some terrestrial impact structures could be asteroidal in origin, caused by objects not having originated in the Oort Cloud. Generally, asteroids might be slower moving than comets and could strike the Earth at less regular intervals. Analysis of Randall's and Reece's impact-cratering data, however, indicated that their "double dark disc model" was statistically three times more likely to be correct than a set of uniform (randomized) impact occurrences.

Thus, as the authors stated, their "double dark disk" scenario seemed a "fascinating possibility worthy of further exploration" because it might "play a significant role in explaining the observed pattern of craters, and possibly mass extinctions ... ultimately reveal[ing] a strong dark matter influence on the history of our solar system and even of life here on Earth."

In a 2014 interview, Randall added diplomatically, "Even if it's a remote possibility that dark matter can affect the local environment in ways that have noticeable consequences over long periods of time, it's still incredibly interesting."[8] Interesting indeed—yet far from proven.

Ah, yes—*proof*. The buzz and popularity of Nemesis faded once no such object was found lurking in Earth's vicinity of the heavens (although absence of evidence is not proof of absence).[9] Still, a 3-D scan of millions of Milky Way stars being conducted by the European Space Agency's Gaia space telescope (launched in 2013) might shed starry light on whether extra mass density exists in the galactic mid-plane, possibly enabling Randall and Reece to find that "ham" and either refine or reject their theory of periodic extinctions.

Especially since the 1980s, physics has presented intriguing information for paleontologists to ponder, but it is anyone's guess whether the picture will be definitively clarified before the inevitable extinction of life on Earth. Not to worry, though. If Randall and Reece are correct, the next Chicxulub-level event shouldn't happen for another ~30 million years. Give or take.

Notes: (1.) "Dark Matter as a Trigger for Periodic Comet Impacts" (2014, April 21). *Physical Review Letters* vol. 112(16): 161301-161305. (2.) See for example, Allen and Diane Debus, (2002), "Portraying Paleocatastrophe" in *Paleoimagery: The Evolution of Dinosaurs in Art* (Jefferson, NC: McFarland Publishers). (3.) "Periodicity of Extinctions in the Geologic Past" (1984). *Proceedings of the National Academy of Sciences* vol. 81(3): 801-805. (4.) Although Earth's solar system is commonly considered to end with Pluto's orbit, the vastness of the entire solar system, let alone the immensity of the single galaxy of which Earth is a part, remains widely unknown. No less mysterious is the question of how the Sun may interact with its stellar neighbors in Earth's region of the Milky Way (which is 130,000 light-years in diameter). The relatively thin, flattened galactic mid-plane is actually quite thick—2,000 light-years, according to one estimate, while the thinner dark disc within would be a mere tenth of that. (In contrast, Earth is only 8.3 light-*minutes* from the Sun.) Once at its periphery, the solar system crosses the galactic plane in about one million years. (5.) The Oort Cloud is the outermost component of the solar system and is composed of trillions of comets; it has never been observed, but its presence is inferred. The innermost edge of the spherical halo of comets that make up the Oort Cloud is thought to begin 0.8 light-years from the Sun, with its outer edge terminating at around 3.2 light-years, or more than halfway to Alpha Centauri—the Sun's closest known stellar neighbor. (6.) "Terrestrial Mass Extinctions, Cometary Impacts and the Sun's Motion Perpendicular to the Galactic Plane" (1"84, April 19). *Nature* vol. 308: 709-712. (7.) "Was it Nemesis That Killed the Dinosaur?" (1985, April 14) *New York Times*. (8.) Choi, Charles Q. (2014, April 28). "Dark Matter Could Send Asteroids Crashing Into Earth: New Theory." Space.com. Retrieved from: http://www.space.com/25657-dark-matter-asteroid-impacts-earth-theory.html. (9.) In jest, authors Marc Davis, Piet Hut, and Richard A. Muller stated that their Nemesis model could be their own "nemesis" if their hypothetical "companion" star was never found. Unfortunately, this has so far been the case. In the most recent effort, NASA's Wide-Field Infrared Survey Explorer, launched in 2009 and in service until February 2011, failed to detect a Nemesis star.

TEN

Topps Dinosaurs Attack! collectible trading cards. See Chapter Thirty-three for more. (copyright 1988, The Topps Company, Inc.)

Bellicose Behemoths

Dinosaurs and those derivative dino-monsters of fiction and film often aren't central to storyline or plot. They are often just 'there' as 'paleo-props,' 'proving' or reminding viewers and readers that the story has some manner of prehistoric bent. That isn't quite the case though when these creatures fight each other—when they are present in particular 'effects' scenes, or if they're imbued with metaphorical purpose. And when dinosaurs and dino-monsters exude metaphor—especially as they have since the early 1950s, it usually doesn't bode well for what's happening in mankind's real world.

The next three chapters have little interrelation except the theme of fighting. Chapter Thirty-three was an earlier submittal of mine to J. D. Lees at *G-Fan* magazine, as I was intent on exploring the not-so-hidden meaning behind these movies and stories. It was somewhat inspired after I won an Ebay bid for a set of Topps *Dinosaurs Attack!* trading cards. So much bloodshed, mayhem and destruction caused by gigantic invading dino-monsters was featured on those cards, which is probably why 1980s-generation kids loved to collect them. Artists who drew the images for those cards must have been raking it in.

By the mid-1990s I'd decided there were just too many real dinosaurs to keep track of (avocationally speaking) in the news and in journals, and so I elected to focus as-time-permits interests on an old favorite genus of mine—*Stegosaurus*. Not only did I begin collecting 'stego' things, and sculpt many kinds of stegosauri, but I also wrote about them too. Chapter Thirty-four is one such manuscript printed in an issue of *Prehistoric Times* during a period when several authors in that magazine tended to write about their 'top ten best' listings of certain dino-things.

And Chapter Thirty-five was somewhat of a nostalgic journey back into my long ago past, at a time when I was considering sci-fi films that captivated me during my most impressionable, youthful years. The chain of events leading me down this path perhaps began while I was serving on Martin Arlt's panel discussion at the 2015 G-Fest concerning *Godzilla Raids Again*. (Another article concerning this experience appeared in an issue of Martin's magazine—*Mad Scientist*.)

The older I get, the more nostalgic my dino-writings become, which is what happened in this book. I would offer my apologies, but—wait ... you'd better put up your puny dukes, pronto! Just *look* at what's coming this way!

> These chapters originally appeared as articles in the following publications.
> Chapter Thirty-three—*G-Fan* no. 72, Summer 2005, pp.8-13.
> Chapter Thirty-four—*Prehistoric Times* no.95, Fall 2010, pp.29-31.
> Chapter Thirty-five—*G-Fan* no.111, Winter 2015, pp.18-23.

Chapter Thirty-three—*When Dinosaurs Attack!*

General Zod: *"My god, that prehistoric monster might get loose! Just think of all the chaos, death & destruction. Do you want bloodshed on your conscience?"*

Dr. Brane: *"The creature feels lost in our modern, civilized world. We must study it before returning it to the wild, an isolated place where it won't be harmed by Man."*

Ms. Drama Queen: *"The extinct were not meant to be resurrected and coexist with Man."*

Mr. Madison: *"We could make a fortune!"*

Man and living dinosaurs are separated by a gulf in time of 66-million years, except, of course, when intermingled through the imaginations of science fiction and horror writers who project grim messages that are often neglected or conflated today. For in fiction, dinosaurs and other prehistoric animals reflect often reflect Man's folly, our aspirations, contemporary understanding of nature, and, especially perhaps, our fears. In the earliest period of fantasy dino-fiction, prior to 1900, fictional human heroes or adventurers were awestruck by living remnants of 'prehistoric' life, without need for aggression. By the 1910s and 1920s, however, humans began haphazardly using weapons and war engines against dinosaurs and their brethren. Ever since, both in movie-land and in the pulps, we killed them shamelessly whenever or wherever they were encountered or appeared, especially when they invaded our cities and shores.

Why such bloodshed? Can't we just, well, get along with science fictional dino-monsters and other prehistoria? And what are the origins of man's apparent militaristic conflict with dinosaurs in literature and film?

One of the most curious products reflecting our fascination for antagonistic dinosaurs is a series of 55 trading cards produced by Topps in 1988, titled *Dinosaurs Attack!*—an inspiration for this chapter. The cards tell the story of Dr. Elias Thorne's attempt to project the past, using a device known as TimeScan. But the technology backfires, instead permitting a full-scale dinosaur assault into the present day. Like *daikaiju*, the dinosaurs are way too enormous. Pictures on the trading cards are often gross, with blood, guts and gore splattering everywhere. Even dinosaurs known to be herbivorous ravenously devour human prey. Yet while we are appalled, we somehow cannot look away. We marvel at the audacity of the artists, who show us disgusting scenes of dino-monster carnage, violence and mayhem. And we take for granted that man simply must wage war with dinosaurs, because … well, *because* … hmmmm.

In Jules Verne's 1864 (and 1867 revised ed.) novel *Journey to the Center of the Earth*, explorers of Earth's deep interior encounter living 'prehistoric' organisms presented in a 'life through time' sequence. In one passage, they're caught adrift on the Central (or Lidenbrock) Sea in the midst of a titanic battle between an *Ichthyosaurus* and *Plesiosaurus*—two Mesozoic marine monsters, and along their journey witness mastodons, mammoths and other Ice Age creatures. But their observations are purely of a scientific nature, as they're led by Lidenbrock, an esteemed Professor of geology. They don't kill the prehistoric creatures for defensive purposes, sport, food or pleasure. The adventurers are merely observing Earth's past natural state through a 'time portal' gained by descent into a volcano. By the 1890s, however, we find that mankind had become less tolerant of dino-monsters, as projected in then contemporary paleoimagery and literature.

In an 1892 restoration of *Stegosaurus*, artist Carl Dahlgren depicted a band of spear-wielding humans anachronistically confronting the beast. This restoration captured the popular belief that prehistoric ('antediluvian') men co-existed with dinosaurs and other prehistoric animals in a primeval, savage state. Then in an 1897 issue of *McClure's Magazine,* the great paleoartist Charles R. Knight painted a Neandertal hunting party attacking a Woolly Mammoth. Please consider—for if prehistoric man couldn't manage to co-exist peacefully with dinosaurs and other prehistoric animals during an age when savagery was the natural order, then how could dinosaurs, if resurrected today, ever manage to get along with modern man, in our present?

Not content to simply make geological observations, take notes, and shoot only photographs when in the presence of fictionalized prehistoric life, by 1900, modern man became the aggressor. Decades before substantive, concerted efforts had been made to conserve rare species and fragile ecosystems, prehistoric fauna discovered in 'lost places' of the globe were truly becoming endangered. In "The Killing of the Mammoth"

(*McClure's Magazine*, 1899) for example, author Henry Tukeman related how modern hunters tracked and killed the very last mammoth, which had survived in the Alaskan wilderness. In Tukeman's short story, mistaken by many readers as a factual account, at least the protagonist expresses remorse after cruelly dispatching the last 'prehistoric monster': "A feeling of pity and shame crept over me as I watched the failing strength of this mighty prehistoric monarch ... It was as though I were robbing nature, and old Mother Earth herself of a child born to her younger days, in the dawn of Time."

In Frank Saville's *Beyond the Great Wall* (1901), it is a monstrous brontosaur which must be killed if the intrepid band of explorers are ever to escape from Antarctica. Another contemporary news story penned by Franz Hermann Schmidt and printed in the Jan. 11, 1911 *New York Herald* related a horrific encounter with a "bullet-proof monster" described as a plesiosaur haunting Colombia's Solimoes River in South America. Dramatic accounts such as Tukeman's and Schmidt's are in the same thematic vein as a rather famous 1912 novel by Arthur Conan Doyle.

But Conan Doyle had already spun a tale of evolutionary forces at play, creating a fearsome beast derived from the Ice Age Cave Bear. In his little known 1910 story, "The Terror of Blue John Gap," which some believe may be a thematic precursor to *The Lost World*, protagonist Dr. James Hardcastle eventually kills the cave monster in its lair before it can further menace the British countryside.

In today's culture, Doyle's 'Lost World' novel would be politically incorrect. Doyle's heroes eradicate one crypto-species of 'primitive' hominids, as if they don't deserve to survive. They also bag dinosaurs and pterodactyls with elephant guns. I guess the only good dinosaur (or prehistoric ape) is a dead (or extinct) dinosaur. At the climax scene of the 1925 silent movie, *The Lost World*, a brontosaur tramples London. A contemporary movie advertisement for the classic film—showing an imagined yet unfilmed scene, preys visually on our minds, revealing an enormous, four-fingered theropod dinosaur clutching a train car with its right foot talons as it rages at fleeing people below. In the background we see city lights. The dinosaur invasion had begun, provocatively in forms of paleoimagery.

While many regard Doyle's 'first strikes' as mere skirmishes against ancient natural fauna, during the World War I era, two other writers, Edgar Rice Burroughs and Vladimir Obruchev, waged an all-out assault on prehistoria found in other lost world settings, respectively, in Caspak and Plutonia. In *The Land That Time Forgot* and its two sequels, first published in 1918, Burroughs' adventurers slaughter Caspak's dinosaurs and other indigenous prehistoric fauna with reckless abandon. And while Obruchev's adventurers penetrate into the Arctic's inner hollowed cavity, a setting named Plutonia, they leave a trail of bloodshed, as if they owned the place. Furthermore, Burroughs' heroes waged war on the mighty pterodactyl race—the Mahars, and a host of other prehistoria deep within the Earth, at its hollowed core named Pellucidar, in a series of popular stories beginning in 1914. 'Terrorist actions' such as these could only prompt vengeful retaliation … eventually!

In his 1929 story, "The Greatest Adventure," writer (and mathematician) John Taine's explorers obliterated a prehistoric pseudo-dinosaurian fauna discovered thriving in Antarctica. True, the scientist-led expedition was ostensibly on a noble scientific quest to discover the mysteries of life's origins there, but along the way, they wiped out as many of the great beasts as they were able, before they could escape and conquer the world! And so—with each passing fictional encounter, mankind seemed to 'drop' increasing numbers of prehistoria dead in their tracks, and in *their* natural habitat, no less. Ultimately, inevitably, matters went too far, one might say. And in 1933, during the time of Nazi oppression, man (or was it beauty?) killed the majestic, prehistoric ape named Kong!

Forever more, the parade of life through the ages would march to the steady pounding beat of war drums.

By the 1930s, movieland dinosaurs were catching up to the prehistoria of fantastic literature, largely due to the successes of special effects animators. Using dino-creations made of clay, plasticine, rubber and fabric, film producers were learning how to suspend disbelief effectively, much as artists and masters of speculative fiction already had honed their craft with pen, paper, and their fertile imaginations. And the world was in upheaval once again, as nations led the march to a second World War, climaxing with detonation of two atomic bombs. Perhaps it just felt safer to read about or observe from a safe vantage point the war on dinosaurs, creatures that were commonly (but not only) emblematic of prehistoric savagery in nature, although now projected into modernity.

Kong, discovered on his native island home, Skull Island (or 'Mountain'), cohabitated with reptiles surviving from the Jurassic and Cretaceous periods which both he and Ann Darrow's rescuers battled to the death prior to Kong's capture. When Kong—best viewed as a great prehistoric ape, rampaged New York City, he is defeated by U.S. biplane pilots firing machine guns. Now denizens from prehistory were engaging mankind's military forces. Shortly after Kong's demise, Ruth Plumly

Thompson weaved both a living dinosaur skeleton as well as maritime war waged by two island countries into the plot of a children's book, *Speedy In Oz* (1933). And Czech author Karel Capek (who in 1921 coined the term 'robot'), penned *War With the Newts* (1936), chronicling mankind's discovery and eventual conflict with a sentient race of amphibious salamander species. These direct evolutionary descendants of a Miocene genus, *Andrias scheuchzeri*, develop technological prowess. Mankind helplessly loses the war versus *Andrias*! Soon after, in the pages of the pulps, we read of man's exploits on other 'prehistoric worlds'—in outer space, where 'exo-dinosaur' fauna fall at the hands of human plunderers.

By 1940, both a Venusian 'dinosaur,' in Henry Kuttner's short story "Beauty and the Beast," and an enormous, Godzilloid-tyrannosaur from the Arctic which, when thawed, attacks Superman's Metropolis had menaced our cities. Although a pair of atomic bombs swiftly concluded the second World War in late 1945, the resulting dinosaurs of film and literature were secretly and silently planning a devastating counterattack. And by 1950, through Isaac Asimov's grim tale—"Day of the Hunters" (1950), we had learned something too about the dinosaurs' destructive tendencies and the science fictional possibility that they may have (nearly?) annihilated themselves, 66-million years ago. (A metaphor for mankind.)

Perhaps unwittingly, Ray Bradbury touched off a firestorm with a pair of his most beloved short stories, "The Fog Horn" (1951) (originally titled "The Beast From 20,000 Fathoms"), and "A Sound of Thunder" (1952). In the former, Bradbury told the mournful tale of the last dinosaur who, in mistaking a modern lighthouse foghorn for the cry of another of its own kind, mounts the beachhead of modern civilization. Pausing before waging unilateral attack, the monster returns to the sea. This story would influence (however indirectly) a number of fantastic films through the 1950s Cold War period. For in the wake of Hiroshima and Nagasaki, and during a heightened period of nuclear testing in America and the USSR, dinosaurs represented nuclear power and radioactive holocaust, man's ultimate folly.

Bradbury's second pivotal story, "A Sound of Thunder," introduced what is still known today as the "butterfly effect." Harping on man's techno-folly, Bradbury cautioned how when science meddles in the unknown, Nature may come back with roaring vengeance. In this case, hunters traveling to the Cretaceous period to seize a Tyrannosaur trophy get more than what they bargained for. To be sure, the primeval hunt comes with its fearful moments, but dispatching the King of the Tyrant Lizards proves to be less problematic than the life of one butterfly, inadvertently extinguished by the careless hunters. The future history of our planet is profoundly altered in consequence.

Are men, guns and dinosaurs perhaps an even more lethal combination than guns and alcohol? Fifties writers of fantastic fiction certainly seemed to think so. For besides Bradbury's classics, Brian Aldiss brought us "Poor Little Warrior (1958), Arthur C. Clarke contributed "Time's Arrow" (1952) to the furor, while L. Sprague de Camp penned "A Gun for Dinosaur" (1956) and Poul Anderson published "Wildcat" (1958 … all of which involved humans traveling backward through time, only to be slain by vicious dinosaurs. 'Don't mess with nature,' the dinos seemed to be telling us! But we didn't heed their warning. Although the government and military forces often felt pressured into using a nuclear deterrent against the dinosaurian invasion of the 1950s and 1960s, I suspect that nuclear weapons wouldn't have repelled the scaly, fire-breathing horde. Against the new breed of highly 'evolved' prehistoric super-monsters, which defied the laws of physics, chemistry and most known biological principles, what good would an atomic or hydrogen bomb have done? Several of these ferocious beasts, like the Rhedosaurus, Godzilla, Rodan, Anguirus, and the Paleosaurus were already symbolically 'radioactive' anyway! Even Kong came back in 1962, albeit in mutated form—larger than ever, roaring with vengeance! And so they came, trampling cities in their stead: New York City, Tokyo, Osaka, London, inspiring fear and cosmic angst—presaging doom and our ultimate oblivion.

But not every prehistorian was huge. After all, the amphibious Gillman featured in 1954's *Creature From the Black Lagoon* was a descendant from the Devonian period, which menaced scientists bearing malice toward its kind in the Amazon. In movie sequels, the Gillman is captured; however, attempts to humanize it prove deadly. Lord knows what threats *that* crypto-species will pose to mankind someday, reminiscent of those pesky 'Newts'! It should also be said that not all dinosaurians penetrating our outer defense networks proved antagonistic: such as the museum display dinosaur which came to life for a joyful romp through an imaginary city in Syd Hoff's children's book *Danny and the Dinosaur* (1958), yet 'Danny's' dino-pal was an exception.

The full frontal super-dino-monster assault has been well chronicled elsewhere in scores of *kaiju* books and fanzines. In pulp fiction as well as on the silver screen, more cities toppled under Godzilla's massive clodhoppers than any other dino-menace. However, during the 1960s, Japan's major 'ironical icon,' Godzilla was soon joined in the fray by 'Ogra' (i.e. 'Gorgo'), Varan, Reptilicus, Yongary, Ghidorah, Gwangi, Gamera, Gappa,

Gorosaurus, a *T. rex* featured in the 1960s film, *Dinosaurus!*, and, in time, elegant marine Titanosaurus. These and other *kaiju*-prehistoria launched a concerted effort, storming through the world's major defenses. Mankind barely survived these horrific ordeals, yet I am here to tell the tale of what followed, as humanity was brought to the brink of global disaster.

By the 1970s, following scientific description of the lively North American genus, *Deinonychus*, the new wave of 'stealth' dinosaurian fighters had already been 'bio-engineered' in the fertile minds of sci-fi and horror writers. By the 1970s, *Deinonychus* was becoming a familiar fiend, appearing in popular dinosaur books, comics and the pulps, where its growing menace to man, both in the remote past as well as in our present, became evident. Another 'secret weapon' from prehistory had already been known for decades however, as the genus *Velociraptor* was first described back in 1924, although then its cunning, gruesome killing tendencies (as ritualized in novels and film) could not have been guessed. *Deinonychus* and its evolutionary cousin *Velociraptor* were not capable of destroying entire cities like *daikaiju*, yet they joined the ranks of Godzilla and fellow *kaiju* in the fateful war on civilization by committing 'terrorist attacks.' The 'rise of the raptor' had begun!

While *kaiju* traditionally were resurrected from the emanations of fiery volcanoes, raptors and a host of other dinosaurians now were being spawned through biochemical means, usually from DNA preserved in amber or other kinds of fossils. At first the vicious raptors were encountered in stories such as Harry Adam Knight's novel *Carnosaur* (1984), or Allen Steele's 1990 novella, "Trembling Earth," but soon Michael Crichton employed them to fuller effect in his blockbuster 1990 novel *Jurassic Park*. Furthermore, intrigue over the intelligent genus *Velociraptor* led to a host of tales concerning technologically advanced 'raptor' evolutionary descendants, known as dinosauroids, which threaten mankind as in Frederick D. Gottfried's 1980 short story, "Hermes to the Ages," in which a dinosaur 'fossil,' named genus & species "Homosaurus mercer," is found on the Moon, or as in Thomas Hopp's pair of exciting *Dinosaur Wars* novels, where 'pteronychus' dinosaurs launch an all-out attack on our planet from their lunar base to reclaim Earth for themselves. The plausibility of intelligent 'dinosauroids' enlivened the possibilities for man's (fictional) encounters with extraterrestrial beings, some of which could be viewed as hostile dinosaurs—descendants of Earth's dino-raptors. At the opposite end, however, perhaps reflecting our 'softer side,' we find Piers Anthony's 1990 novel, *Balook*, concerning a peaceful, genetically engineered family of *Indricotherium* hounded by irritating, senseless human beings.

It was a short leap from the pages of the raptor-infested pulps and contemporary literature to the silver screen and direct-to-video release, where unleashed raptors soon had a field day punishing our troops, unscrupulous business people and misguided scientists with cruel, sickle-shaped talons. Blood was splattered throughout movies such as *Jurassic Park* (1993), *The Lost World: Jurassic Park* (1997)—both scripted from Crichton's 1995 novel, *Jurassic Park III* (2001), and more recent fare such as *Raptor Island* (2004). (Look closely and you'll notice even how the vicious 'Alien' species, starring in the series of movies beginning with *Alien* (1979) through *AVP: Alien vs. Predator* (2004) do resemble dinosaurs—even though they are decidedly not of that ilk!) And as if we hadn't noticed already, in a dinosaur cartoon produced by Steven Spielberg advertised in its catchy title *We're Back: A Dinosaur's Story* (1993), based on Hudson Talbott's 1987 children's tale, police and military are called in, just as the friendly but confused dinos retreat to a museum's dinosaur 'sanctuary.'

A difficulty with war is that often people tend to forget what the triggering factors originally were—which are inherently complicated anyway. Over time then, war can become a self-perpetuating, even 'traditionalized' ritual, like what's happened in our war with dinosaurs and other prehistoria. True, dinosaurs have symbolized many things through the decades, and many an apt metaphor has been plied upon their immense carriages. But there is something at large, beneath it all, which most of us wouldn't recognize in this day and age. Why did this war begin? First, I'll refer you to Paul Semonin's invaluable *American Monster: How the Nation's First Prehistoric Creature Became a Symbol of National Identity* (2000). Although I don't wish to digress, I gleaned the following from Semonin's scholarly treatment of the American Mastodon.

Since the 1850s, man's perception of nature and prehistory were shaped by two philosophies: (1.) the 'myth of wild nature,' where all species are considered to be at war with one another in an intense 'Darwinian' struggle for existence, and 2.) *prehistoric* nature was also represented as a terrible, savage ordeal ruled via nature red in tooth and claw. As noted by Semonin, embracing such symbolism and metaphor justified white Europeans' usurpation of Native Americans, for example, or even other races in other places of the globe, because of a belief in genetic superiority. And because the 19th century's (highly misconstrued) 'ladder of evolution' inexorably (and as some thought, atheistically) led to the pinnacle—creation of a white European race, then all that could have only been accomplished through countless millennia of prehistoric *conflict*.

In an evolutionary sense, through geological time, not only had man's ancestors ascended beyond the extinct dinosaurs, we had also conquered other extinct races of man, such as Neandertal. For example, in his 1920 short story, "The Grisly Folk," H. G. Wells speculated how the Neandertals were out-competed by the contemporary Cro-Magnon. So, by the early 20th century, it was socially acceptable to fantasize about dinosaurs and other prehistoric (or—read 'primitive') species as being somehow inferior and in conflict with modern man who would symbolically fire guns upon them, thus celebrating our supposed superiority over prehistoria—broadcasting the silly message, "we beat them before and we still can today … bang, bang—you're (still) extinct and we're not!" (Here I am reminded of that jolly ride that patrons took at the "Prehistoric Forest" in the Irish Hills, Michigan, where visitors were handed toy guns to shoot the dinosaur statues partially concealed along the winding trail.)

If Charles Darwin were alive today, perhaps he'd be astonished at the myriad, strange distortions of his theory. Dino-symbolism has certainly evolved through 170 years to its present form. To Japanese, however, derivation of dino-monsters and other prehistoria favored in the western world may seem less relevant than nuclear holocaust signified by Godzilla and other *kaiju*. But now, neglecting their evolving cultural symbolism and despite all associated destruction, blood and gore, doesn't the prospect of fantasy dinosaurs engaging our modern military seem, well, somehow 'cool'?

Can we really hold out against the prehistoria? Our forces remain paralyzed, while the dino-monsters attacking on all fronts, multiply.

And then, almost miraculously, something happens to "TimeScan."
Is there salvation for mankind? Do giant dino-monsters have 'souls'?

Chapter Thirty-four—*Plated Paragon: Stegosaurus' Greatest Movie Monster Battles*

It should be no surprise for you that many of our most famous movie monsters, proffered as anachronistically 'prehistoric,' are modeled after dimensions and features of real prehistoric animals. Often, many artistic liberties are taken. Think of blood-thirsty *Tyrannosaurus*, for example. Just add some dorsal plates along its spine, lengthen the arms proportionally, scale its size up by 20 times, and *voila* - you have a Godzilla! *Jurassic Park's "Velociraptors"* were not as huge as portrayed in film. Yet each special, scientifically impossible case (e.g. Rodan, Gorgo, Reptilicus, Rhedosaurus, etc.) emblazons its own set of endearing qualities and anatomical configurations. Alternatively in genre films, we find the prehistoric monsters as known to science appearing rightfully as 'themselves' (although often inaccurately posed), such as arguably the third most famous (real) dinosaur of them all—plated *Stegosaurus*. Herein, we shall consider *Stegosaurus'* greatest movie monster battles, of all-time. Why am I doing this? Well it's simple, "Stego" happens to be my favorite dinosaur.

The idea of the giant prehistoric monster battle was accentuated by artist Charles R. Knight's century-old, influential portrayals of the first and second most famous dinosaurs, respectively—*Tyrannosaurus* vs. *Triceratops*, yes the famed three-horn! The most awe-inspiring of these mural paintings is on display at Chicago's Field Museum of Natural History. Many movie dinosaurs appearing as stop-motion effects were based on Knight's artistry. Many scenes scripted for films, sci-fi novels and short stories have been inspired by Knight's 'Rex vs. Tops' depictions. Think of scenes in *The Ghost of Slumber Mountain* (1919), *The Lost World* (1925) or *The Last Dinosaur* (Tsuburaya Productions, 1977) for example.Knight's artistry has mesmerized me for a lifetime. However, in case you're unfamiliar with Knight, read what two 'rays' of light—Ray Harryhausen and Ray Bradbury said of him:

> "Knight was not only one of the first to reconstruct prehistoric life in a romantic way ... (his) influence ... extended into other fields of endeavor. ...In the world of motion pictures Willis O'Brien and myself were enormously stimulated by Knight's visions of the world of the past as we created new adventures in the world of fantasy and imagination. O'Brien was the first filmmaker to realize prehistoric animals on film making them the stars of *The Lost World* and *King Kong*—their believability made all the more by the groundwork done by Knight ..." (Ray Harryhausen)

> "My history is very similar to Ray Harryhausen's. We both saw the Willis O'Brien film, *The Lost World* in 1925 and we both fell in love with his Knight-inspired dinosaurs." (Ray Bradbury)

As suggested in my *Prehistoric Monsters* (McFarland, 2010), of the earliest 30 or so restorations of the famous 'plated puzzle' completed between circa 1884 and 1926, Charles R. Knight did five of these, which were the best, most original, accurate and influential of the lot. Many artist competitors published knock-off restorations stylized after Knight's work, a 'copy cat' practice that continued through the mid 20th century. I won't go into the traditional discussion here emphasizing whether *Stegosaurus* had a single row, versus double-paired or staggered rows of plates, versus other type of dorsal osteoderm and tail-spike arrangements, but you can find this story recounted in a chapter of another of my books, *Paleoimagery* (McFarland, 2000).

Fortified and fueled by Knight's illuminating prehistoric visions, film artists such as O'Brien and Marcel Delgado soon found themselves designing sculptured 'puppets' for stop-motion use in projected movies such as *The Lost World,* the never completed "Creation," and *King Kong* where, despite its supposed gentle or docile demeanor, *Stegosaurus* appeared menacingly. In the very early filmic years, "Stego" did not battle another prehistoric monster. Such one-on-one fighting (albeit of the defensive variety) would have to wait until Walt Disney Studio's "Rite of Spring" segment released in *Fantasia* (1940). By then according to paleontologists, although *Stegosaurus*' conventional adversary was thought to be horned carnivore contemporary *Ceratosaurus*, in *Fantasia Stegosaurus* battled what appears to be a *Tyrannosaurus*.

And so this will be my aim—to critique this battle versus three other great battles, waged by our "plated paragon" against other dino-monsters.

While *Fantasia* was a cartoon, the other three battles were projected using stop-motion animation.

Let the contestants step forward! Stego's brief yet survival of the fittest performances in *Planet of Dinosaurs* (1978), *Fantasia* (1940), *The Animal World* (1956), and *Journey to the Beginning of Time* (Czech-1955/USA-1966) will be described, compared and herein evaluated. But first, what came 'before' in the movies. A short summary of Stego's supporting roles in its First National Pictures and RKO heyday shall suffice.

First National Pictures' stegosaurs in *The Lost World* were inspired by a passage in Arthur Conan Doyle's original 1912 novel, wherein doomed explorer Maple White claimed to have witnessed a living specimen atop the fabled 'lost' plateau. Stegosaurs animated for the 1925 film are quite visible fleeing in scenes where herds of various dinosaurs are stampeding from a volcanic eruption. According to Neil Pettigrew, however, in a test reel prepared by O'Brien's team, "...two tyrannosaurs attack a group of herbivorous dinosaurs, then each other. The victor is set upon by a triceratops, then flees and unsuccessfully attacks a stegosaurus." (*The Stop-Motion Filmography*, McFarland, 1999) I haven't witnessed this scene, although perhaps chopped versions of the film that I've seen omitted this sequence. Or, it could be also that the Stego-Rex battle was never intended for released final footage, but in any case the scene isn't widely known.

A *Stegosaurus* (originally designed by Delgado for RKO's abandoned "Creation") happens to be the first dinosaur encountered by Skull Island's jungle explorers in *King Kong*. And this specimen is a scaly detailed dandy—much larger than in real life. First appearing off in the misty distance, it meanders back in, closer this time. Then it charges toward Carl Denham and the *Venture's* crew - right towards us! A gas bomb, followed by rifle shots brings down the magnificent beast. Stego's spiked tail twitches in death. Mark Berry notes, "...the idea of the 'dead' animal suddenly reviving for one final thrust has since become a genre cliche'."

Dorsally-plated (non-stegosaur) 'dinosaurian' varieties and abominations would multiply in prehistoric monster films (e.g. *One Million B.C.* (1940), *Gojira* (1954), *The Lost World* (1960). John Taine's 1929 novel, *The Greatest Adventure*, incorporated gigantic Antarctic faux-dinosaurians, one variety of which was plated—illustrated by Lawrence Sterne Stevens for the cover of *Famous Fantastic Mysteries* (June 1944) where it was reprinted. However, there were only several cases of genuine stegosaurs appearing in the movies. The most memorable of these were scripted into dramatic fighting scenes with hungry carnosaurs. (A *Stegosaurus* skeleton model was shown briefly in *Gojira*.)

Which leads us to the first real battle fought by *Stegosaurus,* featured in Disney's *Fantasia*. In the case of individual stegosaur combatants, it's not whether they won or lost; what matters most is whether they fought the good fight.

The "Rite of Spring" segment in *Fantasia* was titled and somewhat inspired, of course, after music composed by Igor Stravinsky. As Mark Berry states in his most excellent *Dinosaur Filmography* (McFarland 2002), "Yes, it is 'cartoon' animation, but in 1940 no one had seen anything like this. *King Kong* had offered marvelous dinosaurs but the *Fantasia* fauna was in bursting Technicolor, rendered in dramatic highlights and shadows that any of the great cinematographers would have been proud to achieve in live action." *Rite of Spring* is an evolutionary sequence, taking us from primordial earth stages, rapidly through ages of invertebrates and fishes on into the majestic Era of Dinosaurs. Although the ordering of life as portrayed and sequenced in *Fantasia* until this point of the story was more or less accurate, any dinosaur fan knows that *Stegosaurus* went extinct about 90 million years before *Tyrannosaurus* evolved. They emphatically didn't live together.

Fantasia's fight scene was directed by Bill Roberts and Paul Satterfield, and the accuracy of its dinosaurs was facilitated by American Museum paleontologists. As Berry comments, while the original battle was suggested as a *Tyrannosaurus* versus *Triceratops* scene, "...an anonymous story man—who knew his dinosaurs—had another idea. 'No,' he said, apparently with much conviction, 'we want to use the *Stegosaurus* because of (the) action of his tail and the four spikes.'" This wasn't the first nor would it be the last time that artistic license won out over scientific accuracy.

"Ding, Ding": Round One, "In this corner ...."

So in *Fantasia*, herbivorous dinosaurs wallow in their comfy mud holes, eating lush vegetation, slaking their thirst. Suddenly, lightning and stormy thunder announces arrival of a three-fingered, demonic *Tyrannosaurus*! Or as Berry aptly characterizes, "the embodiment of merciless nature." All the dinosaurs flee but *Stegosaurus*, being slowest, can't escape Rex's cruel jaws. Rex sinks sharp, six-inch saber-teeth into Stego's tail, forcing the plated wonder to turn and stand his ground—futilely we sense. Stego swings its massive tail three times through the pounding rain, only wounding the unstoppable Rex. Face to face, Rex encroaches upon retreating Stego, the latter glaring, the former baring teeth that have brought down enormous prey time and again. Rex bites tortuously into Stego's neck. Sinking into soft mud, light fades from Stego's weary eyes. It's all over in 45 seconds—Rex growls triumphantly toward the heavens above.

For the 1964 New York World's Fair, Walt Disney, with WED Enterprises, created a "Magic Skyway" for the Ford Wonder Rotunda. Here, many life-like animatronic recreations of prehistoric animals were displayed, including *Stegosaurus* and a 22 foot high *Tyrannosaurus rex*. The tableau—recalling *Fantasia's* dramatic dino-battle scene—was later transported to Disneyland in Anaheim, forming part of its "Primeval World" display, where I was able to enjoy it in 1976.

Next, for Irwin Allen's documentary film *The Animal World* (1956), Ray Harryhausen (facilitated by Willis O'Brien in a 'supervisory' role), created a short "prehistoric sequence" comprising several stop-motion animated dinosaurs, featuring a battle royal between *Stegosaurus* and its traditional adversary (according to the vertebrate paleontologists of the past), horned *Ceratosaurus*. If you're a sixties kid like me, you might remember seeing scenes of *Animal World's* dinosaurs on those old Viewmaster reels.

"Ding, Ding": Round Two

Leading off *The Animal World*, following colorful glimpses of *Allosaurus, "Brontosaurus"* and its hatchling, *Stegosaurus* appears near a cliff. The diorama is a simple tabletop surrounded by painted background scenery. Stego enters, introduced as a "... mild mannered fellow (an) introvert by heredity." Stego realistically chews leaves, until 20 foot high, bipedal *Ceratosaurus* approaches. The two square off, tails swishing apprehensively. To no avail, Cerato bites into a rubbery looking dermal plate. Stego waves its tail and they circle each other. Then the great carnivore seizes an opening, going for Stego's vulnerable neck. The neck bite is forceful and deadly. Stego swings its tail spikes once, wounding Cerato. Blood issues from both creatures now, yet it is Stego who topples over, lifeless. A second Cerato leaps into the scene and they struggle for the cadaverous 'prize.' Finally, as so often happens in dino-monster films, both Ceratos, embraced in a death struggle, tumble over the cliff. The

Stego/Cerato fight scene, described by Neil Pettigrew as "protracted," only lasts 48 seconds.

Additional footage was filmed showing hungry carnivores ripping out bloody hunks of flesh from Stego, but Allen decided to have these excessive scenes excised. The Prehistoric Sequence in general has received a range of critique through the years, from Bill Warren's "...unconvincing but lively dinosaurs...", and Pettigrew's "...one of Harryhausen's and O'Brien's less interesting films..." to printed Press accolades. Harryhausen recollected, "Unfortunately the close-ups looked exactly what they were—mechanical models. Neither Obie nor I really approved but we realized it was out of our hands. The dinosaur sequence was the most talked about part of the picture."

To me, however, the stop-motion is quite lively and smooth. Fully cognizant that these are mere mechanical models, regardless, the result is impressive (even though the scene shows us dino-monsters as they were perceived, half a century ago). Individual segments in the nine minute prehistoric sequence aren't accentuated by artistic glass shots, suggesting a lush, misty humid jungle such as marvelously used in *King Kong*.

For our next 'contestant,' Czech animator Karel Zeman produced one of my favorite films, chock-full of stop-motion animated scenes, *Journey to the Beginning of Time*" (1955, 1966-USA). This gem appeared in serial format on Frazier Thomas' *Garfield Goose* children's Chicago WGN broadcast half a century ago. *Journey to the Beginning of Time* isn't exactly regarded as a true sci-fi film, per Bill Warren's strict definition, yet its fantastic elements truly seized my youthful attentions during the early 1960s. This is one of my childhood 'guilty pleasures,' a sci fi-ish sort of tale about 4 boys rowing backward through many geological ages down the river of time—from the present toward Earth's murky "beginning." In between ends of their mysterious journey, loads of recreated prehistoric animals are featured, spanning the breadth of geological time. Zeman's masterpiece is the first movie featuring stop-motion dinosaurs projected in color. And, oh yes - there's a marvelous, tension-filled stop-motion animation battle between *Stegosaurus* and *Ceratosaurus* featured in this film too!

"Ding, Ding": Round Three

The four adventurers have rowed themselves back into the Mesozoic Era, where they confront a blend of Jurassic and Cretaceous dinosaurs. A *Stegosaurus* plods along the rocky shoreline. The boys also see herbivorous hadrosaurs chomping vegetation and two giant brontosaurs, one on the shore clearly based on Czech artist Z. Burian's 1950 painting. (Burian, a magnificent artist in his own right, was a devotee of Knight.) An older boy nicknamed "Doc" takes many notes while another boy snaps a photo. They moor the rowboat as dusk approaches. On the other side of the river, a *Stegosaurus* ambles in from left of screen. A horned *Ceratosaurus* strides from the right. Uh oh!

Silhouetted in the dimming sky, the two dino-monsters prepare for battle! *Ceratosaurus* wastes no time, going in for Stego's vulnerable neck, clamping down hard. The great carnivore unsuccessfully attempts to turn Stego over on its side. Stego hangs in there but wisely retreats from its deadly foe. They circle, and Stego's impotently wagging tail finally lands two blows upon Cerato's chest, repelling the attacker. Blood issues both from Cerato's chest as well as Stego's neck. Stego limps along toward the setting sun ... before expiring. Total time of this battle—a record, ladies and gentleman, one and a half minutes! The next morning, the boys walk beside a life-sized prop sculpture of the dead plated dinosaur.

While puppets for this well animated scene were also clearly inspired by Burian's painted restorations of these two dino-monsters, in closeup view the models don't look either too detailed or convincing. Yet, spied from afar, as in 'across the river,' their movement is quite good. Also, the *Ceratosaurus'* roaring isn't well synchronized with its mouth movements. While in all three cases considered so far, Stego still hasn't won a bout, this is the first time that our plated paragon inflicts a potentially mortal tail strike upon its attacker. This is the longest Stego battle sequence so far witnessed. Yes, Stego dies, but may this fight be judged a tie, that is if Cerato also died off camera, resulting from Stego's meager offensive?

The most recent entry to be considered here is the battle scene between an alien *Stegosaurus* and *Tyrannosaurus rex* in *Planet of Dinosaurs* (1978). These scenes were created by Stephen Czerkas who sculpted the puppets, and Jim Aupperle with Doug Beswick. Critics often dismiss this sci-fi film as lackluster, but this is due to lackluster acting and uninspired dialogue, not because of the genuinely impressive stop-motion footage created for several animation sequences, collectively featuring more dino-monsters than are usually witnessed in such arduously completed fare. Astronauts pilot spaceship *Odyssey* on its mission, but due to engine failure they crash-land on a life-supporting planet. Unfortunately, most of the inhabitants are monsters—that is, alien dinosaurs—yet such as formerly lived on Earth and even a big, hairy spider to boot!

"Ding, Ding": Round Four

In *Planet of Dinosaurs*, a roaring Stego surmounts a sand dune, causing screaming human actors to scramble for safe higher ground, from where they observe the terrifying scene about to unfold. A 'deathbeast' *Tyrannosaurus rex* then enters from behind rocks at right of frame. Both monsters are extremely warty-looking, yet imposing. They face off. Stego feints right while Rex closes in. Stego incessantly waves its spiky tail but this menacing motion is poorly aimed. The tail never seems to connect—but then again, maybe it does, once! Rex, also swishing its tail, goes for Stego's side and nibbles into several rubbery-looking dermal plates. Next, Rex bites fiercely into Stego's right hind limb. Suddenly Rex loosens his toothy grip. It's hard to tell but Stego may have wielded a 'phantom' non-lethal blow with its tail upon Rex's jaw! On the instant 'replay' we spy a blood streak on the left side of Rex's face. Rex goes for a jugular kill, raising the doomed Stego in its cruel jaws off the ground, then flipping its victim over onto its right side. Rex realistically devours gooey flesh from Stego's abdomen. As you may surmise, this battle sequence follows the 'traditional' pattern—carnivore fiercely biting into spinal plates, then ultimately delivering a death blow to Stego's neck or head. You can see *Fantasia's* influence here.

Total time of this fight scene is one minute. But actually, much of that one minute sequence is intercut with footage of a foolish human character retrieving a laser gun that lies in between the battling dino-monsters. So, reducing the intercut scenes, the stop-motion footage between Stego and Rex in *Planet of Dinosaurs* only lasts 30 seconds.

Planet of Dinosaurs was poorly received and not released theatrically until 1981, as a second feature drive-in billing. As Pettigrew commented, "In 1978, this seemed like a last gasp for dinosaur movies and for stop-motion effects (two genres heading rapidly toward extinction)." Pettigrew also notes that the Stego-Rex battle "...animated by Beswick lacks real energy and is weakened by live-action cutaways of disinterested actors.... Faults aside, this is still one of the great dinosaur pictures."

Okay - so what?

First, how did Stego fare overall in its title bouts? Second, what's the significance of this abbreviated study?

We've considered two of Stego's classic battles staged with a tyrannosaur, and two versus its geological contemporary *Ceratosaurus*. Stego clearly lost both battles versus Rex, but *may* have tied one time with Cerato in *Journey to the Beginning of Time*. Which is the most realistic battle, or the most artistically portrayed? For realism, both Stego/Rex battles must be disqualified simply on the basis that these dino-monsters were not contemporaneous. And dinosaurs on another planet, or having human eyewitnesses? Equally implausible. In my opinion, *Journey to the Beginning of Time's* stop-motion seems a little more fluid, however, and so that's the best of the four title fights! (You are welcome to disagree. I didn't intend to start a 'rumble' over this.)

What is the significance of such scenes in sci-fi and fantasy films? These dino-monster battles are in a large sense an offshoot of Charles R. Knight's century-old paintings, especially his revered *Tyrannosaurus* versus *Triceratops* restorations. These dramatic visual recreations deified the idea of the giant prehistoric monster battle—which lent itself so keenly to science fictional themes. While Knight's most famous compositions featured *terrestrial* combatants, analogous science fictional themed implications had already been explored previously half a century earlier by Jules Verne (with artist E. Riou), who entertained us with "imagetext" depiction of a giant *marine* reptile battle in *Journey to the Center of the Earth*. In the years following Verne, Riou and Knight, major motion picture studios seized the notion of huge battling reptilians for stupendous special effect portrayals. The theme persists to the present day. Think of *The Lost World* (1960), Toho's *Gigantis the Fire Monster* (1959), *One Million Years B. C.* (1966), and *The Last Dinosaur* (1977) for example. Many of the huge animals portrayed in such films are not truly 'prehistoric' in a scientific sense, but at least in filmic context are viewed as such. The popular theme persists to the present day.

Stego continued to make appearances such as in *The People that Time Forgot* (1977), or (using CGI rather than of the stop-motion variety) in *The Lost World: Jurassic Park* (1997). There's a brief Stegosaurus vs. *Allosaurus* encounter in Don Glut's *Dinosaur Valley Girls* (1996), in which Stego fends off *Allosaurus* with a tail swipe. (And, yes, for brevity's sake, here I am omitting from consideration those cgi-bolstered, science pseudo-docu-dramas—beginning with BBC's *Walking With Dinosaurs*, 1999, which include stegosaur scenes.)

Perhaps someday that "mild mannered fellow" *Stegosaurus* will score a decisive knock-out victory over a predatory dino-monster yet in an epical sci-fi/horror flick!

Chapter Thirty-five—*IMHO: Very Best of the pre-1960s GIANT Pseudo Dino-Monster Battles*

In my humble opinion there are three.

Without further ado, I shall name the triad of films in which the subject monster battles appeared, although otherwise not in any particular order (yet): Irwin Allen's *The Lost World* (1960), Hal Roach's (both Jr. and Sr.) *One Million B.C.* (1940), and Toho's *Godzilla Raids Again* (or 1959's *Gigantis the Fire Monster)*. Yes, I know *The Lost World* was released in 1960 but it was in production prior to then, and because it carries special relevance to my psyche it is included here. I should lay out my criteria, ground rules and rationale for my selections. And to further clarify, I do mean battles waged by dino-monsters against fellow dino-monsters .... (*not* versus humans). After all, how we so love when GIANT dino-monsters fight. (Incidentally, I can think of no more fitting example of the abject filmic-inspired *terror* caused by sheer astonishment at sight of *gigantic* dino-monsters, than the cave sequence in Toho's *Rodan, The Flying Monster*, wherein Kenji Sahara's character "Shigeru" temporarily loses his mental faculties after seeing the monstrously-sized enormity of a pterosaur *hatchling*.)

First, while I deeply admire stop-motion animation, particularly when it comes to dinosaur portrayals in movies (e.g. excellency achieved in RKO's *King Kong* or Ray Harryhausen's *One Million Years B.C.*), unlike many fans, I'm not a purist when it comes to animating prehistoric monsters (or pseudo-dinosauria). After all, there *are* other ways of effectively conveying the look, feel and gravity of such fantasy beasts, then and now. Notably today, we're attuned to computerized animated effects ("cgi"), and in my youth we marveled at cartoon animation, suitmation as well as use of live lizard 'actors' which had been adorned with prosthetic horns, fins and neck frills. While some of the latter techniques wouldn't pass muster anymore among the modern generation of monster movie fans, to emerging baby boomers decades ago, such monsters could certainly elevate the psyche. And so, for purposes of this article, you will note that I'm not delving into films involving cartoons or stop-motion. (Readers can peruse

Neil Pettigrew's 1999 book concerning the latter category of films, *The Stop-Motion Filmography*.)

One Million B.C. is a marvelous film, rarely shown today, although in the early 1960s it regularly appeared on Chicagoland afternoon science fiction programs.Notably, Donald F. Glut cites the personal significance of his initial viewing of *One Million B.C.* toward his ensuing burgeoning career of the 1960s and beyond. In fact, his own 1996 movie, *Dinosaur Valley Girls* (also employing live lizard 'giant' dino-action), was inspired by *B.C.* As related in a chapter of Glut's *Jurassic Classics* (2001) ("One Million B.C. (the Original)"), *B.C.* was most likely the first nighttime televised movie he ever watched during the early 1950s. Furthermore, his article about the movie, appearing in the August 1966 issue of *Modern Monsters* became his first published (and paid for) piece concerning dinosaurs of the movies. In *JC*, Don mentions also that during "… those days before home video, the text of the article was largely based upon my own memories of *One Million B.C.*, seen over the years in various television reruns." (p.176) Today, in utter contrast, besides Glut's synopsis in *JC* there are other sources both in books and available online that summarize the movie. I also have fond reminiscences of Don showing *One Million B.C.* in its entirety during a two day course concerning dinosaurs of the movies, of which I was a participant, in September 1985, held at the Field Museum of Natural History in Chicago. Of the three subject films, today *B.C.* may seem most shrouded in mystery, therefore necessitating a short plot summary. Whoa, "Shrouded in mystery?" Yes – my hunch is, at least relatively so. That's because most younger fans may have never seen (or heard?) of *B.C.* because it's in glorious black and white, and also because reliance on live lizards (and other animals, but also employing one *Tyrannosaurus* suitmation effect) for the monsters wouldn't seem convincing, say, to the millennial generation, who would probably prefer watching other sci-fi and fantasy films instead (and who may prefer Harryhausen's similarly titled 1966 entry instead). So hopefully a very brief summary will suffice for our purposes.

The film concerns two tribes of cave people (who resemble modern humans or Cro-Magnon), those from the Rock Tribe and others belonging from the Shell tribe, as their story is deciphered from cave markings by a modern archaeologist and related to a group of mountaineers. And so, one million years ago, as the ancient tale unfolds, virile young hunter Tumak is banished from his people after challenging his father and leader of the Rock Tribe, Akhoba, over fresh meat from a dinosaur carcass slayed by Tumak. Tumak leaves in a rage, encountering several prehistoric animals along his journey, and soon encounters beautiful Loana of the Shell Tribe, who invites him to the cave dwelling of her spear-fishing Shell Tribe. Here it is apparent that while members of the Rock Tribe are crude in their ways and manners, the more socially adept Shell Tribe people are far more caring and sharing of their food, supplies and possessions.

Tumak (played by Victor Mature) and Loana (Carole Landis) are two highly attractive specimens and, as Turner Classic Movies host Robert Osborne whimsically commented in a late 2005 television broadcast, both "big-chested." There is mutual attraction between the two cave people, and after Tumak is thrown out of the Shell Tribe's lair, she follows him through the prehistoric jungle, infested with monsters—each looking absolutely enormous thanks to the special effects team—including an armored glyptodont (really an armadillo upon which rubber horns had been glued). Then approximately 50 minutes into the film, the first great huge pseudo dino-monster battle ever filmed, a full 2 minutes and 10 seconds of vigorous tumbling and thrashing, begins. Of course, *B.C.*'s dino-monsters (i.e. a live dwarf alligator sporting an impressive flexible rubber sail had been affixed over its spine—mimicking a Permian *Dimetrodon*, matched against an unadorned tegu lizard) do not resemble real dinosaurs in the least, but they truly are 'saurian.' (Note – in RKO's *King Kong*, Kong—a mammal, always fought the dinosaurians of Skull Island, but the dinosaurians are never seen combating each other.[1]) In "*B.C.*" such creatures are collectively referred to by Loanna's people as "neecha."

Don Glut summarized the ensuing battle in his *Dinosaur Scrapbook* (1980):

> "As the cave people take refuge in an earth fissure, the two reptiles become engaged in a furious battle that, thanks to the Society for the Prevention of Cruelty to Animals, could never be filmed today. Over and over the two animals roll, jaws locked upon scaly necks, blood flowing profusely, until the *Dimetrodon* lumbers off victorious. This sequence has become almost a classic in itself and frequently finds itself spliced into later films." (p.124)

The use of live lizards was even advertised as a 'plus,' rather than a deficiency, as mentioned in a *B.C.* press kit: "There is no animation whatsoever in the action of *1,000,000 B.C.* (sic). All the animals are 100 percent real: miniature descendants in many instances of the picture book variety we have come to associate with early times." (quoted from Stuart Galbreath IV's *FilmFax* article)

In his 1995 *FilmFax* article, Stuart Galbreath IV notes that while RKO's *King Kong* re-release in 1952 stimulated "a whole slew of giant monster movies," *B.C.* generated a rash of low-budget productions to not-so-ingeniously incorporate footage and outtakes from Roach's picture into their own." Although you'd probably almost need to be of the baby boomer generation to realize it, between 1943 (beginning with *Tarzan's Desert Mystery*) through 1989, segments of the *B.C.* movie sometimes including the famous battle sequence footage made it into several other 'prehistoric' "B" films, television shows, and even commercials, as well. (Strangely, Galbreath mentions Toho's *Rodan* as one of those beneficiaries of *B.C.*'s stock dino-monster footage.) *B.C.* was retitled *Battle of the Giants,* a selection of excerpts from the parent film (or should I call them 'Roach *clips*'? … ah, sigh) and released by Castle Films for the 8mm and 16mm home movie market during heyday of the 1960s monster boom.

Eventually, Tumak, with Loanna in tow, is reunited with his people, but this time Tumak is in charge (by this time usurping his father Akhoba's authority who is played impeccably by Lon Chaney). Loanna instructs the Rock Tribe in polite ways, table manners and social graces, but shortly after a nearby volcano explodes, and then the lucky survivors are summoned to combat another giant neecha threatening the Shell Tribe. After defeating this dino-monster, members of both tribes unite and live happily ever after, inscribing their remarkable story on cave walls for future archaeologists to decipher.

Analogous to how Glut's fond regard for "*B.C.*', Irwin Allen's *The Lost World* holds a very special, nostalgic place in my heart. I've written about my early experiences seeing this film in an article titled "Back to the Volcano," my first 'monster' article and written for *Scary Monsters* magazine. Allen's "Irwinosaurs" indelibly imprinted themselves within the fiber of my being at a highly impressionable time.That is, when my grandparents took me to see it for the first time at age 5 1/2 during the summer of 1960 at a theater in downtown Chicago. (For many years thereafter it was also shown in Chicagoland on "Family Classics" a Sunday evening program hosted by Frazier Thomas.)

I am presuming that more of you are familiar with this production, which is 20 years more recent than *B.C.* (and to delight of younger audiences also was filmed in color). So I won't summarize the entire movie, loosely founded upon Sir Arthur Conan Doyle's magnificent 1912 novel which most of us generally know.

As Glut states in his *Dinosaur Scrapbook*:

> "Perhaps if the reptiles had gone unidentified as in *One Million B.C.*, the picture might not have been so ludicrous. But no one, not even the youngest dinosaur buff in the audience, could rightfully accept it when Professor Challenger (Claude Rains) declares that a monitor lizard with a rubbery ceratopsian frill attached to its neck was really a *Brontosaurus*. ... The highlight of *The Lost World* was a furious battle between the sail-backed crocodilian and the pseudo-brontosaur, though a similar fight scene had been done better in *One Million B.C.*" (pp. 126, 134)

Well, I was one of those young "buffs" in the audience and while I inherently knew the on-screen, gigantic-looking Irwinosaurs didn't resemble any real dinosaurs I was familiar with then, nonetheless their appearances (and roaring sounds!) thrilled me to the core. And in retrospect, contrary to Glut's opinion, *Lost World's* battle scene does seem better to me than that in *B.C.* It's not a stretch that 1960's *The Lost World* shaped me into the dino-monster loving fan I am today.

In sync with Glut, Mark Berry further chides the *Lost World's* battle scene and the prosthetic 'costumes' worn by Irwin Allen's dino-monsters, stating, "The big dino fight pitted one of the six-foot-long monitors, outfitted with a ceratopsianlike neck frill, against a caiman alligator wearing the obligatory back fin." (p.252) And continuing:

> "Not content with mere back fins, the lizard dressers glued horns, plates, barbels, and even a floppy rubber neck frill on the critters, but the more prosthetics they added, the goofier the animals looked. The frill, for instance, makes its wearer look rather like some sort of ancient idol one might find inside a pyramid. We hear sound effects of great trees falling every time one of the behemoths trudges through the forest, as if large animals never use the same trail twice. The specter of possible animal abuse always hangs over scenes like the big dino battle, which is not only distasteful but also a blatant copy of the famous fight in *One Million B.C.* " (p.249)

James Van Hise adds that these two (real) reptiles are "natural enemies." (p.152)

It is hard to objectively disagree with Berry's subjective remarks, although …. All I can say is that at age 5 ½ - witnessing all this saurian splendor and the larger than life combatants on a giant movie screen – was simply stupendous and terrifying. I found myself quickly not caring that Professor Challenger's "dinosaurs" really couldn't be dinosaurs. They looked too cool to be so easily dismissed. And the nearly 3-minute long battle sequence proved an unforgettable, harrowing experience. So what did I notice that was so captivating?

Just over the one hour movie mark, the alligator, bedecked with Angilas-style recurved head horns, and replete with a sail, as well as upward curving spines protruding along the side of the sail, strides in for the kill, *versus* the frill-adorned ceratopsian "brontosaur" (replete with two rows of stegosaurian plates along its spine) and which had been thrashing quite noisily, roaring loudly, through the lush jungle. David Hedison (as Ed Mallone), protecting Jill St. John's character (Jennifer Holmes), fires his gun at them in utter desperation, crying "It's like a toy against them." The combatants cause them to flee toward a ledge alongside a rocky cliff. With monsters' jaws savagely interlocked, the 'brontosaur's' tail whisks Hedison and St. John off onto a lower ledge. The alligator-creature lunges for its adversary's belly. Then with the brontosaur gripped in an apparent neck hold, bodies intertwined, the pair of giant dino-monsters topples over the cliff, off the edge of the mysterious plateau. Wow! Chills. Keep in mind this was 'merely' a science fiction/fantasy film, not unlike so many others that we enjoy.

As in the case of *B.C.*, Irwin Allen shamelessly re-used outtakes such as the famous dino-battle sequence, from *Lost World* for his several television mid to late 1960s sci-fi productions, including my then absolute favorite show – *Voyage to the Bottom of the Sea*. But occasionally there was something apparently new to see as well. For instance, in VTBS' "Night of Terror" first aired on Oct. 9, 1966, a rubber-frilled ceratopsian reappeared – ruling the prehistoric island. In his *Carbon Dates,* Glut states that Irwin Allen not only used stock scenes from his version of *The Lost World* for this episode, "… but apparently also new footage utilizing a similar lizard wearing the same left-over frill." (p.109) Another episode, "Turn Back the Clock" (1964), used *Lost World* stock footage so blatantly, that the television audience complained. Irwin Allen became so wedded to his 1960 20th Century-Fox production that, likewise as what happened in the case of the *Voyage to the Bottom of the Sea* movie followed by his titular television series, he also considered a television program series centered on *The Lost World*.

Finally, there was *Gigantis, the Fire Monster's* outstanding dino-monster battle, of which I won't say as much here because G-fans know this one inherently. I participated on Martin Arlt's "Gigantis" panel at G-Fest XXII and for further information concerning my panel experience, you'll need to also read my article on this movie and (printed in Martin's *Mad Scientist* magazine no.31). But permit me to summarize why I so enjoyed this 9-minute thrillerama ride of a *battle* sequence (albeit interspersed with footage not showing the monsters). I first witnessed "Gigantis" on Chicagoland television during the early 1960s. I saw this at a very impressionable age, when popular fascination over monsters (generally speaking), and my personal love for paleontology was rapidly escalating.

The "Gigantis" (or the 2nd Godzilla beast played by Haruo Nakajima) vs. Anguirus (played by Katsumi Tezuki) battle begins near the 36-minute mark, and just seems to go on forever (one of the blessings of well-orchestrated suitmation staging, versus the vagaries of using live reptiles). The fighting takes place during an imposed city blackout, and the darkened eerie scenes with flames licking upward toward an unforgiving sky, coupled with the strange monster combatants' roaring, buildings toppling, fire-breathing from gaping maw, vigorous tussling to the death ….to my 6 year-old mind then, even if this fiercely fought, filmed action was all impossible - geez - wasn't this exactly how 'real' dinosaurs were supposed to live, fight and die. Toho actually intended their two monsters to mimic real dinosaurs known to science, with Godzilla's suit design stemming from paleoartistic perspectives of *Tyrannosaurus, Iguanodon,* and *Stegosaurus*. Meanwhile, Anguirus's costume seems derived from both *Ankylosaurus,* and an 1897 Charles R. Knight restoration of the (ceratopsian) horned "*Agathaumas*." Even then I could see and comprehend these important analogies (i.e. between select 'real' dinosaurs and Toho's chimeric 'reel' dino-monster designs). Yet, these dino-monsters weren't in the least bit saurian; instead, they were realistically and convincingly played by human actors. Bravo!

Other contestants for the crown? Well, of course, some of those other B-films of the late 1940s, 50s and beyond containing footage of faux-prehistoric dino-monsters battling other pseudo dino-monsters (that are not cgi- or stop-motion animation effects), are outtakes from either "B.C." or Irwin Allen's *Lost World*. In my humble opinion (IMHO), no others quite compare with the three stated here.

And so which are the best and worst of the three? I'd place the fright night fight scene in "Gigantis" (or *Godzilla Raids Again*) at the top. It helps that both the Gigantis (i.e. 2nd Godzilla) and Anguirus monsters, although impossibly huge, at least vaguely resembled genuine dinosaurs. They certainly appeared far more 'dinosaurian' that did the *Lost World's* "Irwinosaurs" or the battling combo of *B.C.* beastiary. Irwin Allen's *Lost World* battle scene then comes in second, and I'm assigning *B.C.*'s iconic dino-monster sequence third place.

Stop-motion animation is often considered the most exalted form of achieving dino-monster effects in film, especially by lovers of classic movies. Today, most such effects are created using computer technology. But when we consider the three genre movies mentioned here, we see that stop-motion puppet animation or cgi isn't the only way of profoundly stirring our imaginations, especially among the youthful-minded, or those who wax nostalgic.

Note: (1.) Also, the excellent sail-backed *Dimetrodons* featured in 20th Century-Fox's 1959 movie, *Journey to the Center of the Earth* are not seen battling other dino-monsters, or even each other.

Key References:

Mark F. Berry, *The Dinosaur Filmography*, (McFarland Publishing), 2002.

Allen A. Debus, "Get Real! Dinosaur Masquerade," *G-Fan*, vol. 1, no. 65, Nov/Dec 2003, pp.28-34.

Allen A. Debus, "Two-Maligned Dino-Monster Films," *Mad Scientist* no.31, 2016 (in press).

Stuart Galbreath IV, "Long Long Ago Before Jurassic Park: The Making of One Million B.C.," *FilmFax* no.48, Jan./Feb.1995, pp.32-36.

Donald F. Glut, *Jurassic Classics: A Collection of Saurian Essays and Mesozoic Musings*, (McFarland Publishing, 2001), Chapter titled "One Million B.C.: The Original," pp. 174-183.

Donald F. Glut, *Carbon Dates: A Day by Day Almanac of Paleo Anniversaries and Dino Events*, (McFarland Publishing), 1999.

Donald F. Glut, *The Dinosaur Scrapbook*, (Citadel Press), 1980.

James Van Hise, *Hot Blooded Dinosaur Movies*, (Pioneer Books), 1993.

ELEVEN

Front cover of the original dinosaur scrapbook as drawn and designed by the author's father Allen G. Debus in January 1935. (unpublished volume I) See Chapter Thirty-eight for m

Personal Petroglyphs

As recounted in my Foreword to Steve Brusatte's 2002 book, *Stately Fossils: A Comprehensive Look at the State Fossils and Other Official Fossils* ... so there I was, sitting on the curb in front of our Park Forest, Illinois home one late summer afternoon, back in the year Maris hit 61. That day I found my first fossil, a cylindrically-shaped bit of whitish stone encircled with grooves, reminding me of a broken screw. My avocational interests in 'real' paleontology stemmed from what happened next—a simple thing really, just asking my Dad what it was. He not only identified it, but also explained its significance. This became a major turning point in my life.

In explaining it was something called a "crinoid," my Dad opened a remarkable window into the deep past, a portal into prehistory that began long before the time of Stonehenge or the age of Ancient Egypt with its dynasties and pyramids, spanning periods of time that seemed incomprehensibly long ago. I realized that fossils such as my crinoid stem permitted curious-minded souls to catch fleeting glimpses of a world vastly different from today's, a world before the time of modern humans and our inventions. I listened to words that day which I could barely comprehend, but which 'stuck' solidly, such as "Paleozoic," "Silurian," and "echinoderm." Yet I thirsted for more. As I still do.

Fossil collecting became a passion of mine in the 1960s. Though I was still in grade school and had never collected fossils *in situ* or in the field, I had become a self-made "expert" on fossils after absorbing my favorite book of the time devoted to fossils—*Fossils: A Guide to Prehistoric Life* (A Golden Nature Guide, New York: Golden Press, 1962, by Frank H. T. Rhodes, Herbert Zim, and Paul R. Shaffer), which advertised on the front cover its "481 illustrations in color." That book is still nestled on my bookshelf, its pages worn from the many hours of youthful fossil hunting excursions it was carried on.

We were lucky to have fossiliferous rocks in the immediate vicinity of places we lived, especially during the years between 1964 and 1968. Initially, my younger brother Rick (who later became a professional geologist a decade later) and I would sit in the grade school gravel parking lot (in the *summer* of 1966, so there were no cars parked there), turning over small limestone rocks one at a time, searching for fossils nearly every day. What hauls we made! After a couple hours of picking over stones, we would cross the grassy field to our house, sunburned and sweaty, our pockets bursting with valuable "finds." Developing an uncanny sixth sense for when our best discoveries were about to be made, we found gastropods, brachiopods, cephalopod fragments, coral fragments, crinoid stems and calyxes, and trilobite fragments. Many of our treasures were in beautiful condition. These hunks of rock from a prehistoric age fascinated us. It amazed us that that nobody else had thought to comb through these fossiliferous fields. I later discovered that the gravel had most likely come from the nearby Thornton Quarry, a Niagara limestone Silurian reef where we collected several years later.

We spent a year (1966-67) and two subsequent summers (1969 & 1971) in England, and found more gravel deposits laden with Mesozoic treasures right outside our apartments, just ripe for the picking! Now we ambitiously added ammonites, belemnites, clams, sponges, sea urchins and worm tubes to our collection. Between 1970 through 1972, my parents had been cajoled into taking us to quarries, exposures, and road cuts to collect fossils both in Illinois (participating in field trips sponsored by the Illinois State Geological Survey—for whom I was employed briefly in 1977 to 1978), and also in England (Cambridge area). But we learned of the hazards of field collecting too. While we happily hammered away on limestone, our Mom was regularly and savagely attacked by bees that were attracted to her evidently 'alluring' hairspray.

As our collections grew, so did our library of paleontology books. Our knowledge of this fascinating subject flowered. We made every effort to identify what we found using references we accumulated.

There were also late October college geology and paleontology field trips to sites in Iowa, Arkansas and Wisconsin. Throughout, I labeled my fossils dutifully and filled two notebooks with relevant notes, even illustrating the best fossils for posterity. I also learned the "pleasures" of camping from these experiences, after camping in freezing conditions, as one such example, in 18°F snowy weather, and stomping about with frozen feet. (Perhaps partly due to "calming" beverages we heartily consumed the night before to 'thin' the blood in our veins.)

My fossil collection still resides in our garage and basement, spared from numerous spring-cleaning 'downsizings.' Into adulthood, interests naturally shift, however my enthusiasm for paleontology, dinosaurs, and, yes, fossils has remained a constant.

It was sometime in 6th grade when I announced that I wanted to become a paleontologist when I grew up. Immediately concern was raised over inevitable starvation, as this was not known to be a lucrative field. Paleontology was viewed then as a "dead" science, unexciting because there was so little happening in the profession. There would be few jobs for degreed paleontologists and in the end, this was the rationale to selecting a safer college major. Oh, the inhumanity of it all! If only I'd known how substantially circumstances would all change in such a short time. For in short succession we experienced: the plate tectonics revolution with its reliance on fossils; the enlightening punctuated equilibrium theory of organic evolution; a dinosaur renaissance; a lively debate over the nature and causes of mass extinctions; and a host of fossil discoveries that would make our heads spin with joy and astonishment. Paleontology was enlivened, and thanks to my Dad's encouraging description of what a strange screw-shaped stony fragment was, long ago, I found it impossible to let go.

I then embarked on a professional career in the field of chemistry (evolving into environmental chemistry), while avidly maintaining avocational paleontology interests. After the mid-1980s, when it became virtually impossible to find the time to collect fossils, other paleo-pursuits more readily accommodated my work-a-day family/suburban lifestyle.

It is possible that the now long lost, aforementioned crinoid fossil may not have been the very first 'stone relic' I ever found, recognized as such. My (paternal) grandparents took care of me for extended periods during summer months in the late 1950s to 1961. During one of those occasions, while digging with a toy 'army' shovel in a wooded section of their back yard in Northbrook, Illinois, I happened to notice in the shallow hole what was described for me immediately afterward as a stone Indian arrow or spearhead, 2.5 inches long—still retained in my home office. An old Paleoindian encampment on their property, perhaps? Or was it instead 'planted' there by one or both of my prankstering grandparents for me to suddenly find (not unlike the fossils which fooled discoverers at Piltdown)? The weird thing about it is that I never even considered the latter possibility until I reached my 60s.

And so we've reached the final mashup of chapters and stories for this volume delving into my paleo-past.

These chapters originally appeared as articles in the following publications:
Chapter Thirty-six—*Prehistoric Times* no. 121, Spring 2017, pp.47-49.
Chapter Thirty-seven—*G-Fan* no.106, Summer 2014, pp. 32-35.
Chapter Thirty-eight—*Prehistoric Times* no.96, Winter 2011, pp.29-31.

Chapter Thirty-six—*One Earth For Us*

In July 1928, astronomer William Maxwell Reed (1871 - 1962)—known as W. Maxwell Reed ("WMR") - finished the manuscript of a book primarily intended for youthful readers concerning the history of life on Planet Earth. His *The Earth For Sam* (1930) became a genre favorite, maintaining popularity for at least two generations of dedicated paleo-philes. Fathers like mine would share their copies with children during the 1960s, thus indoctrinating us with the thrill of comprehending life through an immensity of geological time. Decades later, now older folks like me fondly recall Reed's masterpiece. But, of all the popular books published on dinosaurs and those 'other' prehistoric animals then, why does this particularly well-fortified example (still) stand out? And, hey, while we're at it—who was this titular Sam character anyway?

First, "Sam" was Samuel McCobb Reed, the book author's then young nephew who must have been several years older than my father was in 1928. As W. Maxwell Reed states in his Preface, the lad:

> "…used to ask me to tell him about the formation of rivers, mountains, and clouds and about the earth and stars. Since obviously such questions cannot be answered briefly and since my time was limited on week-end visits, I told Sam I would write him letters. Some of the chapters in this book are these letters almost unmodified. My friends suggested that I complete the series of letters and publish them in book form."

Pretty cool idea! Many of us have fond memories, having taught our own kids about paleontology, dinosaurs and other wonders of the deep mysterious past. W. Maxwell Reed (hereafter "WMR") then went on to write other books in the "Sam" series as well, delving into other, non-paleontological topics, including *The Stars For Sam* (1931), and *The Sea For Sam* (1935). According to a short Wikipedia entry, WMR was educated at Harvard, and taught astronomy at Harvard and Princeton universities. A smart and impressive guy.

So for those of you who haven't yet had the pleasure, what's inside this gem?

Mainly, this is a volume concerning the most majestic and distinctive fauna and flora representing Earth's successive ages. Since we're all familiar with such geo-historical outlines, fairly standard even then, it is unnecessary to delve fully into those familiar details. *The Earth For Sam* remains an exquisite melding of paleo-images with text; it was one of the first truly popular life through geological time books published in America, tailored for youthful enthusiasts. The book *is* rather formulaic, much like many popular books printed in Europe and America particularly from the 1870s on, presenting fundamental geological principles interwoven with, or prefacing the 'meat'—a dramatic unfolding of the evolution of life. Overall, *Earth For Sam* bears a distinctive modernistic ring, while still enticingly reflecting that 'golden age' of "dinosauriana."

WMR clearly borrowed an approach established by 19th century authors in presenting the history of life through time on Earth, while offering geology lessons along the journey. So we read about salt mines, how glaciers form, erosive power of river systems and the origin of geological features such as mountains, folded rocks and volcanoes. But these discussions are interwoven with discussions concerning the main theme—how life originated and key fossil forms from each of the major, successive geological Eras and Periods. And like so many of the early books for non-specialists of the time, WMR opens his age-by-age coverage with the Silurian Period following several short, leading chapters on how life originated and began to evolve, nature of the fossil record and key geological processes (e.g. volcanoes and chemical weathering of rocks).

Earth For Sam lacks a Bibliography, yet its primary references are relatively easy to glean. References consulted or cited by WRM include: Henry F. Osborn's *The Origin and Evolution of Life* as well as several of his other monographs and books, Alfred Wegener's *The Origin of Continents and Oceans*, F. A. Lucas's *Animals of the Past*, Gerhard Heilmann's *The Origin of Birds*, H. G. Seeley's *Dragons of the Air,* Henry R. Knipe's *Nebula to Man* as well as his *Evolution In the Past*, William Berryman Scott's *History of Land Mammals in the Western Hemisphere*, F. H. Knowlton's *Plants of the Past*, Charles Schuchert's *Outlines of Historical Geology*, and *Water Reptiles of the Past and Present* by Samuel W. Williston. Important paleo-references of the time!

The book is sumptuously enriched with over 200 figures, more than one per every two pages. As we know, the ranks of those who would be qualified "paleoartists" today has swollen considerably since the 1920s. But counted among *Earth For Sam's* many figures are black and white reproductions from a veritable 'Who's Who in Paleoart' of the time, including Charles R. Knight, Erwin S. Christman, Alice B. Woodward, E. B. Seeley, Gerhard Heilmann, S. W. Williston, Richard Deckert, J. Smit, E. P. Bucknall, Bruce Horsfall, C. Whymper (etc.), with Knight's fine work taking pride of place. There are also numerous, quaint cartoons by Karl Moseley and, of course, many photographs of museum fossil specimens, geological diagrams and natural geographical features. Bountiful images: there's an 'art' in their proper selection. A lavish affair for youngsters then and now.

The unadulterated joy and pleasure one derives from these dusty old books perhaps principally stems from their outlay of Life's mysterious 'progression' from and through the dawn ages to the present day. For *Earth For Sam* is a delightful life through geological time affair, taking us from the dawn of Earth time and life's origin (then stated surely as a then mind-staggering one *billion* years ago) to the most recent post-glacial period, while warning of a coming period of global warming and mankind's destructive tendencies. It certainly was 'up to date' for its time, as WMR makes full use of adaptational concepts, without dabbling in once popular then "orthogenesis" ideology, while employing contemporary radioactive ages for geological periods, then a relatively new technological breakthrough.

A highly evolutionarily-themed book, *Earth For Sam* refers to natural variations within species, using phrases "elimination of the unfit" and "survival of the fittest," although without mentioning then, more technical-sounding terms such as "Darwinism" or "evolution." WMR does refer to "survival of the fittest" in the "Oligocene Period" (i.e. he refers to Paleogene "epochs" instead as "Periods"). Also, as a metaphor, he likens the destruction of the Eocene creodonts by carnivores of Asian origin to the "ruthless" treatment of "native Indians" by European white men in the Americas. Using a series of then state-of-the-art diagrams, WRM expertly describes how the fish fin evolved into a primate hand, how birds evolved the capacity of flight from ancestral reptilian genera, evolution of whales from terrestrial mammalian ancestral stock, and how dawn horses became …. horses.

Additionally, maps profess to show how configuration of the North American continent may have appeared in earlier periods, while WMR outlines processes and factors that led to subsequent stages of geographical revolution, with concomitant climate changes. A wintry ice age is held responsible for K-Pg boundary changes. Remember that even though several scientifically minded men had by then already speculated that asteroid and comet impacts would be devastating, such possible causes (of extinction) were usually dismissed or mocked in the early 20th century (because they weren't 'uniformitarian' or 'Lyellian' in nature). Also, although Wegener had written of 'continental drift,' there was as yet no convincing geophysical mechanism for causing such movement on a "fixed" continental scale.

Book chapters imply or rely either on gradualistic causes or terrestrial 'revolutionary' shifts leading from one Era to the next. For example, WMR comments cryptically at the opening of the Triassic that, "Then another of the earth's revolutions came and utterly destroyed them." He continues, philosophizing, "It is odd, having been born ... in one revolution, they couldn't stand another." (p.161) At the terminal Cretaceous extinctions, we learn that glacial conditions may have been one root cause. (pp.239-241) And at the dawn of Eocene time, WMR resorts to Knipe's by then 'oldtimey' 1912 theatrical metaphor, namely that "The passage of time from the Cretaceous to the succeeding Eocene is shrouded in darkness. It is as if the lights in a playhouse had been abruptly extinguished, and after a lapse had been restored, disclosing a stage crowded with new characters." (p.243) And of course, in the style of a pictorial essay, there's ample coverage of the Mesozoic Age. (Two quibbles: he refers to *Ceratosaurus*—my dad loved Smit's life restoration here—as a member of the horned dinosaur group, and suggests that *T. rex* co-existed with *Stegosaurus* during the Cretaceous; well, WMR was an astronomer.)

Throughout, WMR flashes arresting writing skills, such as when during the recent Ice Ages, "Thick furs were very fashionable garments for the fauna in the Pleistocene days." (p.343) Or, in referring to walking, winged "...*Dimorphodon* ... in a hurry he looked something like an old-fashioned cavalry officer with spurs and long saber and a fierce expression." (p.198) We also read how *Archaeopteryx's* voice may have sounded and why. (p.189)

Later, WMR quotes from a short fiction story written by W. D. Matthew, published in *Natural History*, about "Pithecanthropus." (pp.330-333) Since its riddle hadn't been solved yet, WMR also includes Piltdown Man—the "Pliocene human"—in this chapter. He recognizes that Greenland may become warm and ice-free once again, but forecasts this environmental change to ~ 30,000 years hence (rather than anticipating the far more rapid warming that has been occurring within only the past few decades).

Perhaps inspired by then recent, impressive prehistoric life exhibits in Chicago, the Field Museum's 'dinosaur hall' adorned by Knight's murals displaying geological history, and both the Sinclair dinosaurs and The World A Million Years Ago on display at the 1933-34 "Century of Progress" Chicago World's Fair, and given *The Earth For Sam's* immediate success, a flurry of similarly themed paleo-books were published. In this vein, two rarely cited contemporary additional popular entries, intended for younger readers, merit mention here: W. W. Robinson's and Irene B. Robinson's *Ancient Animals* (1934), and Leon Morgan's pamphlet, *The World A Million Years Ago* (presumably 1933 or 1934). Artists for each were, respectively, Irene Robinson and H. G. Arbo. Although the Robinsons' book would seem a stylistic precursor to Knight's, the advent of prehistoric man is not explored therein, concluding ominously: "The story of man is not for these pages. While he rules, however, new monsters cannot live on this earth that was born of the sun." Meanwhile, Leon Morgan's booklet adopted a similar paleo-perspective to WMR's.

Earth For Sam is rife with ideas, it's interesting to mention a few of WMR's flights of fancy. WMR is prone to intriguing speculation, barking up the tree of accidental, historical 'contingency.' Such as, if ants (insects originating in the Mesozoic Era) had evolved to the size of horses, WRM speculates, their "… wonderful capacity and their instinct to help each other would have made them … rulers of the earth. They probably would have killed all animals that were their enemies, just as we are doing today. Under such conditions … cousins of the apes would never have had an opportunity to become men." (p.179) He even concludes with an intriguing stream of paleontological "what-ifs," discussing what if Earth's 'experiment' could be rerun on 100 different planet Earths, instead of how evolution actually proceeded. Then who or what might be standing here today, instead of us? Today, the foundational value in *Earth For Sam* isn't as much as in its presentation of eye-catching, extraordinary imagetext, but more so in teaching us that wise men, decades ago, appreciated the quirky and randomized factors leading to our present planetary circumstance—now grasped so dearly yet precariously by an apex primate.

While many of you may have heard of *The Earth For Sam*, let us move on briefly to WMR's *second* paleo-oriented book, which very few of you may know of – *Animals On the March* (1937, co-written with Jannette M. Lucas, edited by Edwin H. Colbert). This is possibly the third book published in America dedicated to and emphasizing prehistoric mammals, intended primarily for non-specialists. *Animals On the March*, a kind of 'sequel' to *Earth For Sam*, was still published in that early period prior to the advent of punctuated equilibria evolution theory. Besides Knight's magnificent restorations, WMR also employed paleoart by Margaret Flinsch, Mrs. E. M. Fulda, John W. Hope and Bruce Horsfall.

While following ample coverage of "land-bridge" theories, WMR describes how populations of extinct mammals may have roamed as sea levels and land areas both rose and fell, affecting oceanic currents and paleoclimate, WMR delves into a synopsis of vertebrate evolution – quickly striding toward his main subject: mammals, ancestral to modern. One of the first such chapters is a point of view description 'told' in the words of an equine "scholar" (a modern horse, no less!) about the genealogical history of its own noble race. From there he goes on to describe the rise of other mammalian groups (felines, dogs, rhinos, camels, deer, elephants, etc.), since the Cretaceous. Charles Darwin's *Origin of Species* is mentioned, yet in cases of shovel-tusked proboscidea and the Irish Deer, this time he ponders the (then contagious) theory of orthogenesis leading to their extinctions.

Despite their universal, expansive and conservationist outlook, WMR's paleo-themed books were written a decade following World War I, during that dark period leading toward World War II and invention of atomic warfare. WMR was evidently a pacifist, weary of man's politically fueled, worst tendencies trending toward global war and eradication of species. His mindset reminds me of modern socio-biologist E. O Wilson's. In modernity, as we reflect upon how things used to be, we tend not to recognize that scientific men, science popularizers, pondered deep ramifications of paleontology, including extinction, decades ago. And long before Stephen J. Gould began writing about such matters as evolutionary what-if, "contingency," luckily, a little known astronomer strove to put everything into perspective for "Sam" … and us.

Chapter Thirty-seven—*Godzilla, the Reawakening*

It's difficult to believe that, like Godzilla, I'm in my 60th year (i.e. in 2014)! I don't feel truly old, but like it or not I'm clearly entering 'old codger' status. So the other day I was musing how I first became acquainted with Godzilla in childhood and on into early high school years, then also later in adulthood when I became more fully (re-)introduced to the monster and his Japanese cohorts. For those of you who might be interested, here's my story in less than 2,500 words!

Already by the early 1960s, I was already primed and predisposed to dinosaurian things, thanks to my father. As a young lad, I became a fan of *King Kong*, and had enjoyed museum displays featuring dinosaur bones and skeletons as well as many paleontology books, toys and models. Also, I had seen the Irwin Allen film, *The Lost World* at a theater (which scared the pants off me!!), but felt cheated by never having seen *Gorgo* unlike all my first-grade pals who had. When we moved to a Chicago suburb in 1961, I became attuned to particular television stations that featured monster movies during the afternoons on certain days of the week, and here it was that I saw *Godzilla, King of the Monsters* for the first time. Around that time as well, I was thrilled to see another suspiciously similar monster battling what I recognized as a hybrid horned/ankylosaur, although for some reason the bipedal, titular monster was named "Gigantis." I was hooked.

The Japanese military also fought a sort of super-cool, winged pterodactyl (two of them, no less!), each much larger than the one Kong battled atop Skull Mountain, featured in a film named "Rodan," which mesmerized me on the same channel. And there also was another spiny-backed creature, Varan, who, to me with my vivid imagination, was really not so "unbelievable." Those miniature sets looked so realistic to my then youthful eyes! I beguiled myself into believing that the 'weighty' camera action of Godzilla slowly rumbling through Tokyo was created by use of small models and stop-motion animation (as in 1933's *King Kong*), because I had no understanding of how they achieved the "slow motion" photographic 'illusion of mass' effect with a man in a monster costume. As much as I loved Godzilla, I knew infuriatingly little about this fascinating dinosaur-monster.

Of course there were many other stimuli ongoing in my environment then, but it's important to note that this truly was the inaugural heyday of Monster fandom, the Era of *Famous Monsters of Filmland*, Aurora monster models, and many monstrous things that began appearing regularly on television (e.g. *The Outer Limits*) and in stores. Monsters of every persuasion permeated all conceivable media. A trendy fad surely, but despite my early devotion to the dino-monster clan, surprisingly there was very little available information concerning those giant Japanese dinosaurian movieland creatures. *Famous Monsters* ran numerous articles about Boris, Fay, the two Lons and Bela, or the latest one-star horror flicks from Hollywood such as *The Horror of Party Beach*, but there was one *FM* issue purporting to have reliable background on Godzilla and his brethren, and this was issue # 30, with its enlightening feature, "Giants From Japan" that included many revealing, behind-the-scenes photos.

When our Dad took my younger brother, Rick, and I to see *King Kong vs. Godzilla* in 1963, well, my mind was forever warped! I believe summer vacation was just about underway when I spied an advertisement in the *Park Forest Star* for this 1963 blockbuster, emblazoned with a great illustration of Godzilla. What can I say, but – from the perspective of my then 9 –year old brain, *greatest movie ever!!!* Subsequently, I staged epic monster battles between Kong and Godzilla using our Aurora models to smash our Marklin railroad train setup, and to the dismay of my parents and fourth-grade teacher, wrote numerous stories about my original "*kaiju*" (even though I didn't know that word then), instead of diligently doing homework. "King Collidese" was one of my giant monsters that grasped the world in fear, battling other loathsome creatures of my own dark invention. I recall around this time often playing "The Godzilla Game" with Rick on our bedroom floor, while listening to Zacherle's "Monster Mash" album.

Perhaps an academic slump at the time resulted in my father deciding *not* to take me to see *Godzilla vs. the Thing* when it came out in 1964. So I remember having to ask a classmate to *draw* on notebook paper for me his recollected scenes from the movie so I would have a better idea what the plot was about and who this mysterious "Thing" was. Later, buoyed by a successful fifth-grade academic campaign, my father took me (and Rick) to see *Ghidrah, the Three-headed Monster* upon its American release. (Geez—did my too-much-protesting father actually also enjoy watching these movies too?) Well, this 'adverse influence' had the side effect of once more curbing scholastic enthusiasm, resulting in my generally lackluster sixth-grade performance. With so many cool monster movies and science fiction shows out there to dwell upon, it was difficult to concentrate on school.

Before high school began, there was *King Kong Escapes* with its interesting "Gorosaurus," that our dad took us to see at a local drive-in. There were 'casualties' too. The Debus family moved from one suburb to another during August of 1968, and my parents junked all my old comics and magazines and monster models & other cool horror stuff ... except for my Aurora Godzilla model. I don't recall whether I salvaged it from the garbage, or whether my parents consented in allowing me to keep it. But I still have it today.

Then during the best Christmas vacation ever, my brother and I watched repeated viewings during a marathon broadcast of a new movie, *Godzilla vs. the Sea Monster.* (During the afternoons I read Tolkien and Burroughs' *Land That Time Forgot*.) Then later that following spring, like a breath of fresh air, along came *Destroy All Monsters*. Around this time too, Chicagoland began its "Creature Features" program, which is where I finally got to see *Gorgo*! But by the time of *Monster Zero*, I found myself seriously, yet sadly questioning whether I had outgrown such fare.

Fortunately, by the time high school came along, I was a little more mentally engaged with studies, although my passion for Godzilla didn't fizzle until late in my sophomore year, that is, after *Monster Zero's* 1970 release. Throughout this period, however, the assault of Toho pictures continued (although I still didn't get to see them all). But those which we did see wielded ever lessening impact; after all, when it came to extracurricular affairs, girls and baseball were then competing for my attention, gaining an edge.

Competitive classes, plus the impact of having less leisure time forced me to abandon Godzilla and his pals. I became more selective about sci-fi shows, ensuring that these were a little more "high brow" than those featuring men in scaly suits knocking over cardboard buildings. (By then, for example, *Lost in Space* had yielded to *Star Trek*.) When college years began in 1972, realistically, there was no time for monster movies. Chemistry, physics and calculus classes ruled the day. Plus I didn't own a television. The campus library beckoned instead. So while I *heard* about Godzilla's exertions against the Smog Monster (a title certainly relevant for the time), I never went to see any of the newer Toho films of the 1970s. Graduate school was even more involved, and so by the time I was finally out in the 9-to-5 workaday world, there were no more new Godzilla movies on the horizon. And if there was any new information about how they made all those formerly beloved films, say – published in fanzines of the 1970s, I remained completely out of touch. Despite my great memories, Godzilla was essentially dead to me. But then a curious thing happened, as I was destined to reestablish my nerd-like persona.

During the time of my first (doomed) marriage, there was more time, especially on the weekends, available for watching monster movies and rediscovering my avocational love of the science of paleontology. As disparate as these generalized topics may seem, for me, they somehow intertwined in a manner that was difficult to explain. And through time the bond grew.

Now I realize there were a number of articles about Godzilla published during the 1970s, although with my science blinders on, I never took note of them. I also never read the Godzilla comic books and paid no attention to the *Godzilla Power Hour* cartoon. And so I never knew about the best Godzilla reference stemming from that period, a chapter titled "Godzilla, a new king," in Donald F. Glut's wonderful *Classic Movie Monsters* (1978); I procured my copy in early 2000. But by then (as we'll see) I had already befriended Don.

So when did I reacquaint myself with Godzilla? It was during the fall of 1982. My parents had already bought me a new book for my birthday in August, *Dinosaurs., Mammoths and Cavemen: The Art of Charles R. Knight*, coauthored by Glut. I had heard of that author's name before, yet couldn't place exactly how, when or why – (maybe in association with movie monsters or Dracula, or perhaps in a *FM* or *Star Wars* context?). But, coincidentally, I had just published a long article (one of my very first) for a local Earth Science club newsletter titled "They Painted the Dinosaurs," delving into all the classic paleoartists who had restored dinosaurs and other prehistoric animals. So Knight was of great common interest.

By then, I was going through a divorce and found myself traveling a few weeks later through southwestern Wisconsin at the time on a vacation. I took the Charles Knight book along to read during the evenings in hotels. I decided to write Glut's publisher when I returned home, hoping they would forward my letter and that Glut would reply. But then in one small town, on a bookstore shelf I spied a large, thick reddish book titled *The Dinosaur Scrapbook*, also written by Donald F. Glut. As if the Knight volume wasn't spectacular enough, this second book jettisoned me out of my funk, and for a few minutes as I paged through it, I felt like a kid again!

Dinosaurs of Worlds Fairs, movieland dinosaurs, science fictional dinosaurians of every possible persuasion, dino-paleoart, dinosaur parks and museums, pictures of dino-things which I had enjoyed since the early days of my youth, but had felt isolated from for so long were all there on view, page after page of saurian splendor in *The Dinosaur Scrapbook* (1980). And there was even an entire chapter on "Prehistoric Monsters Japanese Style"! This was the first book I'd ever come across that cleverly and successfully *comingled* information concerning those fanciful Toho giants with facts about 'real' dinosaurs (instead of only combining with other movie monsters). But while I was familiar with, or had actually seen firsthand much of the real dinosaur material presented therein, thanks to Glut, I also realized there was so much about the Japanese dino-monster industry that I had never imagined.

Why I never purchased this second book right then and there, which was *so* subliminally calling out for, nay—ordering me to take it home, I'll never know. I must have been too strapped for cash at the time. I still needed to buy gasoline for the long drive. I didn't want to run up my credit card tab. But then, voila!, as if my dad had some remote, extrasensory grasp of my re-awakening, for Christmas that year I was astonished to receive—yes, *The Dinosaur Scrapbook* as a present!! And so now I was really able to tear into the book at leisure.

By then, of course, Glut's publisher had forwarded my letter and Don had replied. Amazingly, as things turned out, while he lived in Burbank, every Christmas he returned to Chicago to visit his mother, friends and family. And his mother lived less than 3 miles—as the crow flies—from me. We agreed to get together over the holiday, and indeed we met at my apartment on Dec. 28, 1982, where we talked about dinosaurs, sci-fi & monsters, and he autographed my books. In the *Dinosaur Scrapbook* he wrote "…And thanks for showing me your scrapbooks!"

Glut's informative chapter on Prehistoric Monsters Japanese Style was enlightening, yet as I realized decades later, merely a synopsis of what was provided in *Classic Movie Monsters*. For one thing, he nicely summarized all of Godzilla's adventures occurring since the last one I'd seen in 1970, and also he whetted my appetite with news of other Asian monsters I'd never heard of …. "Gappa," "Gamera," "Yongary" & others. And incredibly, Don had made his own stop-motion films including a Godzillean creature and other giant dino-monsters resembling those I'd known and loved. This book was a feast for 'saur' eyes. I was no longer living in a fool's paradise, or so it seemed.

In 2012, Don published his memories about writing *The Dinosaur Scrapbook*, which by now has attained status as a sort of minor classic, although focusing on its dinosaur content. (See *Prehistoric Times*, 2012, no.100, pp.49, 53.) No, his "Scrapbook" isn't perfect; there are scattered errors (see *Prehistoric Times*, 2002, no.51, p.27). But thanks to Don's Godzilla writings of the 1970s, and perhaps especially to his *Dinosaur Scrapbook*, he's still regarded as a Godzilla expert.

For me, I suppose there was yet a 3rd Godzilla 'enlightenment,' following my *Dinosaur Scrapbook* reawakening. I remarried and we had two daughters, which preoccupied much of my time after work and on the weekends. I read voraciously (mainly science fiction and paleontology books) on train commutes to and from work, but for many years during the 1980s increasingly dormant Godzilla was rarely in my sights. That is, until I began subscribing to *Prehistoric Times* magazine in 1993, wherein I noticed an ad for a peculiar reference to something named "G-Fan" — evidently a zine devoted entirely to Godzilla. If only there had been such a magazine decades before! I was intrigued and subscribed. There really was much about Godzilla that I didn't know or realize, which is why I treasure this magazine today.

Astonishingly, more movies had come out that I'd never heard of since the 1984 remake. People associated with the zine had begun contacting the old stars of those early classic films. And who would have ever predicted half a century ago that eventually, every year thousands of Godzilla fans would flock to a location not far from our current residence to commune in appreciation over the 'Big Guy.' What an odd turn of events for this 60- year romp.

Chapter Thirty-eight—*The Original Dinosaur Scrapbooks*

For those among the ranks of the older dinophile generation, perhaps, didn't it seem like a dream come true when you received your meaty copy of Don Glut's *The Dinosaur Scrapbook* (Citadel Press, 1980)? So many pictures of dinosaurs conveyed in pop-cultural splendor! *The Dinosaur Scrapbook* was jam-packed with information concerning so many facets of dinosaurian and prehistoric animal 'culture'- showcasing how they've appeared through the decades, categorized in such a wide variety of media! A veritable treasure trove of information, published just as paleontology surged into what was arguably its greatest decade. I received my copy for Christmas in1982 and couldn't put it down. But Glut's supremely entertaining book wasn't the 'original' dinosaur scrapbook, after all. In fact, my father had already compiled five genuine, unpublished scrapbooks decades before. And so, with his recent passing (in 2009), I hope to share with you the essence of what I'd already assimilated from him about dinosaurs and prehistoric life from what he entitled "Monsters of the Prehistoric Past," (scrapbook Volumes I through V), compiled from January 1935 to circa 1941.

My father (Allen G. Debus) was eight years old when he began filling his first scrapbook. Actually, the impetus behind the scrapbooks was his mother, Edna, who had gained some prior fascination for prehistory. (I later discovered that during the early 1920s she'd taken several courses in Geology during her 2 ½ years at Northwestern University.) The Scrapbooks reflect a tensional imagetext. In the case of Volume I, for example, what gets preferentially saved through pasting onto a finite number of scrapbook pages? Pictures or words; restoration portrayals and photos versus text in the form of articles conveying information beyond the visual? Typically, understandably—given a young boy's growing fascination, the visual element predominates. And for those of you who are intrigued about my father's interests in Depression era popular paleontology, you can read his recollections titled "A Dinosaur Hunt in Chicagoland in the 1930s," chapter 33 in *Dinosaur Memories* (Authors Choice Press, 2002).

Scrapbook Volume I begins with imagery reflecting paleontology's chief contemporary pop-cultural theme—life through time. My dad had crudely sketched postage stamp-sized vignettes of the major geological Periods, from the Earth's formation to Neandertal times, which he labeled a "Prehistoric Chart." Much of this volume and the rest mirror this reigning interest in pictures of prehistoric life, arranged wherever possible within the pages/confines of a scrapbook into chronological order. The next page shows a delightful color drawing clipped from a magazine named "A Scene in the Late Permian Period." Today, there's no telling where it came from or who the artist was. Then on page 3 begins the first in a series of V. T. Hamlin's "Dinny's Family Album: Monsters of the Prehistoric Past" cartoons, pasted in order as they were collected. Here, leading off, we see the Woolly Mammoth, although without researching I can't say whether there were already others in the series that appeared in newspapers prior to the Mammoth that weren't noticed or saved, prior to my dad's and nurturing grandmother's compilation of Scrapbook Volume 1. According to my dad, "Dinny's Family Album" was printed alongside the comic strip "Alley Oop" in the Sunday *Chicago Times*. Obviously this series carried significance for my father, who, in his own hand subtitled his Scrapbook series, "Monsters of the Prehistoric Past." *Tyrannosaurus* appears on page 5 of his first Scrapbook. Sixty-three of Dinny's 'family' are represented within Volume I's 103 album pages.

I lack room here to describe the contents of each Scrapbook in detail, as opposed to outlining general contents and themes explored. Volume no. I continues with newspaper articles about one of the (then) ubiquitous Barnum Brown's latest incredible discoveries, pasted-in without dates or proper reference - (a 'malpractice' which my father, who eventually became the consummate scholar and historian later chided his children for). Charles R. Knight's museum restorations were often printed in periodicals of the time to illustrate news of startling discoveries made, such as by the prevalent Barnum Brown. A cartoon on page 18, titled "The End of the Dinosaurs," had lasting influence on my father. It shows a drawing of three Jurassic dinosaurs dying in a freezing glacial age! (This particular illustration appears in one of my dad's favorite books, W. Maxwell Reed's *The Earth For Sam*, 1930 ed., and this with another adjacent scrapbook picture may be part of Reed's book jacket.) Extinct, unfit dinosaurs were also considered as emblematic of outmoded utility and uselessness, as exemplified in one of my favorite old cartoons, titled "The Penalty of Size," where a dirigible was viewed as no more useful in 'modern times,' than an 'extinct prehistoric dinosaur.'

My grandparents also took my father to see the exhibits at the (then) Chicago Natural History Museum, as recorded by the appearance of black and white museum postcards showing Frederick A. Blaschke's sculptural restorations of Tertiary perissodactyl mammals ('titanotheres' and 'Mesohippus.'). A curious postcard showing only the rear of an *Apatosaurus* skeleton records that the so-called 'dinosaur halls' were then very much in embryonic stages of completion. Another quaint pictorially arranged geological time chart with scrawled-in title, "The Ages of the Earth," appearing on p. 20, reinforces where we've been so far (and foreshadows what will be featured) within Scrapbook contents.

Much of the rest of this Scrapbook features more of 'Dinny's Family' and, with increasing frequency, black and white Chicago Natural History Museum postcards—especially showing Charles R. Knight's restorations. At a later time, my dad realized it was best not to affix post cards with glue, but to instead use postcard hinges. Below each museum postcard, my dad typed (on an antique Royal typewriter) the words from the captions onto sheets of paper—no longer readable on the flip side of each postcard—which were then glued onto scrapbook pages next to each postcard. My dad's growing fascination for paleoanthropology is evident on page 34, where he self-illustrated the words to Laura E. Richards's poem "The Cave-Boy." But there are also newspaper articles interspersed between images, such as one titled "Reconstructing the Biggest Beast that Ever Lived," concerning erection and display of the Berlin *Brachiosaurus* in pre-World War II Germany. Yet, according to another journalist covering the Barnum Brown 'beat,' instead, the "World's Most Gigantic Prehistoric Beast Lived in the Rocky Mountains. Scientists Find." This creature could walk a mile in 500 steps! Near the end of Volume I, amidst other imagery, there are pasted-in photos showing prehistoric scenes from RKO's *King Kong* and First National's "The Lost World." My dad astutely listed his short Bibliography on the final page, which included references by Knight, W. Maxwell Reed, W. W. Robinson and Leon Morgan.

Compilation of "Monsters of the Prehistoric Past Volume II," began during the summer of 1937. This time the book is a small 3-ring binder, which my grandfather—then president of Chicago's Modern Boxes situated on Pulaski—brought home for him to continue documenting his explorations of the prehistoric realm. From the get-go, we're welcomed by more of Dinny's extended 'family,' beginning with *Hypsilophodon*. On page 5, is the first in this series not devoted to a particular prehistoric animal, but instead, 'The Age of the Earth,' proclaiming that our planet was then thought "as nearly as science has been able to determine it roughly 2.25 billion years old." Increasingly, we see articles and snippets on early Man and the evolution of early vertebrates 'ascending to' modern *Homo*. On page 11 is an interesting drawing obviously inspired by Sinclair's dinosaur display at the 1933/34 Chicago World's Fair. My dad began pasting in Sinclair dinosaur stamps into Volume II around this time (as well as into the booklets Sinclair provided). And Charles R. Knight's famous *T. rex* vs. *Triceratops* Chicago Natural History Museum mural appeared on page 1 of the Sunday *Chicago Times* on April 11, 1937. My dad's young artistry is showcased on several pages too. Quite interesting are several Associated Press photographs showing Rapid City South Dakota's famous Dinosaur Park statues at 'in-progress' stages of near completion.

Of course, as I know from my own paleo-scrapbooking days of the 1980s and 1990s, one typically pastes in the material as it comes - whenever it's published and available. So in the middle of Volume II, reflecting contemporary popular interests in the evolution of modern man and his ancestral tree, articles on "Peking Man" and another pasted-in publication bearing the strange caption, "Modern Man got his face from an old fish," (the author's arrangement is rather racial in tone when perceived in modern light), also appear. There was considerable interest even then concerning the later falsified 'Piltdown Man' who was then thought to have lived half a million years ago. Naturally, these articles were illustrated using photographs of then newly instituted Chicago Natural History Museum paleoanthropological dioramas and sculptural displays. Later in Volume II, is John A. Menaugh's article devoted to where "Science Places Man Among the Primates."

But following these explorations of human prehistory, life through time rings true again, as my dad evidently procured a comic strip or coloring book concerning the history of life on Earth. These pictures are arranged from the "before prehistory" stage "when the Earth was a flaming ball") through the Ice Age, ending with the Mastodon. Dinosaurs and Mesozoic life predominate in this sequenced pictorial. A glossy - paged informative magazine article by William K. Gregory, titled "Building a Super-Giant Rhinoceros," describes how American Museum workers reconstructed its "Baluchitherium" skeleton. It seems my dad may have had a green colored cardboard lunch box adorned with dinosaur figures that eventually made its way to his scrapbook 'chop shop' too. Volume II concludes with several magazine restorations of Cretaceous herbivorous dinosaurs, and many tracings of prehistoric animals, which I've 'traced' to their origin—

introductory chapters of H. G. Wells' *The Outline of History*. A handwritten page lists "The Earth's Calendar," stating the age of a curious assortment of geological "periods," (several of which are instead stages or epochs, yet apparently in common paleo-parlance) beginning with the Archean (1.5 billion years ago).

"Monsters of the Prehistoric Past - Volume III," another small 3-ring binder, was begun during the summer of 1938, arresting our attention with its first headline "A Prehistoric Continent Found," concerning fossiliferous deposits discovered in northern Mexico. Next, my dad sliced a copy of the Sinclair Dinosaur booklet, the one showcasing James E. Allen's colorful dinosaur restorations. This display ends with another geological time chart containing small portraits of life characteristic to each geological period taken from the Sinclair booklet. This is followed by another stylistic chart representing Earth's geological ages, beginning at "3 billion BC." More paleo-themed cartoons follow, including those in color titled "Fragments," alongside black and white dinosaur restorations now essentially unique to these Scrapbooks—meaning I've never seen them reproduced anywhere else. Articles on the Dodo and the Okapi were inserted; in the latter case my dad apologetically typed an explanatory heading, "I am putting this article in because it tells of the ancestry of the little known Okapi." (p. 49) By this time, to save room within scrapbook pages, my dad occasionally folding article pages several times over, so several (large) magazine pages could be more easily compacted into one scrapbook page, thus conserving precious space.

A newspaper clipping photo on page 61 containing the caption, "Young paleontologists—These Oklahoma City youths found this perfect specimen of a 250,000 (sic - should be 'million' instead) year old 'cotylorhynchus' recently. This is the most perfect specimen of such an ancient lizard ever unearthed, say envious scientists." must have caught my dad's envious eye. Although my grandparents frequently took my dad to the Field Museum (as it was later named) on numerous occasions, never during his formative years did Dad participate on fossil collecting field trips (until he took us kids along for such ventures decades later). As a college undergraduate he double-majored in chemistry and history, and so never did geological field work. While compiling scrapbooks was certainly compelling for him, *when*, he may have wondered, would he begin to make his own discoveries and contributions to this fascinating, unfolding field known as paleontology?

One of several articles by Roy Chapman Andrews pasted into pp. 86-87 titled "How to Find Fossils," probably didn't fully satisfy my dad's yearning. Andrews later wrote about the recent discovery of the Komodo Dragons (Vol. III, pp. 106-107); a picture caption questions, "Are they related to the King of Tyrants?" Following several original drawings of prehistoric men, Volume III continues with announcement of discovery of a living Coelacanth! On page 97 my dad wrote all in capital letters, "Next page see the skull of the largest monster that ever lived." So then on page 98 the monster is revealed. There's a rare Associated Press photo showing Dr. T. E. White posing adjacent to the Harvard Museum of Comparative Zoology's grinning *Kronosaurus* skull. Yes—important discoveries were being made all over the world but my dad was simply too young to contribute to the growing wealth of knowledge.

Decades before Gary Larson's "Far Side" cartoons, comic strip artists were poking fun at dinosaurs. One amusing entry preserved in Volume III shows a stampede of large and rather anxious, if not amorous looking dinosaurs and a woolly mammoth passing by two perplexed and frightened hunters. The caption reads, "Gad, Winston, what sort of mating call did you blow?" And during a winter snowstorm on April 6, 1938, photographer (Joseph Malacina) captured a wispy brontosaur shape among snow-covered branches in foliage along North Pulaski Road. And there's an apparent 'Blondie' cartoon where the titular character inadvertently causes a huge dino skeleton to collapse, despite a conspicuous "Do Not Touch" sign.

Two articles follow in succession. First, Robert J. Casey in a 'Special to The Chicago Daily News,' reported how workers who built the Fort Peck reservoir dam across the Missouri River had become quite familiar with dinosaurs, bones of which they'd often encountered during construction phases. Then on p. 102 we read how "Heat, not cold, killed off Dinosaur..." Interestingly, Professor Raymond B. Cowles of UCLA even suggested that large reptiles surviving the Cretaceous-Tertiary transition "... developed hair at this time to reflect the heat of the sun, thus keeping their body temperatures closer to optimum...." Clearly, even then, there was disparity in the popular press as to how dinosaurs died.

"Monsters of the Prehistoric Past Book IV" began on February 3, 1940. On that date my dad was 13 ½ years old yet still very much a paleo-phile. With high school years fast approaching that coming fall, Volume IV is the shortest. Following photos of Barnum Brown standing adjacent to an amazing *Triceratops* sculpture and another Roy Chapman Andrews article, we're treated to Greer Williams' article on "The Origin of Bird Life," illustrated with outstanding life-sized sculptures of prehistoric avians, including the *Phororhacos*. The next major entry, written by George H.

Eckhardt and illustrated using Charles R. Knight's Field Museum mural showing the 'Cradle of Life,' concerns chemist Dr. Edgar T. Wherry's 'new' theory of how life may have originated from the "Marriage of volcanic gases (i.e. methane and ammonia) and quartz crystals in the warm steaming pools of Earth's youth 1 million years ago." Ideology such as this led to the famous experiment of 1953 conducted at the University of Chicago by Stanley L. Miller. Also, mentioned in this article is the 16th century 'alchemist' and physician Paracelsus, the historical figure whom my father would later complete his doctoral dissertation on at Harvard. It seems my dad's interests, as reflected in this last scrapbook (although sprinkled with paleo-snippets from the newspapers), were broadening, even into the realm of archaeology as in the case of another late entry.

There's one more (undated) scrapbook, a small binder (formerly) filled with old black and white museum postcards; no news or magazine clippings. So, Volume IV was essentially his scrap-booking 'swan song.' The 1940s world was rapidly shifting and my father moved on into other fields of personal and academic endeavor—beyond the dinosaur scrapbooks.

Forty years later, I began compiling *my own* scrapbooks. True—I was inspired somewhat by Glut's *Dinosaur Scrapbook*, but I know Dad's old originals were also on the forefront of my mind. Starting with news clippings from 1979, my 15 thick (paste-in) scrapbooks spanned that amazing secondary period of the Dinosaur Renaissance—a bit beyond the 1996 discovery of feathered Chinese dinosaurs. Maybe if *Prehistoric Times* lasts another half a century, these would seem sufficiently 'historical' to describe here…by someone.

INDEX

A

Abbott, L.B. 30–31, 143–45
"A Century of Progress" (exposition) 62
"A Day In the Life of a Dinosaur" 163
Agathaumas 23, 25, 83, 90, 229, 233–37, 353
AGON, the Atomic Dragon 158–61
Albertosaurus 80
Alen, P.G. 73
Allen, Irwin ... 28–31, 115, 144–47, 170, 341, 347, 352–54, 368
Allen, James E. 73, 378
Altispinax 240, 243
Alvarez, Walter 319
American Museum of Natural History 73, 90, 190, 192, 221, 252, 291, 295, 297
Andrews, Roy Chapman 68, 378, 379
Angier, Natalie 319
Anguirus 261, 268, 283–84, 288, 305, 333, 353
Animal World, The 144, 339, 341
Anytime Rings, The 128, 171
Ape Gigans 275, 278–81
Arata, Jack .... 227, 256, 275, 280, *See* Preface
Archaeopteryx ... 90, 101, 113, 131, 151, 173, 222, 245–53, 255, 365
Arctic Giant, The 28, 193, 306
Arlt, Martin .... 154–55, 157, 277, 327, 353, *See* Preface
Asimov, Isaac 38, 42–43, 126, 168, 198, 332
"A Sound of Thunder" ... 2, 11, 39, 41, 43, 45, 136, 165, 167, 185, 332
At the Earth's Core 2, 128, 131, 269
At the Mountains of Madness 128
Augusta, Josef 79, 86, 91, 115

B

"Backyard Terrors" 206
Bailey, Jack Bowman 238, 243
Bakker, Robert T. 54, 92, 94, 101–2, 108, 128, 173–75, 195
Barnes, Arthur K. 35–36
Barshofsky, Philip 176–77
Baryonyx 238
Bear, Greg 20
Beast From 20,000 Fathoms, The ... 21, 28, 45, 47, 147, 197, 262, 332
"Beast of the Yungas, The" 181
"Beauty and the Beast" 33, 34, 195, 332
Beil, Charlie 211, 214
Berry, Mark .... *See* Preface

Berry, Mark F 28, 31, 46, 49–50, 127, 143–47, 286, 303, 339–40, 351
Bissette, Stephen R 60, 306
Blade, Alexander .... *See* E. Hamilton
Blish, James 182, 264
Bogue, Mike 1, 128, 156
Bradbury, Ray 2, 11, 39–45, 136, 160, 165–67, 185, 332–33, 337
Breithaupt, Brent H. 237
British Museum (Natural History) .. 15, 79, 83, 92, 115–16, 120–21, 223, 239, 243
Brontosaurus ..... 49, 108, 116, 145, 207, 213, 231, 341, 351, *See* Chapter Eighteen, *See* Chapter Six, *See* Chapter One
Brothers, Peter H. 154, 157
Brown, Barnum ... 28, 68, 73, 76, 80, 210–11, 220, 235–36, 375, 379
Brown, Lewis S. 172–73
Brusatte, Stephen L. 106, 127, 358
Buck, Frank 35, 43
Burian, Zdenek 53, 78, 84–86, 91, 94, 115, 237, 343
Burpee Museum 54, 85
Burroughs, Edgar Rice 2, 13, 39, 48, 126, 128, 131–36, 178–79, 269, 306, 330, 370
Burton, Maurice 92, 116–17, 121

C

Camarasaurus 91
Camerascopes, Ltd. 223–24
Capek, Karel 263–64, 332
Carcharodontosaurus 103
Carey, Diane 44
Carroll, Earl 66, 75
Cassiday, Bruce 128, 171
Ceratosaurus .. 139, 218, 222, 224, 238, 240, 242–43, 338, 341–43, 345, 365
Chicago Natural History Museum .... *See* Field Museum, *See* Field Museum, *See* Field Museum, *See* Field Museum, *See* Field Museum, *See* Field Museum
Chicago Sunday Tribune, The 56, 195
Chicago World's Fair *See* "A Century of Progress"
Christman, Erwin S. 63, 78, 91, 236–37, 364
Clarke, Arthur C. 42, 176, 333
Clos, Lynne .... 228, *See* Preface
Colbert, Edwin H. 4, 108, 120, 207, 239, 366
Conan Doyle, Arthur .2, 10, 40, 46–47, 48–50, 63, 130, 142, 156, 165, 182, 219–20, 238, 264, 283, 285, 288, 330, 339, 350, *See* Chapter One
Cooper, Merian C. 20, 191, 276, 288, 291, 308

Cope, Edward D..... 90, 110, 130, 233–35, 237–38
Corrigan, Ray "Crash" 285–88
Cosmic Landscapes.................................. 126, 129
Creature From the Black Lagoon............ 178, 333
Crichton, Michael.. 20, 45, 128, 185, 195, 263–65, 296, 298, 334–35
Croghan, Jason .... 259, *See* Preface
Crowther, Peter 120–22
Csotoni, Julius 104
Czerkas, Stephen 93, 105, 233, 343

D

Daikaiju... 6, 38
Damon, Joseph 63–75
Danger Dinosaurs!... 165
Dark matter .... *See* Chapter Thirty-two
Darwin, Charles. 89, 110, 111, 133, 254, 280, 285, 367
Daspletosaurus....................................... 80–81, 105
David, James F. 169–70
Davidson, Jane P. 96–98, 104
De Camp, Sprague L. 44, 168, 182, 333
De Paolo, Charles.. 306
Debus, Allen G..................... 67, 75, 128, 357, 374
Debus, Diane 5, 93, 104, 321
Debus, Edna P. 6, 375
Debus, George W.W. 208, 215, 313, 376
Debus, Karl 225
Debus, Lisa..*See* Preface
Debus, Richard................... 86, 163, 213, 276, 359
del Rey, Lester 170
Delgado, Marcel.. 22–24, 27, 28, 31, 63, 138, 234, 293, 298, 338
Denham, Carl 308, 339, *See* Chapter Twenty-nine
Dennis, Kristen....................................*See* Preface
Dennis, Ryan ....*See* Preface
Dent, Lester 185
Desmond, Adrian J. 78, 86, 102, 128, 198
DeVito, Joe................ 284, 290, 295, 299, 303, 304
Diane, Debus... 111
Dimetrodon 66–67, 71, 142–46, 215, 220, 224, 280, 286, 349
Dinah.. 65–66
Dinny 66, 207, 211–15, 236, 375–76, 389
Dinosaur Art 95–96, 98, 103
Dinosaur Filmography, The 29, 31, 46, 49, 139, 143, 340, 354
Dinosaur Memories........... 3, 5, 225, 276, 315, 375
Dinosaur renaissance... 84, 95, 101, 104, 150, 173, 188, 195, 200, 237, 262, 298, 312, 360
Dinosaur revolution.. 102
Dinosaur Summer.. 20
Dinosaur Wars 334
Dinosaur World... 3, 11, 69, 75, 81, 204, 209, 229, 238, 256
Dinosauriana 97
Dinosauroid.. 186, 263–64

Dinosaurs Attack!.. 59, 195, 325, 326, *See* Chapter Thirty-three
Dinosaur Scrapbook, The....... 3, 75, 139, 142, 145, 158, 208, 210, 234, 349, 351, 371–73, *See* Chapter Thirty-eight
Dinosaur Sculpting: A Complete Guide .. 5, 75, 97
Dinosaur Valley Girls (film) 31, 139, 140, 345, 348
Dinosaur Valley Girls (novel) 140
Dinosaur Valley Girls (novel) 175
Diplodocus 28, 57–58, 181, 189–90, 195, 218, 221–22, 224, 233, 261
Dixon, Dougal.. 98
Doc Savage: The Time Terror......................... 184
Drake, David.. 44
Druktenis, Dennis......................... 10, *See* Preface
Dufault, Danielle... 100

E

Earth For Sam, The ....375, *See* Chapter Thirty-six
Edwards, Vernon......................... 63, 223–25, 242
ERBANIA.. 126–27
"Exiles of the Stratosphere" 180
Extinct Animals 16, 22, 242
Extinct Monsters 14, 196, 227, 235, 240–42

F

Fantasia (1940).................. 193, 286, 338–41, 344
Fantastic Adventures........... 9, 36, 38, 311, 313–15
Faraday, Robert.. 128, 171
Fearn, John Russell 178
Field Museum of Natural History... 30, 53, 54, 76, 190, 206, 337, 348
Field, Robert Scott 306
Figuier, Louis................................. 16, 90, 256, 280
Flammarion, Camille.................. 27, 189, 194, 389
Flash Gordon 144, 146
Fleischer, Max................................. 28, 193, 223
Fogg, William 46, 50
"The Fog Horn" 42, 45, 332
Footprints of Thunder 169
Ford, Tracy... 243
Fossil birds (Mesozoic) 227, *See* Chapter Twenty-three
Fossil Spirit, The 174
Foster, Phil.. 163
Fredericks, Mike .... 53, 69, 75, 84, 86, 229, *See* Preface

G

Gaw.. 284, 297, 299–302
"Gertie On Tour"... 191
G-Fan 4, 12, 38, 55, 127, 129, 151, 154, 157, 195, 198, 199, 215, 260, 277, 289, 306, 326, 354, 361, 373

Ghidrah, the Three-headed Monster 149, 153, 156, 195, 270, 272, 369
Ghost of Slumber Mountain, The 22, 283, 337
Giant Ground Sloth 68, 174, 189, 285–88
Gigantis, the Fire Monster 25, 31, 74, 149, 195, 268, 283–84, 287, 300, 306, 345, 347, 353–54
Gigantosaurus .... 189–91, *See* Frontispiece
Gillman 14, 333
Gilmore, Charles Whitney 63, 211, 224, 242
Glut, Donald .... 3, 5, 14, 20, 49, 50, 71–75, 79–81, 85, 139–42, 145, 158, 175, 208–9, 210–13, 216, 229, 233–40, 276, 285, 290, 302, 345, 348–54, 371–72, 374, 380, *See* Preface
Godzilla 1, 6, 10, 14, 21, 38, 46, 47, 54, 56, 57, 68, 98, 127, 137, 149, 158, 161, 178, 188, 191, 198–200, 260, 261, 268, 288, 290, 300–303, 309, 304–9, 333–36, 337, 353–54, 368–73
Godzilla (1954) 147
Godzilla (1998) 197
Godzilla (2014) 186, 307
Godzilla no Gyakushu 283–84
Godzilla Raids Again 195, 304, 306, 327, 347, 354
Godzilla vs. King Ghidorah 304–5
Godzilla vs. the Smog Monster 197, 262
Godzilla, King of the Monsters ... 21, 199–200, 368
Godzillasaurus 300, 305
Gojira .. 12, 28, 38, 47, 50, 159, 161, 191, 197–98, 199, 262, 304, 339, 340
Gorgo 21, 35, 38, 79, 158, 178, 261, 268, 337
Gorgo, (1961) 272
Gorgo' 333
Gorgosaurus .... 31, 53, *See* Chapter Seven
Gould, Stephen J. 99, 110, 136, 188, 196, 201, 312, 367
Graziano, Lisa M. 167
Graziano, Michael S. A. 167
Green, Johnathan 183
Guyot, Arnold H. 109–10, 113
Gwangi, The Valley of 138

H

Hadrosaurus 89–90
Hagenbeck, Carl 15, 182, 217, 218–20, 222
Hagenbeck, Claus 218
Hamilton, Edmund 38
Harryhausen, Ray 1, 19, 28–29, 63, 98, 138–41, 144–46, 152, 195, 236, 337–38, 341–42, 347, 348
Hawkins, Benjamin Waterhouse 63, 70, 85, 89–90, 94, 108–11, 116, 155, 207, 211, 217–21, 225, 233, 238–40, *See* Chapter Eleven
Hawley, Russell 11
Heilmann, Gerhard 242, 246, 256, 363–64
Hell Creek 167
Hot-Blooded Dinosaurs, The 78, 86, 102, 114, 128
"Hothouse Planet, The" 35–36
Houdini, Harry 28
Howgate, Mike 85, 114, 224

Hoyt, Harry O. 21, 23, 29
Hubbell, Will 44
Hunterian Museum 224
Hutchinson, H. N 14, 112, 227, 229, 235–36, 240–43
Hylaeosaurus 154–55, 154–56, 233

I

Iguanodon . 17, 27, 58, 88–89, 130, 154, 189, 197, 203, 218, 219, 221–23, 231, 238, 353
Irvine, Alex 267
Irwin, G. H. 9, 36
Irwinosaurs 28, 144, 147, 350, 351, 354

J

Jefferson, Thomas 285, 287–88
Johnson, Martin 43
Johnson, Osa 43
Jones, Willis Knapp 181–82
Journey to the Beginning of Time 45, 69, 339, 342, 345
Journey to the Center of the Earth (1959 film) 30, 141, 142, 148, 354
Journey to the Center of the Earth (novel) ... 2, 13, 39, 128, 133–34, 174, 278, 283, 329, 345
Jurassic Classics 3, 5, 71, 142, 348, 354
Jurassic Park ... 13, 21, 28, 31, 39, 45, 64, 98, 104, 128, 138, 140, 167, 184–86, 195–97, 262–64, 269, 296, 298, 334–35, 337

K

Kalt, Rhoda Knight 107, 110
Kanerva, John 211–15
Kastner, Chris 206, 209
Keep Watching the Skies 47, 199, 308
King Kong ... 20, 22, 24, 32, 37, 47, 49, 54, 63–64, 68, 78, 90, 115, 138, 145–47, 190–95, 273, 284, 299–300, 303, 308, 338–40, 342, 347–50, 368, 376
King Kong vs. Godzilla . 1, 195, 276, 284–85, 306, 369, *See* Chapter Twenty-nine
King, Stephen .... *See* Chapter Twenty-six
Kirkland, James I. 44
Knight, Charles R. 2, 21, 22, 24–25, 62–63, 73–74, 77, 80, 83, 84, 90–91, 94, 99, 107–10, 115–16, 193–94, 219, 229, 233–36, 337–38, 256, 283, 286, 288, 294, 329, 337–38, 345, 353, 364–67, 371, 375–77, 380
Kuttner, Henry 33–34, 37, 195, 332

L

La Brea Tar Pits 208, 292
Laelaps 89–90, 194, 238, 294

"Lair of the Grimalkin" 9, 36
Lakes, Arthur 233, 244
Lambe, Lawrence 77–80
Lambeosaurus 53, 76–77, 80, 105
Land That Time Forgot, The ... 2, 13, 48, 128, 135, 306, 330, 370
Landis, Carole 141, 349
Landis, Geoffrey A. 44
Lankester, E. Ray 16–19, 22, 242
Lanzendorf, John ... 81, 100, 105, 229, *See* Preface
Lavas, John 16–18, 46, 86, 94
Le Monde Avant La Creation de L'Homme 194, 389
Lees J. D. .... *See* Preface
Lermina, Jules 189, 195, 198
Long Jr., Frank Belknap 180
LoRusso, Don 94, 98
Lost Atlantis, The 47, 49–50, 236
Lost Continent, The (film) .... 12, 262, *See* Chapter Four
Lost Continent, The (novel) 50, 60
Lost World, The (1925 film) ... 2, 13–29, 147, 194, 262, 330
Lost World, The (1960 film) 2, 29–32, 144–46, *See* Chapter Thirty-five
Lost World, The (novel) 17–20, 46, 264, 330
Lovecraft, H. P. 128
Lucas, Frederik A. 94, 363
Lull, Richard Swann 63, 236–37
Lydekker, Richard 238, 241
Lynch, Vincent .... 191, *See* Frontispiece

M

Mammoth 10, 65, 66–68, 141, 167, 172, 184, 189, 198, 217, 223, 278, 287, 329–30, 375, 379
Mammoth (novel) 167
Mantell, Gideon 17, 88–89, 174, 197
Marsh, Othniel .. 111, 112, 130, 221, 229, 232–33, 235, 238, 244
Marsten, Richard 165, 176
Mason, James 30, 143
Mass extinctions ... 6, 200, 312, 313, 315, 360, *See* Chapter Thirty-two
Masthay, Carl 229
Matthew, W. D. ... 78, 190–91, 193, 201, 218, 365, *See* Frontispiece
McCarthy, Steve 85
McCay, Winsor 3, 191–95, 308
McCormick, Maureen 109
Megalonyx 285–88
Megalosaurus . 18, 89, 120, 189, 227, 237, 238–44
Megatherium 224, 285, 286–87
Merkl, Ulrich 3, 60, 191, 194–95, 201, 308
Messmore and Damon 64–73, 125, 192, 216
Messmore, Francis B. 69, 70–71, 74–75
Messmore, George Harold 63–66, 75
Meteor Crater (Arizona) . 311, *See* Chapter Thirty-one

Meyer, A. 19
Mill, John 174
Miller, C. R. 220
Miller, P. Schuyler 176
Mitchell, W. J. T. 96, 97, 101, 188, 196, 201
Mokele mbembe 62, 75
Moment Out of Time 166
Monger, Gareth 99
Monoclonius 220, 229, 233–37
Morgan, Leon 72, 366, 376
Morris, A. 314–15
Muller, Richard A. 319, 322
Mullis, Justin 276

N

Nakajima, Haruo 152–55, 353
Neecha 31, 139–40
New York World's Fair (1964) 206, 208, 213, 341
Nicholls, Robert 101, 103
Night Shapes, The 182
Nye, Nicholas 20

O

O'Brien, Willis 20, 22–29, 31, 49, 63, 138, 144, 234, 293, 298, 338–42
Oakley, Kenneth 119, 131
Ogden, D. Peter 126–27
"One Prehistoric Night" 176–77
Osborn, Henry F. 28, 62, 79–80, 110, 181, 201, 235, 244, 363
Otachi 266, 268
Owen, Richard 89, 112–13, 155, 165, 239–40, 245, 285, 302

P

Pacific Rim (film) ... 129, *See* Chapter Twenty-five
Pacific Rim (novel) .... *See* Chapter Twenty-five
Palaeozoic Museum 89, 108–10, 112, 207, 211, 217
Paleoart ..... 2, 53, 62, 84–86, 88, 91, 115–16, 150, 154–55, 164, 168, 197, 200, 204, 206, 234, 243, 364, 367, 371, *See* Chapter Nine, *See* Preface
Paleoimagery. 2, 5, 62, 84–86, 116, 164, 200, 204, 209, 211, 232, 329–30, 338, *See* Chapter Nine
Pallenberg, Josef .. 15, 63, 203, 207, 211, 236, *See* Chapter Twenty-one
Panic in Paris 189, 195, 198
Parker, Neave .. 2, 83, 84, 86, 91–92, 94, 239, 243, *See* Chapter Twelve
Paul, Frank R. 38
Paul, Gregory S. 93, 94, 239, 256
Periodicity .... *See* Chapter Thirty-two
Petitpas, Dan 126

Pettigrew, Neil .. 23–24, 26–28, 48, 339, 342, 344, 348
Piltdown Man . 15, 19, 69, 120, 131, 295, 365, 377
Planet of Dinosaurs 45, 339, 343–44
Pohl, Frederick 42, 179
Porter, Elbert 77, 208
Powers, Peggy 163
Prehistoric Times . 2, 4, 11, 50, 53, 67, 69, 75, 84–87, 94, 96, 98, 102, 104, 106, 127, 129, 168, 204, 209, 224, 228–30, 236, 243, 302, 312, 326, 361, 372, 380
"Prehistoric Forest" 207, 336
"Prehistoric Park" 210–13
"Prehistoric Park" 214–15
Primeval World, (Disneyland) 204, 208, 341
Psittacosaurus 101, 103
Pterosaur . 25, 57, 99, 113, 149–53, 178, 245, 250, 272

R

Rains, Claude 29, 144, 351
Randall, Lisa 316, 318–21
Raup, David M. 312, 317–20
Reece, Matthew 316
Reed, William Maxwell 362, 375–76
"The Resurrection of Jimber-Jaw" 184
Return to the Lost World 20
Revolt on Venus 36, 180
Rey, Luis 103–4, 105
Rhedosaurus 1, 21, 47, 154, 160, 261, 306, 333, 337
Roach, Hal 30, 64, 141–42, 146, 175, 347, 350
Robeson, Kenneth .... *See* Lester Dent
Rockwell, Carey 36, 180
Rodan .. 2, 21, 57, 261, 268, 333, 337, *See* Chapter Fifteen
Rodan, The Flying Monster 347
Roof, Katharine Metcalf 27
Rovin, Jeff 156, 158, 278

S

"Sands of Time, The" 176
Sanz, Luis 139
Sawyer, Robert J. 44
Scientific American ... 35, 101, 154, 191, 201, 203, 222, 229, 231–33, *See* Frontispiece
Sepkoski, Jr., J. John 317–18
Sereno, Paul 54, 106
Serling, Rod 45, 161
Settles, James B. 314–15
Shark-Crocodile .. 275, *See* Chapter Twenty-seven
Sheldon, Roy .... *See* E. C. Tubb
Shepstone, Harold J. 220–23
Shin Godzilla 197, 304, 307
Simak, Clifford D. 44
Sinclair Refining Company 73–74, 215
Skeleton Crew 270, 273

Sky People, The 36, 178–79
Smit, J. 235, 240–42, 256, 364–65
Space Stories and Sounds 279–80, 282
Spinosaurus 104, 196, 237–38, 240, 242–43
Starfire: The Mending, Book 1 186
Starlog 86, 107
Stegosaurus ... 18, 22, 67–68, 72–73, 79, 116, 120, 125, 154, 164, 204, 210, 215, 218, 221–24, 231–33, 243, 267–68, 293, 309, 326, 329, 353, 365, *See* Chapter Thirty-four
Stern, Bill 279–80
Stirling, S. M. 36, 178
Stockton, Stuart Vaughn 186
Stout, William 243
Suitmation . 1, 68, 125, 127, 137–38, 141, 147–48, 153, 264, 268, 305, 347–48, 353
Swinton, William E. 91, 100, 116–22, 211, 214, 224, 242
Symmes, John C. 132

T

Tennant, Todd .... 389, *See* Preface
The Lost World: Jurassic Park 148, 335, 345
The Mist (film) 270
The Mist (novella) 270, 272
The Parasaurians 185
Thunder of Time 170
Time Machine, The ..... 18, 165, *See* Chapter Three
Tom Corbett 38, 71, 180
Topps 59, 195, 325, 326, 328
Triceratops .. 21–28, 39, 48, 49–50, 57, 67, 69–74, 77–80, 108, 110, 120–22, 120–22, 142, 167, 171, 186, 192–94, 207–9, 220, 222–24, 235–37, 244, 253, 292–96, 337, 340, 345, 377, 379
Tsuburaya, Eiji 1, 28, 151, 288
Tubb, E. C. 166
Tunnel Through Time 170
Twenty Million Miles to Earth 35, 195
Tyrannosaurus . 21, 22, 27, 36, 39–41, 57–58, 67–68, 70, 72–74, 76–80, 116, 120, 135, 180–81, 186, 192–96, 209, 215, 256, 283, 294, 337–41, 343–45, 348, 353, 375

U

Unknown Island 147, 275, 284–88
Unnatural History 183
Utley, Steven 44

V

V. rex 296–97, 299
Varan 149–50, 153–57, 333, 368
Varan, the Unbelievable 149, 153
Varley, John 167
Venus .... 1, 11, 70, 178–80, 194, 332, *See* Chapter Two

Verne, Jules..... 2, 13, 16, 19, 23, 39, 128, 132–34, 136, 142, 164, 174, 269, 278–81, 283, 285–86, 288, 329, 345
Vinther, Jakob 103
von Meyer, Hermann 240–42

W

Walking With Dinosaurs (BBC) 98, 345
Warren, Bill 47, 199, 308, 342
Wells, H. G...... 18, 39–42, 45, 132, 165, 167, 261, 263, 336, 378
Wells, Robert 185–86
Whip 35–38
White, Steve 95, 104
Wiebe, Maidi 81
Williams, Gary .... 85, 106, 204, 225, 229, *See* Preface

Winchell, Alexander D 95
Winslow, J. H. 19, 181
Witton, Mark P. 104, 150
Wonders of Geology 88, 174
"World A Million Years Ago, The" 64, 66, 69–73, 125, 206, 208, 366

Y

Ymir 35, 195

Z

Zallinger, Rudolph 77, 92, 94, 102, 106, 116, 151, 198
Zangerl, Rainer 77
Zeman, Karol 45, 69, 342

ABOUT THE AUTHOR

Allen A. Debus has written numerous articles, several for journals and many more for a variety of fanzines—the latter including *Prehistoric Times, G-Fan, Mad Scientist, FilmFax* and *Scary Monsters*—usually intertwining the subjects of science fiction and paleontology. He was a contributing editor of *Fossil News: Journal of Avocational Paleontology*. Allen has also written seven other paleo-related books: two of the self-published variety, and five for McFarland Publishers emphasizing prehistoric animal popular culture & imagery. He is a self-taught dinosaur sculptor and a retired environmental chemist.

ABOUT THE COVER

The cover was designed by Todd Tennant in 2017 for this book, for hire. (**Front**) This image, based on the 1884 'Old-stego' *Scientific American* restoration described here in Chapter Twenty-two, is from C. Flammarion's *Le Monde Avant La Creation de L' Homme* (1886). (**Back**) Images are (clockwise from top); photo by A. Debus of two dinosaur toys—*Therizinosaurus vs Deinonychus*; a portion of a *Lost World* movie still (1925, copyright First National Pictures); postcard personally addressed to author showing Calgary's Zoo sauropod "Dinny" received via mail in 1963; "King Komodo"—an original dino-monster invention conceived and created by Todd Tennant; a *Lost World* movie ad, 1925—copyright First National Pictures. Central image showing movie stills from *Agon, The Atomic Dragon* (copyright 1965, Tsuburaya Productions).

www.ingramcontent.com/pod-product-compliance
Lightning Source LLC
Chambersburg PA
CBHW082201220526
45470CB00010B/3007